The Role of Women in Cultivating Sustainable Societies Through Millets

Raghvendra Kumar
GIET University, India

Ishaani Priyadarshini
University of California, Berkeley, USA

IGI Global
PUBLISHER of TIMELY KNOWLEDGE

A volume in the Practice, Progress, and Proficiency in Sustainability (PPPS) Book Series

Published in the United States of America by
IGI Global
Engineering Science Reference (an imprint of IGI Global)
701 E. Chocolate Avenue
Hershey PA, USA 17033
Tel: 717-533-8845
Fax: 717-533-8661
E-mail: cust@igi-global.com
Web site: http://www.igi-global.com

Copyright © 2024 by IGI Global. All rights reserved. No part of this publication may be reproduced, stored or distributed in any form or by any means, electronic or mechanical, including photocopying, without written permission from the publisher. Product or company names used in this set are for identification purposes only. Inclusion of the names of the products or companies does not indicate a claim of ownership by IGI Global of the trademark or registered trademark.

Library of Congress Cataloging-in-Publication Data

Names: Kumar, Raghvendra, 1987- editor. | Priyadarshini, Ishaani, 1992- editor.
Title: The role of women in cultivating sustainable societies through millets / edited by: Raghvendra Kumar, Ishaani Priyadarshini.
Description: Hershey, PA : Engineering Science Reference, [2024] | Includes bibliographical references and index. | Summary: "The objective of this book is to promote the cultivation and consumption of millets as a means of building sustainable societies, while also recognizing and enhancing the critical role that women play in this process"-- Provided by publisher.
Identifiers: LCCN 2023034584 (print) | LCCN 2023034585 (ebook) | ISBN 9781668498194 (hardcover) | ISBN 9781668498200 (ebook)
Subjects: LCSH: Women farmers--India--Hyderabad. | Millets--India--Hyderabad. | Sustainable agriculture--India--Hyderabad.
Classification: LCC HD6073.F32 H84 2024 (print) | LCC HD6073.F32 (ebook) | DDC 331.4830954/84--dc23/eng/20231121
LC record available at https://lccn.loc.gov/2023034584
LC ebook record available at https://lccn.loc.gov/2023034585

This book is published in the IGI Global book series Practice, Progress, and Proficiency in Sustainability (PPPS) (ISSN: 2330-3271; eISSN: 2330-328X)

British Cataloguing in Publication Data
A Cataloguing in Publication record for this book is available from the British Library.

All work contributed to this book is new, previously-unpublished material. The views expressed in this book are those of the authors, but not necessarily of the publisher.

For electronic access to this publication, please contact: eresources@igi-global.com.

Practice, Progress, and Proficiency in Sustainability (PPPS) Book Series

Ayman Batisha
International Sustainability Institute, Egypt

ISSN:2330-3271
EISSN:2330-328X

Mission

In a world where traditional business practices are reconsidered and economic activity is performed in a global context, new areas of economic developments are recognized as the key enablers of wealth and income production. This knowledge of information technologies provides infrastructures, systems, and services towards sustainable development.

The **Practices, Progress, and Proficiency in Sustainability (PPPS) Book Series** focuses on the local and global challenges, business opportunities, and societal needs surrounding international collaboration and sustainable development of technology. This series brings together academics, researchers, entrepreneurs, policy makers and government officers aiming to contribute to the progress and proficiency in sustainability.

Coverage

- Global Business
- Global Content and Knowledge Repositories
- Sustainable Development
- Intellectual Capital
- Knowledge clusters
- Innovation Networks
- ICT and knowledge for development
- Strategic Management of IT
- Eco-Innovation
- Technological learning

IGI Global is currently accepting manuscripts for publication within this series. To submit a proposal for a volume in this series, please contact our Acquisition Editors at Acquisitions@igi-global.com or visit: http://www.igi-global.com/publish/.

The Practice, Progress, and Proficiency in Sustainability (PPPS) Book Series (ISSN 2330-3271) is published by IGI Global, 701 E. Chocolate Avenue, Hershey, PA 17033-1240, USA, www.igi-global.com. This series is composed of titles available for purchase individually; each title is edited to be contextually exclusive from any other title within the series. For pricing and ordering information please visit http://www.igi-global.com/book-series/practice-progress-proficiency-sustainability/73810. Postmaster: Send all address changes to above address.
Copyright © 2023 IGI Global. All rights, including translation in other languages reserved by the publisher. No part of this series may be reproduced or used in any form or by any means – graphics, electronic, or mechanical, including photocopying, recording, taping, or information and retrieval systems – without written permission from the publisher, except for non commercial, educational use, including classroom teaching purposes. The views expressed in this series are those of the authors, but not necessarily of IGI Global.

Titles in this Series

For a list of additional titles in this series, please visit: http://www.igi-global.com/book-series/practice-progress-proficiency-sustainability/73810

Global Perspectives on Climate Change, Social Resilience, and SocialInclusion
Aly Abd ElRazek Galaby (Alexandria University, Egypt) Mohammed Abo ElEnein (American University, UAE) and Hassan Mohamed (Alexandria University, Egypt)
Engineering Science Reference • © 2024 • 380pp • H/C (ISBN: 9781668489635) • US $275.00

Achieving the Sustainable Development Goals Through Infrastructure Development
Cristina Raluca Gh. Popescu (University of Bucharest, Romania & The Bucharest University of Economic Studies, Romania) Poshan Yu (Xi'an Jiaotong-Liverpool University, China & European Business University of Luxembourg, Luxembourg) and Yue Wei (Institution of Public Private Partnerships, Hong Kong)
Engineering Science Reference • © 2023 • 303pp • H/C (ISBN: 9798369307946) • US $230.00

Operational Research for Renewable Energy and Sustainable Environments
Joshua Thomas (UOW Malaysia KDU, Malaysia) Gerhard Wilhelm Weber (Poznan University of Technology, Poland) Roman Rodríguez Aguilar (Universidad Panamericana, Mexico) Elias Munapo (North West University, South Africa) and Pandian Vasant (Modeling Evolutionary Algorithms Simulation and Artificial Intelligence (MERLIN), TDTU, Vietnam)
Engineering Science Reference • © 2023 • 300pp • H/C (ISBN: 9781668491300) • US $240.00

Fostering Cross-Industry Sustainability With Intelligent Technologies
Brojo Kishore Mishra (NIST Institute of Science and Technology (Autonomous), India)
Engineering Science Reference • © 2024 • 400pp • H/C (ISBN: 9798369316382) • US $325.00

Sustainable Development in AI, Blockchain, and E-Governance Applications
Rajeev Kumar (Moradabad Institute of Technology, Moradabad, India) Abu Bakar Abdul Hamid (Infrastructure University, Kuala Lumpur, Malaysia) Noor Inayah Binti Ya'akub (Infrastructure University, Kuala Lumpur, Malaysia) Hari Om Sharan (Amity University, Kolkata, India) and Sandeep Kumar (Rightzone Technologies Pvt. Ltd., India)
Engineering Science Reference • © 2023 • 350pp • H/C (ISBN: 9798369317228) • US $285.00

Crafting a Sustainable Future Through Education and Sustainable Development
Javier Martínez-Falcó (University of Alicante, Spain) Bartolomé Marco-Lajara (University of Alicante, Spain) Eduardo Sánchez-García (University of Alicante, Spain) and Luis A. Millan-Tudela (University of Alicante, Spain)
Engineering Science Reference • © 2023 • 302pp • H/C (ISBN: 9781668496015) • US $215.00

IGI Global
PUBLISHER of TIMELY KNOWLEDGE

701 East Chocolate Avenue, Hershey, PA 17033, USA
Tel: 717-533-8845 x100 • Fax: 717-533-8661
E-Mail: cust@igi-global.com • www.igi-global.com

Raghvendra Kumar dedicates this book to his son, Prakhar Pandey, and his wife, Priyanka Pandey.

Table of Contents

Detailed Table of Contents

The General Assembly of the United Nations (UN) has declared 2023 as the 'International Year of Millets' to promote millets' nutritional, health, environmental, and economic advantages. Millets' unique qualities and advantages have made them important contributors to numerous sustainable development goals (SDGs). Prioritizing millets can lead to a sustainable and inclusive future. Millets are small-seeded grasses with great nutritional value and environmental resistance, surviving drought and flood. Millets were the first cereal grain cultivated, making it one of the oldest agricultural cereals. From historical records, millets were the first cereal grain to have been domesticated, making it one of the oldest farmed grains in the world.

Agriculture systems worldwide need to adopt sustainable cropping solutions to deal with existing issues of malnutrition, food insecurity, resource depletion and lifestyle health issues. Millets have special nutritive and agronomic properties. Millets offer one such solution as they have nutritional, environmental, and economic superiority over the most common staple grains in India. The present study deals with women empowerment in agriculture that reflects increased women's decision-making authority related to agricultural resources, management and production, and income, which will enhance the ability of women farmers to strengthen their economic status. The novelty of this study focuses to address challenges in sustainable production and marketing in millets promotion that impede inclusive growth of women farmers. The findings of this study advocate having an integrated approach to enhance millet value chain efficiencies to develop millet promotion, considering economic, environmental, and health sustainability through women empowerment.

Millets are cereal crops, commonly grown in semi-arid regions. Millets are known to be a good source of fiber, magnesium, phosphorous, zinc, serotonin, and gluten-free product that helps to heal Cancer effectively. The active components of Foxtail millet bran (FMBP) are promoting higher levels of reactive oxygen species (ROS) that help block the STAT3 pathway by reducing the growth of malignant cells in intestine. Nowadays in medical colonoscopy, histological image analysis and MRI are some commonly adapted techniques for the identification of polyps in the colon. However, the manual approaches are very time-consuming and less accurate. There is no specific model available that can measure the effectiveness of millet in healing malignant cell growth in colon. The main goal is to design a deep learning based MoblileNet-V3 model for the detection and classification of histological images. This study will help researchers and medical practitioners to compare the malignancy status before and after the consumption of millet.

This chapter explores the impact of millet cultivation and sustainable development on agriculture, focusing on advancements in production and environmental practices. Key factors influencing production quantity, quality, location, etc., are analysed. It presents an overview of food security assessments, modelling, mapping, and monitoring techniques, utilizing data from various satellites, from spectroradiometers, and ground-truthing, the grain yield of different millet varieties is estimated, accounting for regional variations and the influence of soil bunds on crop growth. Through the application of the random forest technique, the chapter demonstrates how crop yields can be identified and predicted using regression models based on rainfall patterns and millet suitability in specific Indian states. The findings showcase how monitoring techniques can empower farmers in various regions to achieve higher yields, sustainable food production, and improved crop quality along with addressing the limitations or challenges encountered during the research process.

This book chapter explores the profound relationship between women, millets, and their impact on health, empowerment, and sustainable agriculture. Millets, ancient crops with exceptional nutrition, have evolved alongside women's roles in cultivation. Throughout history, women have been key guardians of millet farming, preserving traditional knowledge in challenging environments. Women drive the resurgence

Today, millet has been rebranded as "Nutri-cereals." It became a choice for the taste of many consumers. Its nutri-rich sources and benefits are creating a fascination for many consumers and apparently there will be further increase in demand for the future. On the other hand, millet also attracted attention due to lower investments by farmers requirements such as water fertilizers, pesticides, etc., with good income and market potentiality. Moreover, it is increasing minimum support price (MSP), and millets can also be considered as positive cultivation for the farmers to make decisions on agriculture. The eventual declining trend of the natural resource availability also identifies the need for crop shift as of paramount importance. Millets are generally categorized into major and minor millets, ICAR-IIMR has been leading their efforts for a decade now.

Considering the growing market for bakery products, and healthy food products rising in consumer perception, new functional food product development in bakery industry is a worthwhile focus. In search of appropriate nutritional and functional ingredients, the application of millet flours as alternative ingredients is critically assessed by the researchers. The present review highlighted the India's potential (19% productivity share at global level) in millet production, its nutritional goodness, as well as recent traits in bakery product development. Not only the nutrient goodness, but also the phenolic and bioactive components associated with these nutrigrains, could lead the researchers to categorizing it as superfood. Non-gluten nature associated with these arid gain is the worthy source for nutrient enrichment and fortification in various processed bakery products; its low glycemic index (GI) based nature also confirms millets significant role in curing various disorders, making it to categorised as healing grain.

The issue of security is paramount in society, especially to the women who are being illtreated in this prevailing era. Thus, the authors intend to aid security by bringing up a smart security system for women safety. This system involves an EEG sensor, heartbeat sensor, GPS sensor, and GSM module along with an Arduino. This proposed system purely depends on the psychology of a woman which deals with measuring of brainwaves and heartbeat rate within a given limit. And on further exceeding, the data will be detected depending on the level of waves emitted. These values would trigger the GSM to receive the data and thereby it sends an automatic popup alert in the form of message along with the location. The entire system will be activated to safeguard woman in all aspects.

Millets are a group of small-seeded grasses that have been cultivated for thousands of years in different parts of the world. In many cultures, millets have been considered a staple food, and they continue to be a vital source of nutrition for numerous people today. In the context of women's history, millets have played a significant role in many societies. In India, millets have been a traditional food for centuries, and women have played a critical role in their cultivation and consumption. Historically, millets have also been an important source of income for women in many societies. Millets have also played a significant role in women's health. Millets are highly nutritious and excellent sources of protein, fiber, and vitamins. They are also gluten-free, making them a valuable food source for people with celiac disease or gluten intolerance. Millets are a nutritious and healthy food that offers a variety of health benefits. Incorporating millets into your diet can help improve overall health and well-being.

The Indian government, with the support of 72 nations, proposed to the United Nations Food and Agriculture Organisation to declare 2023 as the "International Year of the Millets." Millets have numerous benefits-are gluten free with high nutritional properties, grown in dry-arid regions, drought and disease resistance, environmental sustainability, and empowering women for socio-economic development. The rural women in developing countries are helping to sustain the millet economy, which has high nutritional value and very tolerant to extreme semi-arid and drought weather conditions experienced due to climate change. This chapter will examine the Indian government's policy towards developing a millet economy amidst the impacts of polycrisis situation experienced around the world; to empower rural women in the agricultural sector, focusing on the economic dimensions; to bring about social transformation and economic significance of millet farming, value adding, improving livelihood, and social transformation for sustainable development through climate smart millet economy in India.

The experiences of women who grow millet in Odisha, India, are investigated in this research. The study identifies the constraints and opportunities for women to participate in millet farming, as well as the effects of millet farming on their lives and means of subsistence, using in-depth interviews, focus groups, and surveys. According to the research, although cultural expectations and a lack of resources present substantial obstacles for women, their participation may be facilitated by establishing women's groups and cooperatives, access to training, and support from NGOs and government initiatives. In addition to proposing steps to increase women's access to resources and support gender-sensitive agricultural extension services, the study highlights the potential of millet cultivation for advancing sustainable agriculture and gender equality.

Preface

Millets are ancient grains that have been cultivated by humans for centuries, and they have recently emerged as a crucial staple food in numerous cultures owing to their abundant essential nutrients and their resilience to changing climates, requiring fewer resources for growth. In the face of the current climate crisis, the spotlight on millets as a cornerstone of sustainable societies has intensified. The Role of Women in Cultivating Sustainable Societies Through Millets delves into the significance of millets in sustainable agriculture and the indispensable role that women play in their cultivation and promotion.

This book seeks to advocate for the cultivation and consumption of millets as a pathway to constructing sustainable societies. However, it goes beyond mere cultivation, as it critically recognizes and highlights the pivotal role of women in this transformative process. Drawing from millennia of historical perspectives, the book examines how women have been key agents of change in millet cultivation, shaping its cultivation techniques, preservation, and the propagation of traditional knowledge across generations.

Chapter 1 discussed about the millets can lead to a sustainable and inclusive future. Millets are small-seeded grasses with great nutritional value and environmental resistance, surviving drought and flood. Millets were the first cereal grain cultivated, making it one of the oldest agricultural cereals. From the historical records, millets were the first cereal grain to have been domesticated, making it one of the oldest farmed grains in the world.

Chapter 2 study deals with Women Empowerment in agriculture that reflects increased women's decision-making authority related to agricultural resources, management and production, and income; which will enhance the ability of women farmers to strengthen their economic status. The novelty of this study focuses to address challenges in sustainable production and marketing in millets promotion that impede inclusive growth of women farmers. The findings of this study advocate having an integrated approach to enhance millet value chain efficiencies to develop millet promotion, considering economic, environmental and health sustainability through women empowerment.

Chapter 3 main goal is to design a Deep learning based MoblileNet-V3 model for the detection and classification of histological images. This study will help the researchers and medical practitioners to compare the malignancy status before and after the consumption of millet.

Chapter 4 demonstrates how crop yields can be identified and predicted using regression models based on rainfall patterns and millet suitability in specific Indian states. The findings showcase how monitoring techniques can empower farmers in various regions to achieve higher yields, sustainable food production, and improved crop quality along with addressing the limitations or challenges encountered during the research process.

Chapter 5 explores the profound relationship between women, millets, and their impact on health, empowerment, and sustainable agriculture. Millets, ancient crops with exceptional nutrition, have evolved alongside women's roles in cultivation. Throughout history, women have been key guardians of millet farming, preserving traditional knowledge in challenging environments. Women drive the resurgence of millet cultivation, advocating for environmental benefits and promoting diverse varieties for food security. Millets' nutritional advantages support digestion, immunity, and health for senior citizens and individuals with dietary restrictions. They are vital for pregnant and lactating women's health and offer significant economic opportunities, fostering women's financial independence. Empowering women with crop monitoring and management improves yields and food security. Millets emerge as potent tools for women's empowerment, health, and economic autonomy, contributing to a sustainable and inclusive future.

Chapter 6 discussed about the millet boosts immunity, detoxifies the body, reduces cancer risk, increases energy, strengthens the brain and muscles, and improves respiratory immunity. Compared to processed grains, millets are more nutrient-dense. Millets contribute to a well-balanced diet and play a significant role in sustainable nutrition. Lack of awareness about the nutritional content of existing dietary patterns and unwillingness to adapt eating habits are two major contributors to poor production and consumption. Millets are no longer extensively grown, and as a result, there are barely any of them on the market. Since 2003, several studies have been performed to assess the millet's increasing demand. This study's objective is to compile marketing mix strategies from those studies and then provide an intelligent perspective on such strategies.

Chapter 7 found that the OMM has remarkable contribution for reviving and encouraging the production, consumption and processing of millet. This is also helpful for improving the employment and livelihood among the tribal in the Koraput district. But Government also need to aware the tribal people reading the benefits of such potential programmes to further enhance the livelihood opportunities.

Chapter 8 are generally categorized into major and minor millets, ICAR-IIMR has been leading their efforts for a decade now, In the development of the value chain through various valuable interventions through food processing techniques, brand they are healthy and convenient products.

Chapter 9 review highlighted the India's potential (19% productivity share at global level) in millet production, its nutritional goodness as well as recent traits in bakery product development is also elaborated to broaden the horizons of thrust research areas of study.Not only nutrient goodness but also phenolic and bioactive components associated with these nutrigrains could made the researchers to actergorized as superfood. non-gluten nature associated with these arid gain is the worthy source for nutrient enrichment and fortification in various processed bakery products; its Low Glycemic Index (GI) based nature also confirms millets significant role in curing various disorders, making it to categorised as healing grain.

Chapter 10 intend to aid security by bringing up a smart security system for women safety. This system involves an EEG sensor, Heartbeat sensor, GPS sensor, GSM module along with an Arduino. This proposed system purely depends on the psychology of a woman which deals with measuring of brainwaves and heartbeat rate within a given limit. And on further exceeding the data's will be detected depending on the level of waves emitted. These values would trigger the GSM to receive the data and thereby it sends an automatic popup alert in the form of message along with the location. The entire system will be activated to safeguard woman in all aspects.

Chapter 11 discussed about the Millets are highly nutritious and excellent source of protein, fiber, and vitamins. They are also gluten-free, making them a valuable food source for people with celiac disease or gluten intolerance millets are a nutritious and healthy food that offers a variety of health benefits. Incorporating millets into your diet can help improve overall health and well-being.

Chapter 12 will examine the Indian government's policy towards developing a millet economy amidst the impacts of polycrisis situation experienced around the world, to empower rural women in the agricultural sector focussing on the economic dimensions, to bring about social transformation and economic significance of millet farming, value adding, improving livelihood, and social transformation for sustainable development through climate smart millet economy in India.

Chapter 13 identifies the constraints and opportunities for women to participate in millet farming, as well as the effects of millet farming on their lives and means of subsistence, using in-depth interviews, focus groups, and surveys. According to the research, although cultural expectations and a lack of resources present substantial obstacles for women, their participation may be facilitated by establishing women's groups and cooperatives, access to training, and support from NGOs and government initiatives. In addition to proposing steps to increase women's access to resources and support gender-sensitive agricultural extension services, the study highlights the potential of millet cultivation for advancing sustainable agriculture and gender equality.

Raghvendra Kumar
GIET University, India

Ishaani Priyadarshini
School of Information, University of California, Berkeley, USA

Chapter 1
Implications of Nutricereals on Consumers

Usha Prabhu
Jain University, India

N. Sathyanarayana
https://orcid.org/0000-0002-4185-7751
Jain University, India

Madireddi S. S. V. Srikumar
Jain University, India

Nazneen Sayyed
Smt. Kamala Mehta VWA College of Commerce, India

ABSTRACT

The General Assembly of the United Nations (UN) has declared 2023 as the 'International Year of Millets' to promote millets' nutritional, health, environmental, and economic advantages. Millets' unique qualities and advantages have made them important contributors to numerous sustainable development goals (SDGs). Prioritizing millets can lead to a sustainable and inclusive future. Millets are small-seeded grasses with great nutritional value and environmental resistance, surviving drought and flood. Millets were the first cereal grain cultivated, making it one of the oldest agricultural cereals. From historical records, millets were the first cereal grain to have been domesticated, making it one of the oldest farmed grains in the world.

INTRODUCTION

The United Nations (UN) General Assembly has declared the year 2023 to be the "International Year of Millets." "Millets are fantastic, nutrient-dense, ancient crops," said Q.U Dongyu, Director of the Food and Agriculture Organization of the United Nations. Millets have the potential to significantly contribute to a wide range of socioeconomic and social goals, including the empowerment of smallholder farm-

DOI: 10.4018/978-1-6684-9819-4.ch001

Copyright © 2023, IGI Global. Copying or distributing in print or electronic forms without written permission of IGI Global is prohibited.

ers, the eradication of poverty, the adaptation to climate change, the promotion of biodiversity, and the transformation of agri-food systems (Banerjee, 2022).

The concern is on the nutritious value that is attached to millets, In recent years, the growth prospects of the millets market have been helped significantly by the rising interest that is being shown in revitalising the consumption of millets across a variety of countries. In addition, several programmes are currently being done to increase millet production and consumption to lower the risk of health complications carried on by diabetes, obesity, and cardiovascular illnesses. It has been essential for the growth of the millet sector that millet suppliers and enterprises that manufacture food and beverages work together to develop unified strategies for meeting the demand for millet products.

Many inhabitants have relied on millets as their primary food source for centuries. Millets are home-grownup crops in many counties of the world- Asia & Africa, although they are believed to have originated in tropical western Africa through evolutionary processes. Millets have been grown in Africa and East Asia for the past 10,000 years. India is one of the biggest producers of millets in the world. Millets are a complex group of cereals that include fonio, sorghum (also called "large millet"), proso, foxtail, barn-yard, teff, Kodo, browntop, pearl, small, finger, and Guinea millets, among others. Millets are resistant to climate change and may thrive in dry environments with little to no additional water or fertiliser. As a result, they are a perfect alternative for nations that wish to enhance their level of self-sufficiency and decrease their dependency on cereal grains imported from other countries. The ability of millets to help farmers adapt to and survive in the face of climate change, malnutrition, and other major challenges is attracting increasing interest from governments and businesses. India has announced its intention to make the International Year of Millets 2023 a "People's Movement" and to make the country the "Global Hub for Millets."(Sahoo & Mahapatra, 2023). Agriculture, forestry, and fisheries contributed 73% growth in real terms to global value-added between 2000 and 2019, which amounted to USD 3.5 trillion in 2018. (*FAO Statistical Yearbook 2021 - World Food and Agriculture*, 2021)

Consumption of millet around the world is down 0.9% from a year ago, but it's forecast to rise be-tween 2019 and 2024. There will be more than $12 billion worth of millets sold by 2025, up from more than $9 billion now. As a result of supportive government measures, the worldwide millets market is expected to expand rapidly between 2019 and 2025. The reason is that there exists a greater propensity among the urban population in Asia and the Pacific region to consume nutritious food.

India has become a permanent representative to the United Nations, Ambassador TS Tirumurti said "To improve production efficiency, get more money for research and development, and make more con-nections in the food industry, it is important to make consumers, manufacturers, and decision-makers aware of the nutritional and environmental value of millets." Now, Millets have been experiencing a resurgence in global production and consumption due to their nutritional benefits, climate resilience, and sustainability.

HISTORY OF MILLETS

(*A brief history of millets*, 2022) Millets may have been the first cereal grain to be domesticated, making it one of the oldest farmed grains in the world. There exists ambiguity on dates of cultivation among thinkers – historians, archaeologists, and agricultural scientists. (*The Story of Millets*, 2018). Millets be-come a globally important staple crop after being farmed in Asia and Africa for over 5,000 years (*About Millets*, 2022). Several ancient Indian texts mention millets, and artefacts from Harappa and Mohenjo-

Daro have shown that millets were grown there. In many regions of India, China, Japan, Korea, Russia, Turkey, Russia, Ethiopia, and other countries, traditional cuisines still include a significant number of millet-based dishes. Some of the oldest Yajurveda manuscripts in India talk about millets. They name Barnyard millet (aanava), foxtail millet (priyangava) and black finger millet (shyaamaka). This indicates that millet eating was quite prevalent, predating the Indian Bronze Age by around 4,500 years B.C. In China, the Han dynasty (206 BC–220 AD) laid the groundwork for the extraordinary development of wheat products. Up until the Han dynasty, the Chinese diet consisted of millet mush and rice that had been boiled. China became known for stuffed and unstuffed dumplings, noodles, breads, and increasingly intricate wheat recipes after the Han era. Millet's drought-resistant growth characteristics made it an essential staple crop in African and Asian cultures until modern irrigation methods were developed. It was crucial even after wheat and rice farming were perfected.

Proso millet, a staple crop eaten all over the world, is said to have been first planted in Asia and Africa about the same time in the Neolithic period (more than 7,000 years ago). Millets have been a mainstay in dry Asia and Africa for millennia. Northern China was dry and frigid, making it an ideal location for millet cultivation. Millet was a staple meal in north-central China by 5000 BCE and all of northern China by 3000 BCE. Among the first and most extensively produced grains is millet-foxtail (Setaria italica). Rice and wheat remain crucial to East Asian semiarid diets notwithstanding their growth (China, Japan, Russia, India, and Korea).

Millet was initially domesticated in China and Mongolia, but it swiftly spread across East Asia, even to high-altitude locations such as the Tibetan Plateau. China and Mongolia are considered to be the birthplaces of millet. The cultivation of proso millet on the Tibetan Plateau appears to have undergone substantial adjustments before finally being abandoned in Eastern Tibet, according to the evidence. On the Tibetan Plateau, wheat and barley eventually came to dominate the market, displacing proso millet.

Millets are exceptionally resilient crops; certain types of millet, such as proso millet, are ready for harvest after only seventy days. This adaptation to short cultivation durations is probably what caused this grain to become the staple food for nomadic societies across Central Asia. As these tribes travelled from one location to another, the grain spread throughout the region. It is one of the earliest cereal grains domesticated by humans, and is making a resurgence because of rising interest in their nutritional benefits, rising concern for the environment, and the urgent need to adapt our food systems to climate change. The nutritional value of these grains is high, and they may be a key factor in our joint efforts to end hunger, adapt to climate change, increase biodiversity, and reform our food production systems.

Millets are among the crops that are the least labour-intensive for farmers. Millets are known as "the crop for the lazy farmer" in many parts of the world. There are only two things that need to be done: the first is to go out to the field and scatter the seeds, and the second is to come back after three months and harvest the grains. Millets can grow in very poor soils, don't need deep ploughing, can be sowed with little equipment or machines, and only need one weeding to be harvested. A successful harvest is likely if there are enough good rains at the right times. Even if the rains are not as plentiful as they should be, farmers can still get a harvestable crop if they use some updated methods and procedures.

According to the writings of Herodotus, a historian from Greece who lived between the years 485 and 420 BC, millet was cultivated in both Greece and Persia during his day. The Romans referred to millet as milium and created a porridge called puls, which was comparable to polenta and was named after them. The explorer Marco Polo reported the following about the food that was available during the reign of Genghis Khan: "They never go hungry since instead of bread they boil rice, foxtail millet, or proso millet for their staple foods." The emperor of Western Europe at the time, Charlemagne, gave orders for the storage and consumption of millet during the season of Lent. Millet, not wheat, was the primary grain used for human consumption in Europe during the Middle Ages, and it was farmed on a much larger scale than wheat.

Archaeological sites in Europe have yielded remnants of proso millet dating back to the latter part of the third millennium B.C., while foxtail (Italian) millet has been farmed in Europe since the Bronze Age, roughly 2000 B.C. During the time of the Roman Empire, about 753 BC to 610 AD, the millets proso and foxtail were utilised.

Millets may live on the earth's leftover moisture and dew, and some can flourish in the third season, winter, given the correct soil and topography. This is possible because millets are able to extract moisture from both the soil and the air. It can play an important role through collective efforts and empower smallholder farmers.

ECONOMIC DIMENSIONS OF AGRICULTURE

The world's population is growing at an exponential rate, thereby growing the need for food, pressing more research into crops like millet. Millets are small-seeded grasses that thrive with little soil fertility and water, millets are considered resilient crops that can withstand high climatic stress, such as drought and flood, and they are perfect crops for the 21st century when we are confronted with depleting natural resources and an era of climate change.

The majority of millet species are original to Africa and were later farmed in other countries in the world. Millets are farmed in 93 countries, however only 7 have over 1,000,000 hectares (Meena et al., 2021). Countries like India, Nigeria, and China account for more than 55% of worldwide millets output through their extensive cultivation of the crop. It is a traditional crop with exceptional nutritional content and health advantages, Millet products are becoming popular with consumers as the prevalence of lifestyle disorders, particularly in urban areas.

Millets were included to the national food security mission in 2018 as "nutri cereals." It was budgeted to get Rs 300 crore for its expansion. Growing millets can help sustain the lives of smallholder farmers and create opportunities for young people and women to find work. The generated revenue has the potential to boost the economy of any country.

TYPES OF MILLETS

Millets come in major and minor kinds. Foxtail millet, finger millet, tiny millet, proso millet, kodo millet, barnyard millet, and brown top millet are small millets. Many African countries cultivate fonio and tef millets.

Sometimes called "famine crops," millets are the only crops that can be reliably harvested even in the worst of food shortages. These crops were sometimes called orphan crops because they are planted last and have lesser market demand and returns. These underused crops are vital to the poor in many places' subsistence, food security, and nutritional security, and their diversity diversifies the food basket.

Figure 1. Different types of millets

Millets are rich in fibre and starch, which minimise the risk of diabetes and other diseases. Millets nourish our gut prebiotic microorganisms.

Sorghum Millet

After rice, sorghum (jowar), wheat, barley and maize is the fifth most extensively produced cereal crop. The major use of sorghum in India is as food, as nutritional animal feed. In addition to various industrial uses, such as the production of drinkable alcohol, the grains are used as feed for poultry. One of the most water- and solar-efficient crops is sorghum, sometimes called the C4 plant. This plant may be used to generate both food and biomass. The crop may be cultivated in a variety of environmental situations due to its innate ability to tolerate drought. Sorghum is a dual-purpose crop in semi-arid settings because its grain and stover are valued for human and animal use.

Figure 2. Sorghum field at seed set stage

Pearl Millet

Bajra, or pearl millet, has the best potential for drought resilience of all of the millets and is the world's sixth-largest crop in terms of both acreage and output. As the third most extensively produced crop in India, pearl millet follows rice and wheat.The current grain production is 9.5 million tonnes, and it is farmed over more than 9.3 million acres. When hybrids of 80-85 days maturity are planted in summer crops under irrigated and high fertility conditions, pearl millet can yield 4-5 t/ha. It reaches heights of 6-15 feet and stands straight up. The inflorescence of this plant looks like a thick panicle of brownish-coloured spikes. Pearl millet, or "Bajra" as it is known in India, is a very nutritious grain that is abundant in protein, fibre, phosphorus, magnesium, and iron. Pearl millet's high mineral and protein content gives it several health benefits.

Figure 3. Pearl millet field at seed set stag

Finger Millet

Finger millet is essential for rural Southern Indians and East and Central Africans. Ragi, or finger millet, may be grown at sea level and in the Himalayas' steep areas. A mature plant can grow to 150 cm tall, and its tiny mustard-colored, light brown, or white seeds are incredibly small. The crop matures in 3–5 months, depending on cultivar and growing circumstances. Finger millet grains provide high-quality protein with tryptophan, cystine, methionine, fibre (10-15%), phytochemicals, calcium, and other minerals. The malting and nourishing grain is excellent. Finger millet is a major crop because it resists storage insect infestations.

Figure 4. Finger millet field at seed set stage

Foxtail Millet

The third-largest millet crop, foxtail millet is farmed for food in Asia's semi-arid tropics and forage in Europe, Australia, North America and North Africa. Foxtail or Italian millet may be underutilised for grain production. It develops a thin, upright, leafy stem that can grow to a height of 1 to 5 feet. The grains' grain structure is quite similar to that of paddy rice. To use them, the outer husk must be removed from them. The Chinese assert incredibly large yields that occasionally approach 11,000 kg per acre. It can withstand drought, thrives at higher altitudes (up to 600 feet), and is frequently cultivated as a crop grown in place of sorghum on soils with black cotton when there is insufficient rainfall. The majority of the crop is grown in China, but some Italian millet is also planted in Russia, Japan, and India. It is mostly sown for fodder in the USA.

Figure 5. Foxtail millet field at seed set

Proso Millet

Arid regions of Asia, Europe, North America, Africa and Australia cultivate proso millet in short seasons. The quick-season, emergency-irrigated crop common millet or broom corn requires minimal moisture. Numerous soil types and climatic conditions are suitable for proso millet. With its drought resistance, proso millet is ideal for places with little water and long dry spells. Proso, which matures 60 to 75 days after sowing, is a short-season crop in comparison to all millets. It's often grown as a summer crop that was late-seeded. Three to four feet is the maximum height of proso millet. The centre of its little panicle droops. Its round seeds have a smooth, shiny shell that is about 1/8 inch across. The colour of the seeds might be white, yellow, orange-red, or brown. This millet was grown in China, Northern India, Russia and the Balkans until rice and other grains replaced it.

Figure 6. Proso millet field at seed set stage

Barnyard Millet

India, China, Japan, and Korea produce barnyard millet for food and fodder. Japanese and Indian millet tolerate various soil and humidity conditions. They grow well throughout the year and at higher altitudes, although maturation takes three to four months. A 60–130 cm tall erect plant, barnyard millet has spikelets that range in colour from brown to purple. In Eastern Asia, sections of Africa, and India, particularly in the mountainous regions of Uttarkhand, it is produced for both grain and fodder and in the Eastern USA, it has been a valuable pasture crop. The native people of that continent use an ancestor, Australian millet (Echinocloa decompositum), as a cereal grain.

Figure 7. Barnyard millet field at flowering stage

Kodo Millet

Kodo millet, native to tropical and subtropical South America, was grown in India 3,000 years ago. Kodo millet is cultivated on the poorest soils in India, although it is unlikely to be grown elsewhere. It is robust, drought-resistant, and can grow in rocky or gravelly soils that other crops cannot. Compared to other millets, which mature in two to four months, it takes four to six months. Grain might be light red or dark grey.

Figure 8. Kodo millet field at flowering stage

Little Millet

Little millet, which is an important component of the diet of tribal people in Sri Lanka, Nepal, and Myanmar, was reclaimed in the Eastern Ghats of India. Little millet develops quickly and can tolerate both waterlogging and drought.

Figure 9. Little millet field at seed set stage

Brown Top Millet

Native to India, brown top millet has only just begun to be cultivated in some areas of Karnataka and Andhra Pradesh, India uses it mostly as a crop for food. Even on poor sandy loam soils, it develops in 60–80 days. Up to 100,000 acres of hay and pasture are planted annually in Georgia, Florida, and Alabama. The seed is used to feed game birds like quail and doves. It grows more quickly than pearl millet and has finer stems as a fodder crop.

Figure 10. Seed-set brown top millet field

Tef Millets

Tef is a cereal that grows mostly in the highlands of Ethiopia and Eritrea (at altitudes of 2700 metres or above). Tef, which is more closely related to grasses than to other cereal grains, is adaptable to many environments. It produces small seeds that are exceptionally healthy due to its high protein content. Tef is ground into flour and used primarily in the production of injera, a type of fermented flat bread that is a staple in the Ethiopian diet. More than 2 million acres are devoted to its cultivation in the African country of Ethiopia.

Figure 11. Tef millet at flowering stage

Fonio Millet

A small amount of fonio is grown, especially in West Africa on plateaus or intermediate elevations with thicker soils. According to an investigation of the protein fractions, fonio grain has a very good nutritional quality.

Figure 12. Fonio crop at flowering stage

This little grass, which grows to 30 to 80 cm, is drought-resistant and matures in 70 to 150 days, depending on the species. Unlike other grains, fonio proteins contain methionine. Among other health benefits, fonio helps prevent and treat diabetes.Both the cereal's gastronomic and nutritional qualities are top-notch. The tiny and challenging-to-peel grains are the only downside, making processing time-consuming.

MILLETS' NUTRITION AND HEALTH BENEFITS

Nutrition serves as a crucial building block for healthy development, yet this fact is often overlooked. Lancet Commissions (Willett et al., 2019) acknowledges the need to identify diets that are both healthy and environmentally sustainable and to increase the use of underutilised plant species like sorghum, teff grains, quinoa and millets, which are both climate resilient and nutritional powerhouses. Millions of people around the world rely on cereal grains millet for sustenance as these are nutritious and energising, these are a good source of several essential nutrients such as protein, fats, vitamins, fibre, minerals and polyphenols. Given that nutritional health is the primary factor in determining the degree to which human potential can be developed and realised (Radhika et al., 2011).

Finger, Kodo, Pearl, Proso, Little, Foxtail, and Barnyard millets feed millions worldwide. Millets are high in protein, minerals, and phytochemicals, like the primary grains.The anti-oxidant concentration and activity are both impacted by the various processing procedures, including soaking, malting, decortications, and cooking (Saleh et al., 2013). Millets, also called "Nutri-cereals," are renowned for their high nutrient content, the millets contain 7-12% protein, 2-5%fat, 65-75% carbohydrates and 15-20% dietary fibre. (*Nutritive value of foods*, 2020)

Pearl millet contains proteins (12-16%), lipids (4-6%), important fatty acids, dietary fibre, B vitamins, minerals including calcium, potassium, iron, zinc and magnesium, and other nutrients. Millet grains are rich in folic acid, riboflavin, niacin and thiamine,. Millets may prevent age-related illnesses. Millet protects heart disease, diabetes, digestive issues, cancer, detoxification, immunity, respiratory health, vitality, muscular and nervous system strength, and degenerative illnesses including Parkinson's and metabolic syndrome (Shull et al., 1990).

Table 1. Comparison of millets and cereals' nutrition (per 100g)

Grains	Sorghum	Pearl millet	Finger millet	Little millet	Maize	Wheat	Rice
Energy (kcal)	334	363	320	329	334	321	353
Protein (g)	10.4	11.6	7.3	8.7	11.5	11.8	6.8
Carbohydrate (g)	67.6	61.7	66.8	65.5	64.7	64.7	74.8
Starch (g)	59	55	62	56	59	56	71
Fat (g)	1.9	5	1.3	5.3	3.6	1.5	0.5
Dietary Fiber(g)	10.2	11.4	11.1	6.3	12.2	11.2	4.4
Minerals (g)	1.6	2.3	2.7	1.7	1.5	1.5	0.6
Ca (mg)	27	27	364	17	8.9	39	10
P (mg)	222	296	283	220	348	306	160

Source: Indian Food Composition Tables and Nutritive Value of Indian Foods

Millets are good for preventing a wide range of diseases, such as high blood pressure, heart disease, cancer, and tumours, because they are rich in nutrients. The millet-consuming population has been linked to lower rates of diabetes (Rajasekaran et al., 2004). Many companies are growing as a result of the rising demand for newborn food and nutrition products made with millets. There has been an alarming rise in the number of people with diabetes, obesity, and heart problems like coronary artery disease, heart attack, arrhythmias, etc. Millets are a great way to keep from getting sick because they are high in protein and minerals like calcium, iron, etc.

MILLETS AND ENVIRONMENT

Millets are eco-friendly, inexpensive, and nutritious. Millets can survive high temperatures, have a low carbon and water impact, and flourish on poor soils without extra inputs. In addition, millets are photo-insensitive, which means they are resistant to the effects of climate change. Millets are eco-friendly crops because they don't need as much water or pesticides as other grains. It has been suggested that increasing millet production and consumption can be a wonderful strategy to improve sustainable agriculture and lessen the environmental impact of farming. One of the reasons why millets are being heralded as the "new age global grain". People today are increasingly concerned about having a healthy lifestyle to combat metabolic disorders and lifestyle diseases.

Environmental Benefits:

1. **Millets Are Drought-Resistant Crops:** Millets can thrive in lower-quality soil. Millet is favoured in India's dry and semi-arid regions because it can withstand conditions that would kill more common crops like wheat and rice.
2. **Millets Require Less Water to Grow:**.According to projections for the year 2040, the industrialised nations will face the greatest threat to their water supply. Once again, millets appear to be a viable solution. The water needed to grow millets is 70-80% less than that needed to grow wheat or rice. For this reason, millets are a viable crop option for areas with limited water resources.
3. **Millets Offer Natural Resistance to Pests:** Studies have linked high levels of pesticide exposure to serious illnesses like leukaemia and non-Hodgkin lymphoma. In this regard, millets also perform admirably. Millets provide inherent protection against insects and pesticides. They may be grown with fewer chemicals, making them a good choice for supporting organic farming.
4. **Millets Are Great for Soil Health:** Millets have a zero carbon footprint since they absorb as much carbon from the environment as they release during production (0.1 to 0.2 CO2eq kg per kg), but rice and wheat both leave behind a larger carbon footprint (0.4 and 0.35 CO2eq kg per kg, respectively). Millets are energy-efficient crops because they require less fertiliser and irrigation, totalling 0.48 to 0.71 kWh per kilogramme of production, whereas rice requires 1.05 kWh per kg.

MILLET BASED BUSINESS

(SolutionBuggy, 2022) Millets, which were once considered to be the food of the lower classes, have recently become a significant component of the diets of the wealthy class. A government report predicts that the international market for millets will expand at a Compound Annual Growth Rate (CAGR) of 4.5% from 2021 to 2026. Businesses are starting to take notice of the growing demand for millets as a result of the government's decision to promote them heavily. The proportion of the global market held by each region, including Asia-Pacific, North America, Africa, and Europe, amongst others, was increasing at a favourable rate. Demand for items made from millet has been growing in metropolitan areas over the past several decades, this presents a significant opportunity for new business owners to enter the market. The value of the millets market is expected to rise from its present level of $9 billion to more than $12 billion by the year 2025, reflecting the annual global production of 25 million tonnes of millets. Encouragement from various governmental bodies to expand the size of the international millets market between 2019 and 2025. Additionally, millets contains calcium, iron, and fibre, all of which contribute to the fortification of vital nutrients that are necessary for the healthy growth of children. In addition to having a higher protein content than rice, millet is an excellent source of essential nutrients such as vitamins A and B, iron, phosphorus, magnesium, and manganese. Millets are being used in an increasing number of infant foods and other nutrition products, and many manufacturers are extending their business operations by purchasing smaller firms to meet the rising demand. Breweries have recently begun offering gluten-free beer varieties to broaden their customer base. It is used in beverages such as beer. Gluten-free beers are brews that have not been contaminated with gluten during the brewing process. These beers were developed specifically for people who have been diagnosed with celiac disease or gluten intolerance.

The processing of organic millet into a variety of food products that are "ready to cook and eat" (RTC, RTE) has been recognised as a marketable possibility by enterprising individuals who have recognised this trend.

The ever-increasing demand in this industry reaches a new high when these products are made available for rapid cooking and are already prepared for consumption.

The distinctive reasons why products made from millet are poised to achieve ever-increasing popularity in the processed food sector.

1. A powerhouse of rich nutrient
2. Gluten-free
3. It has Dietary fiber and resistant starch & micronutrients
4. Easy to cròp and process for the end user
5. It is good for the planet as it is climate-resilient crops
6. It becomes a part of ready to cook & eat- reduces cooking time

A recent study revealed that the combination of millets and legumes produces a complete protein that is easy to digest and rich in a variety of essential elements.

Proteins derived from plants and alternatives to meat are quickly gaining popularity in international markets.

The Indian Institute of Millet Research has recognised 200 new millet industry entrepreneurs. Products with added value are being developed by the startups, and a good number of them have already begun exporting in limited numbers. Millets can assist the FMCG sector amid interruptions in global food grain supply chains, such as those caused by the Ukraine war or severe climate conditions, such as irregular monsoon rains that reduce wheat and rice production.

UNIQUE SELLING PROPOSITION (USP) OF MILLETS

1. Millets are the next smart food because it is packed with nutrients and energy. They are the powerhouse and the Government of India has named it Nutri-Cereals.
2. Nice source of energy as it constantly releases energy.
3. Millets are gluten-free.
4. Millets contain a good profile of micronutrients which is divided into two categories, Type-1 and Type-2.
    - Type-1 which are the functional nutrients that can be stored in the body like iron, calcium, manganese, copper, iodine, and a certain amount of Vitamin A, D, and K which is helpful for cellulose metabolism.
    - The type-2 are the growth nutrients (growth promoters) like Nitrogen, phosphorus, Zinc, Magnesium, and Potassium.
5. Millets contain a balanced amino profile. The protein digestibility can be increased by proper processing and malting.
6. Millets do not contain trans fat.
7. Millets contain fibre and resistant starch which work as pre-biotics.
8. Millets are used in the pharmaceutical industry.
9. Millets are alkaline.
10. Millets come under the therapeutic category.

11. Good for the farmers as it is suitable for dryland agriculture requires fewer inputs that requires less water and is naturally organic.
12. Good for the planet as it is a climate-resilient crops.

The Unique Selling Proposition (USP) of Millets can be produced in the target groups and demographic profiles where we can use millet as the key ingredient in the food-based approach. We all should be committed to bringing about a reduction in malnutrition in the world.

STUDIES ON MILLETS

(Pragya & Rita, 2012) believes that the problem of widespread malnutrition can only be addressed by prioritising the highest possible standard of nutrition for the world's population. Increasing yields of millet and household techniques are important, but it is also important to foster diversification of food production at the national and home levels by the consumers.

(Mal, Padulosi, & Bala Ravi, 2010), thinks that different types of millet can be used to produce a wide range of healthful, high-value food products that appeal to both traditional and non-traditional millet consumers.

Using millets as "food medicine" was demonstrated by (Sarita & Singh, 2016), as antioxidants which include phenolic acids and glycated flavonoids which are found in millet. Millet foods have been labelled as prebiotics because they can boost the survivability of probiotics, which may have positive health effects on humans.

(Coulibaly, Kouakou, & Chen, 2011), undertook a study on millets and concluded that millets are rich in the mineral magnesium, which has been shown to lessen the severity of headaches and heart attacks. Millets have a high concentration of the cholesterol-lowering phytochemical phytic acid.

Millets are rich in vitamins and minerals, with 60–70% carbs, 7–11% protein, 1.5–5% fat, and 2–7% crude fibre. It's rich in B vitamins, magnesium, and antioxidants. Millet contains manganese, phosphorus, iron, and other minerals. Millet proteins include all necessary amino acids except lysine and threonine and are rich in methionine and cysteine (Singh, Mishra, & Mishra, 2012).

(Obilana, 2003) The research states that due to a growing population and worldwide food demand, it is necessary to study plants like millets that may be cultivated locally and eaten by low-income households in India and the Sahel zone.

(Gupta, Srivastava, & Pandey, 2012) discovered that millets' nutraceuticals lower the risk of high blood pressure, heart disease, cancer, and tumours. Adding roughage to the digestive system and delaying gastric emptying are two additional health benefits.

(Sarita & Singh, 2016), focused on millet, which is a high-energy, nutrient-rich meal that can help prevent and treat diseases like obesity, diabetes, Cardio Vascular Disease, and other types of diseases. Millet also helps to reduce malnutrition and nourish the general population. Gluten-free food is millet.

Millets are distinctive due to their high levels of protein, calcium, fibre, and polyphenols (Devi et al., 2014). Millets may be processed due to their high-quality grain. Grain is watered, dehulled, and milled before additional processing (fermentation, malting, extrusion, popping, and roasting). As a staple and household product, its processing must include both traditional and industrial levels, including small and large organisations.

The nutritious richness of millets necessitates research into the unique nutritional and functional qualities of various millet cultivars and the creation of millet-based products with high added value (Amadou, Gounga, & Le, 2013).

Millets are a good source of vitamins and minerals, with 60–71% carbohydrates, 7–11% protein, 1.5–5% fat, and 2–7% crude fibre. They have plenty of vitamin B, magnesium, and antioxidants. Millet has enough manganese, phosphorus, and iron. Millet proteins are rich in sulphur-containing methionine and cysteine, giving them a good source of all necessary amino acids (Singh, Mishra, & Mishra, 2012).

MILLETS TODAY (GLOBAL CONTEXT: PRODUCTION AND CONSUMPTION)

Millet's drought-resistant growth characteristics made it an essential staple food in Africa and Asia long before modern irrigation methods were developed. Millet is still the main staple food for more than a quarter of the world's population. Millets are mostly cultivated in Africa and Asia's dry and semi-arid areas.

The world's 76 million cultivated hectares of millets yield 95 million metric tonnes of food grains in addition to the 300 million metric tonnes of cereal grains. Both sorghum and pearl millet take up a significant amount of land and are responsible for the production of more than 90% of all millets.

More than 40% of the world's millet supply is used by countries in Africa, primarily the nations of Niger, Mali, Nigeria, and Burkina Faso. Millets are particularly common in impoverished countries and regions, such as India and Africa, where ensuring adequate food and nutrition is a primary concern. Millet prices on the international market are notoriously unstable. They are not usually tied to the prices of other significant coarse grains like maize, sorghum, or barley, and are instead largely driven by the volume of available supply.

In Niger, millet makes up about 75% of the total cereal grain intake, while in most other nations in the Sahel region, it makes up more than 30%. In addition, it plays a important role in the economies of Namibia (where it accounts for 25% of total cereal food consumption) and Uganda (where it accounts for 20%).

MILLETS: THE WORLD OUTLOOK

Table 2. Area and production of millets by region (2019)

Regions	AREA(lakh hectares)	Production (lakh tonnes)
India	138	173
Europe	8	20
America	53	193
Africa	489	423
Asia	162	215
Australia & New Zealand	6	12
World	**718**	**863**

• India makes more than 170 million tonnes, which is 80% of Asia's total and 20% of the world's total.
• Average yield around the world: 1229 kg/ha, India: 1239 kg/ha.
(Source: FAO Stat 2021, Ministry of Agriculture & Farmers Welfare)

Table 3. CAGR of millets production in Africa

Area	Element	2007 to 2011 (%)	2012 to 2016 (%)	2017 to 2021 (%)	Overall Result 2007 to 2021 (%)
Central African Republic	Yield	3.13	-19.26	11.28	0.88
Central African Republic	Production	-9.24	-16.22	15.38	-0.34
Eastern Africa	Yield	-0.34	1.00	0.39	-1.04
Eastern Africa	Production	10.06	4.91	0.84	6.39
Middle Africa	Yield	1.07	-0.49	-0.08	1.41
Middle Africa	Production	7.90	2.71	4.77	7.66
Northern Africa	Yield	1.55	0.02	0.08	1.26
Northern Africa	Production	7.37	20.67	7.17	9.75
Western Africa	Yield	10.42	-13.35	3.42	4.18
Western Africa	Production	19.25	-5.57	-0.18	11.18
Africa	**Yield**	**3.01**	**-6.04**	**-0.17**	**0.66**
Africa	**Production**	**10.96**	**2.64**	**2.60**	**8.39**

Source: Authors Computations

Table 3 shows the CAGR of millets production in different regions of Africa from 2007 to 2021. The table can be interpreted as follows:

- The overall result for Africa from 2007 to 2021 shows a CAGR of 0.66% for yield and 8.39% for production.

- The region with the highest CAGR for yield is Western Africa with 4.18%, while the region with the lowest CAGR for yield is Eastern Africa with -1.04%.
- The region with the highest CAGR for production is Western Africa with 11.18%, while the region with the lowest CAGR for production in Central African Republic with -0.34%.
- The period with the highest overall CAGR of millets production in Africa was 2007 to 2011 with 10.96%, followed by 2017 to 2021 with 2.60%, and 2012 to 2016 with 2.64%.
- The period with the lowest overall CAGR of millets yield in Africa was 2012 to 2016 with -6.04%, followed by 2017 to 2021 with -0.17%, and 2007 to 2011 with 3.01%.

Table 4. CAGR of millets production in America

Area	Element	2007 to 2011 (%)	2012 to 2016 (%)	2017 to 2021 (%)	Overall Result 2007 to 2021 (%)
Northern America	Yield	NA	NA	NA	NA
Northern America	Production	NA	NA	NA	NA
Central America	Yield	3.44	-1.57	-0.90	1.18
Central America	Production	3.86	3.39	-1.92	1.52
Caribbean	Yield	0.01	-0.20	0.10	-0.05
Caribbean	Production	-0.39	-0.28	0.21	-0.09
South America	Yield	-5.66	0.57	-1.04	-1.52
South America	Production	-0.25	2.10	8.59	2.81
Americas	**Yield**	**-1.62**	**-0.07**	**-1.72**	**-0.47**
Americas	**Production**	**1.60**	**2.81**	**3.57**	**2.22**

NA – Data Not available
Source: Authors Computations

Table 4 shows the CAGR of millets production in different regions of America from 2007 to 2021. The CAGR is shown for both yield and production. From the table, we can see that:

- In Central America, the CAGR for yield has decreased from 3.44% from 2007 to 2011 to -0.90% from 2017 to 2021, with an overall result of 1.18% from 2007 to 2021. The CAGR for production has fluctuated, with an overall result of 1.52% from 2007 to 2021.
- In the Caribbean, the CAGR for yield has remained relatively stable, with an overall result of -0.05% from 2007 to 2021. The CAGR for production has also remained relatively stable, with an overall result of -0.09% from 2007 to 2021.
- In South America, the CAGR for yield has fluctuated, with an overall result of -1.52% from 2007 to 2021. The CAGR for production has increased from -0.25% from 2007 to 2011 to 8.59% from 2017 to 2021, with an overall result of 2.81% from 2007 to 2021.
- In the Americas as a whole, the CAGR for yield has decreased from -1.62% from 2007 to 2011 to -1.72% from 2017 to 2021, with an overall result of -0.47% from 2007 to 2021. The CAGR for production has increased from 1.60% from 2007 to 2011 to 3.57% from 2017 to 2021, with an overall result of 2.22% from 2007 to 2021.

Table 5. CAGR of millets production in Asia

Area	Element	2007 to 2011 (%)	2012 to 2016 (%)	2017 to 2021 (%)	Overall Result 2007 to 2021 (%)
Central Asia	Yield	4.91	5.20	5.51	5.21
Central Asia	Production	13.85	13.25	19.58	15.37
Eastern Asia	Yield	3.62	-1.01	1.16	2.29
Eastern Asia	Production	1.36	-10.79	4.00	-1.37
Southern Asia	Yield	0.98	-0.89	0.60	0.73
Southern Asia	Production	2.17	0.84	3.64	1.22
South-eastern Asia	Yield	2.33	0.65	-2.09	-0.25
South-eastern Asia	Production	3.59	0.81	-3.03	-0.51
Western Asia	Yield	2.57	-0.19	0.72	0.49
Western Asia	Production	0.21	-3.30	2.04	-1.25
Asia	**Yield**	**1.67**	**-1.76**	**-0.08**	**0.26**
Asia	**Production**	**2.43**	**-1.81**	**1.31**	**0.03**

Source: Authors Computations

Table 5 shows the CAGR of millets production in different regions of Asia from 2007 to 2021. From the table, we can see that:

- The table shows that Central Asia had the highest CAGR for both yield and production in all periods, followed by Southern Asia. Eastern Asia had a negative CAGR for both yield and production from 2012 to 2016, but recovered slightly in the last period. South-eastern Asia and Western Asia had negative CAGRs for both yield and production in the last period, indicating a decline in millets cultivation.
- The table also shows that Asia as a whole had a positive CAGR for yield and production from 2007 to 2011, but a negative CAGR from 2012 to 2016. The last period showed a slight improvement in production, but a slight decline in yield.

Table 6. CAGR of millets production in Europe and the European Union

Area	Element	2007 to 2011 (%)	2012 to 2016 (%)	2017 to 2021 (%)	Overall Result 2007 to 2021 (%)
Eastern Europe	Yield	**NA**	**NA**	**NA**	**NA**
Eastern Europe	Production	**NA**	**NA**	**NA**	**NA**
Southern Europe	Yield	9.16	1.58	-0.12	2.92
Southern Europe	Production	11.37	22.16	-46.66	-10.21
Europe	**Yield**	**9.16**	**1.58**	**-0.12**	**2.92**
Europe	**Production**	**11.37**	**22.16**	**-46.66**	**-10.21**
European Union	Yield	11.55	1.45	-100.00	-100.00
European Union	Production	13.71	24.51	-100.00	-100.00

Source: Authors Computations

Table 6 shows the CAGR of millets production in different regions of Europe and the European Union from 2007 to 2021. From the table, we can see that:

- In Southern Europe, the CAGR for yield has decreased from 9.16% from 2007 to 2011 to -0.12% from 2017 to 2021, with an overall result of 2.92% from 2007 to 2021. The CAGR for production has fluctuated, with an overall result of -10.21% from 2007 to 2021.
- In Europe as a whole, the CAGR for yield has decreased from 9.16% from 2007 to 2011 to -0.12% from 2017 to 2021, with an overall result of 2.92% from 2007 to 2021. The CAGR for production has fluctuated, with an overall result of -10.21% from 2007 to 2021.
- In the European Union, the CAGR for yield has decreased from 11.55% from 2007 to 2011 to -100.00% from 2017 to 2021, with an overall result of -100.00% from 2007 to 2021. The CAGR for production has also decreased, with an overall result of -100.00% from 2007 to 2021.

Table 7. CAGR of Millets production in the world

Area	Element	2007 to 2011 (%)	2012 to 2016 (%)	2017 to 2021 (%)	Overall Result 2007 to 2021 (%)
Africa	Yield	3.01	-6.04	-0.17	0.66
Africa	Production	10.96	2.64	2.60	8.39
Americas	Yield	-1.62	-0.07	-1.72	-0.47
Americas	Production	1.60	2.81	3.57	2.22
Asia	Yield	1.67	-1.76	-0.08	0.26
Asia	Production	2.43	-1.81	1.31	0.03
Europe	Yield	9.16	1.58	-0.12	2.92
Europe	Production	11.37	22.16	-46.66	-10.21
European Union	Yield	11.55	1.45	-100.00	-100.00
European Union	Production	13.71	24.51	-100.00	-100.00
World	**Yield**	**1.64**	**-4.08**	**-0.23**	**-0.02**
World	**Production**	**5.35**	**0.71**	**2.14**	**3.80**

Source: Authors Computations

Table 7 shows the CAGR of millets production in different regions of the world from 2007 to 2021. From the table, we can see that:

- In Africa, the CAGR for yield has decreased from 3.01% from 2007 to 2011 to -0.17% from 2017 to 2021, with an overall result of 0.66% from 2007 to 2021. The CAGR for production has remained relatively stable, with an overall result of 8.39% from 2007 to 2021.
- In the Americas, the CAGR for yield has decreased from -1.62% from 2007 to 2011 to -1.72% from 2017 to 2021, with an overall result of -0.47% from 2007 to 2021. The CAGR for production has increased from 1.60% from 2007 to 2011 to 3.57% from 2017 to 2021, with an overall result of 2.22% from 2007 to 2021.
- In Asia, the CAGR for yield has fluctuated, with an overall result of 0.26% from 2007 to 2021. The CAGR for production has also fluctuated, with an overall result of 0.03% from 2007 to 2021.
- In Europe, the CAGR for yield has decreased from 9.16% from 2007 to 2011 to -0.12% from 2017 to 2021, with an overall result of 2.92% from 2007 to 2021. The CAGR for production has fluctuated, with an overall result of -10.21% from 2007 to 2021.
- In the European Union, the CAGR for yield has decreased from 11.55% from 2007 to 2011 to -100.00% from 2017 to 2021, with an overall result of -100.00% from 2007 to 2021. The CAGR for production has also decreased, with an overall result of -100.00% from 2007 to 2021.
- In the World as a whole, the CAGR for yield has decreased from 1.64% from 2007 to 2011 to -0.23% from 2017 to 2021, with an overall result of -0.02% from 2007 to 2021.

Table 8. CAGR of millets production in developing countries

Area	Element	2007 to 2011 (%)	2012 to 2016 (%)	2017 to 2021 (%)	Overall Result 2007 to 2021 (%)
Least Developed Countries	Yield	1.83	-0.96	-0.82	-0.20
Least Developed Countries	Production	7.45	5.47	2.84	5.58
Land Locked Developing Countries	Yield	1.74	2.84	0.49	-0.63
Land Locked Developing Countries	Production	7.35	3.94	3.01	5.76
Small Island Developing States	Yield	0.01	-0.20	0.10	-0.05
Small Island Developing States	Production	-0.39	-0.28	0.21	-0.09
Low-Income Food Deficit Countries	Yield	1.59	-0.64	-0.20	0.32
Low-Income Food Deficit Countries	Production	6.90	6.01	3.87	5.78
Net Food Importing Developing Countries	Yield	1.63	-1.01	-0.63	-0.19
Net Food Importing Developing Countries	Production	7.12	5.33	3.24	5.51

Source: Authors Computations

Table 8 shows the (CAGR) of millets production in different areas. The table is divided into two elements: Yield and Production. From the table, we can see that the overall production of millets has increased in all areas except for Small Island Developing States, where it has decreased by -0.09%. The highest increase in production is seen in Low-Income Food Deficit Countries with a CAGR of 5.78%, followed by Land Locked Developing Countries with a CAGR of 5.76%. The yield, on the other hand, has decreased in most areas except for Low-Income Food Deficit Countries where it has increased by 0.32%. The highest decrease in yield is seen in Net Food Importing Developing Countries with a CAGR of -0.19%.

In summary, the production of millets has increased in most areas while the yield has decreased. This could mean that more land is being used for millet cultivation or that more efficient farming techniques are being employed to increase production. However, further analysis would be required to determine the exact reasons for these trends.

CHALLENGES IN MILLET PRODUCTION AND USAGE

(Roy, 2023) Regarding millet production, processing, value addition, marketing, and consumption, the millet industry has been confronted with several obstacles. It is anticipated that a significant barrier to entry for the expansion of the company will be the absence of product recognition by a variety of government agencies over the period.

1. **Low productivity of millets:** Millet productivity is very low when compared with other crops like wheat, and rice. Farmers follow traditional methods while growing millets, if the right set of procedures and practises are carried out, the millets yield gap can be reduced to a manageable level. who utilised modern agricultural techniques, such as the System of Millet Intensification (SMI).
2. **Area covered with Millet is Decreasing:** The amount of land that was used for growing millets in the 1960s is less than half of what it was in the present day, according to our comparison. Millet farming has the potential to boost overall output, particularly if abandoned fields and unused land are brought into use. In addition to this, the farmers who grow millet need to be given financial incentives so that they will be motivated to continue growing millets.
3. **Millet Seed Production:** Farmers are having trouble obtaining millet seeds of sufficient quality. There is potential for the introduction of high-quality seeds of minor millets, and one way to realise this potential is through the establishment of seed hubs for the breeding and production of seeds. This seed production needs to be driven by demand, and all of the players in the seed value chain need to work together to improve the chain as a whole.

CONCLUSION

Millets can be a 'Health Brand'-millet food products are known for their nutritional value, but consumers are generally unaware of their nutritional and therapeutic benefits. Millets can play a transformative role in tackling lifestyle diseases, the benefits of it should be brought to the mainstream and promote millet cultivation to consumption. There will be a huge potential market for millet if proper awareness is created.

REFERENCES

A brief history of millets. (2022). The Millet Foundation. https://themillet.org/a-brief-history-of-millets/

About Millets. (2022). SV Millets. https://svmillets.com/millets/#:~:text=Millets%20are%20one%20of%20the,became%20a%20staple%20food%20worldwide

Amadou, I., Gounga, M. E., & Le, G.-W. (2013). Millets: Nutritional composition, some health benefits and processing-A review. *Emirates Journal of Food and Agriculture*, *25*(7), 501–508. doi:10.9755/ejfa.v25i7.12045

Banerjee, S. (2022). *International Year of Millets: Unleashing the potential of millets for the well-being of people and the environment*. Food and Agriculture Organization of the United Nations (FAO). https://www.fao.org/newsroom/detail/international-year-of-millets-unleashing-the-potential-of-millets-for-the-well-being-of-people-and-the-environment/en

Coulibaly, A., Kouakou, B., & Chen, J. (2011). Phytic acid in cereal grains: structure, healthy or harmful ways to reduce phytic acid in cereal grains and their effects on nutritional quality. *American Journal of Plant Nutrition and fertilization technology, 1*(1), 1-22.

Devi, P. B., Vijayabharathi, R., Sathyabama, S., Malleshi, N. G., & Priyadarisini, V. B. (2014). Health benefits of finger millet (Eleusine coracana L.) polyphenols and dietary fiber: A review. *Journal of Food Science and Technology*, *51*(6), 1021–1040. doi:10.100713197-011-0584-9 PMID:24876635

FAO. (2021). *FAO Statistical Yearbook 2021 - World Food and Agriculture*. Food and Agriculture Organization of the United Nations. https://reliefweb.int/report/world/fao-statistical-yearbook-2021-world-food-and-agriculture#:~:text=In%20fact%2C%20the%20global%20value,percent%20of%20the%20global%20workforce

Gupta, N., Srivastava, A., & Pandey, V. (2012). Biodiversity and nutraceutical quality of some Indian millets. *Proceedings of the National Academy of Sciences. India. Section B, Biological Sciences*, *82*(2), 265–273. doi:10.100740011-012-0035-z

Mal, B., Padulosi, S., & Bala Ravi, S. (2010). *Minor millets in South Asia: learnings from IFAD-NUS Project in India and Nepal.*

Meena, R. P., Joshi, D., Bisht, J. K., & Kant, L. (2021). Global scenario of millets cultivation. *Millets and millet technology*, 33-50.

Obilana, A. (2003). Overview: Importance of millets in Africa. [all cultivated millet species]. *WORLD (Oakland, Calif.)*, *38*(2), 28.

Pragya, S., & Rita, S. R. (2012). Finger millet for food and nutritional security. *African Journal of Food Science*, *6*(4), 77–84. doi:10.5897/AJFSX10.010

Radhika, G., Sathya, R. M., Ganesan, A., Saroja, R., Vijayalakshmi, P., Sudha, V., & Mohan, V. (2011). Dietary profile of urban adult population in South India in the context of chronic disease epidemiology (CURES–68). *Public Health Nutrition*, *14*(4), 591–598. doi:10.1017/S136898001000203X PMID:20701818

Rajasekaran, N., Nithya, M., Rose, C., & Chandra, T. (2004). The effect of finger millet feeding on the early responses during the process of wound healing in diabetic rats. *Biochimica et Biophysica Acta (BBA)-. Biochimica et Biophysica Acta. Molecular Basis of Disease*, *1689*(3), 190–201. doi:10.1016/j.bbadis.2004.03.004 PMID:15276645

Roy, T. C. (2023). What are the Opportunities and Challenges in Millet Sector in India? *Millet Advisor.* https://milletadvisor.com/opportunities-and-challenges-in-millet-sector/

Sahoo, J. P., & Mahapatra, M. (2023). *Millets towards a sustainable nutritional security national year of millets.* Revitalization.

Saleh, A. S., Zhang, Q., Chen, J., & Shen, Q. (2013). Millet grains: Nutritional quality, processing, and potential health benefits. *Comprehensive Reviews in Food Science and Food Safety, 12*(3), 281–295. doi:10.1111/1541-4337.12012

Sarita, E. S., & Singh, E. (2016). Potential of millets: Nutrients composition and health benefits. *Journal of Scientific and Innovative Research, 5*(2), 46–50. doi:10.31254/jsir.2016.5204

Shull, J., Chandrashekar, A., Kirleis, A., & Ejeta, G. (1990). Development of sorghum (Sorghum bicolor (L.) Moench) endosperm in varieties of varying hardness. *Food Structure, 9*(3), 8.

Singh, K., Mishra, A., & Mishra, H. (2012). Fuzzy analysis of sensory attributes of bread prepared from millet-based composite flours. *Lebensmittel-Wissenschaft + Technologie, 48*(2), 276–282. doi:10.1016/j.lwt.2012.03.026

SolutionBuggy. (2022). *Millet based business profitable manufacturing msmes solutions.* Linkedin. https://www.linkedin.com/pulse/millet-based-business-profitable-manufacturing-msmes-solutionsbuggy/

The Story of Millets. (2018). Karnataka State Department of Agriculture & ICAR. https://www.millets.res.in/pub/2018/The_Story_of_Millets.pdf

Willett, W., Rockström, J., Loken, B., Springmann, M., Lang, T., Vermeulen, S., Garnett, T., Tilman, D., DeClerck, F., Wood, A., Jonell, M., Clark, M., Gordon, L. J., Fanzo, J., Hawkes, C., Zurayk, R., Rivera, J. A., De Vries, W., Majele Sibanda, L., & Murray, C. J. L. (2019). Food in the Anthropocene: The EAT–Lancet Commission on healthy diets from sustainable food systems. *Lancet, 393*(10170), 447–492. doi:10.1016/S0140-6736(18)31788-4 PMID:30660336

Chapter 2
Millets Promotion in India by Women Farmers for a Sustainable Economy

H. Kalaiarasi
K.S. Rangasamy College of Technology, India

S. Kirubahari
K.S. Rangasamy College of Technology, India

T. Senthil Kumar
Asia Pacific University of Technology and Innovation, Malaysia

ABSTRACT

Agriculture systems worldwide need to adopt sustainable cropping solutions to deal with existing issues of malnutrition, food insecurity, resource depletion and lifestyle health issues. Millets have special nutritive and agronomic properties. Millets offer one such solution as they have nutritional, environmental, and economic superiority over the most common staple grains in India. The present study deals with women empowerment in agriculture that reflects increased women's decision-making authority related to agricultural resources, management and production, and income, which will enhance the ability of women farmers to strengthen their economic status. The novelty of this study focuses to address challenges in sustainable production and marketing in millets promotion that impede inclusive growth of women farmers. The findings of this study advocate having an integrated approach to enhance millet value chain efficiencies to develop millet promotion, considering economic, environmental, and health sustainability through women empowerment.

DOI: 10.4018/978-1-6684-9819-4.ch002

Copyright © 2023, IGI Global. Copying or distributing in print or electronic forms without written permission of IGI Global is prohibited.

INTRODUCTION

Millets are being grown in more than 130 countries across the world and have been considered as an integral part of the healthy diet by half a billion people across Asia and Africa over many years. As per the report of National Investment Promotion and Facilitation Agency (2022), millennials have witnessed a shift in the proportion of cereals and coarse grains such as millets in their daily dietary plan. With a sedentary lifestyle and increasing health awareness, people are making conscious choices on the consumption of millets to curb lifestyle diseases. Millets also provide opportunities for smallholder farmers to improve their livelihoods, and quality of living standards while improving the diversity of the global food supply which ensures food and nutrition security (Clemens et al., 2023). Even the Indian Prime Minister has articulated the importance of enhancing the demand for millets which could improve the income for the growers and provide better nutrition to the consumers.

India is the world's largest producer, and the world's second-largest exporter of Millets and is striving to change the humble millets reputation worldwide. The Government of India had marked 2018 as the national year of millets to promote the production and consumption of millets. Moreover, the United Nations General Assembly (UNGA) unanimously adopted a resolution proposed by India supported by 72 countries to mark 2023 as the 'International Year of Millets' declared on 5th March 2021.

World export of millet has increased from $400 million in 2020 to $470 million in 2021 (ITC trade map). India exported millets worth $75.46 million in the year 2022-23, against $62.95 million in 2021-22. India's export of cereals stood at Rs. 111,062.37 Crore / 13,857.95 USD Millions during the year 2022-23. However, the share of Millet-based value-added products is negligible.

Agriculture sector employs 80% of all economically active women in India, which comprises 33% of the agriculture labor force and 48% of the self-employed women farmers. In India, about 85% of rural women are engaged in agriculture. Yet, only about 13% own land, though women play a significant role in various agricultural activities. According to Periodic Labour Force Survey (2019-20), 59.9 per cent of women in India are agriculture workers. In rural India, this percentage is 75.7. Economic Survey of 2020-21 mentions that with growing rural to urban migration, there is 'feminization' of agriculture sector, with increasing number of women in multiple roles as cultivators, entrepreneurs, and laborers. As per the report of the Institute of Human Development (IHD, 2014), 70% of all women engaged in cultivation are from households witnessing migration and about 60-80% food is produced by rural women.

Women farmers across the world play a key role in agriculture and food security; they contribute in several ways to improve the farming income. Thaker et al., (2016) opined women farmers is also a story of alternative rationalities to large-scale corporatization of Indian agriculture by interlinking local interpretations of food value to broader logics of development, nutrition, and hunger. The women farmers come together in interpreting food as culture, rooted in local cultural understandings of health and nutrition. Gowri et al (2020) stated the millets have been a staple of the Indian diet, especially in rural India, for years and remain prevalent even today for a balanced diet. The Government of India has identified millet as a safe bet to enhance farmers' income, especially for women farmers, and as a reliable grain to ensure India's nutritional and food security. In addition to many health benefits, millets are also good for the environment with low water and input requirement which lead to sustainability in agriculture.

Millets Pave the Way to Achieve Sustainable Development Goals Through Women Farmers

Millets contain numerous health and environmental benefits and that is the main reason for its promotion during the year of millets in India and during the global year of millet. Such promotion is particularly important for all the stakeholders involved in the millet value chain. As per the report of Ministry of Agriculture and Farmers Welfare (2018), declaration on the year of millets and rebranding of millets as "Nutri Cereals" were done for extensive promotion and demand generation. The Department of Agriculture and Farmers Welfare has taken a proactive multi-stakeholder engagement approach to achieve the aim of International Year of Millets 2023 and to promote the Indian millets across the world. Ministries, States, and Indian embassies have been allocated focused months in 2023 to conduct various activities to promote International Year of Millets and increase awareness about the benefits of millet for the Consumer, Cultivator and Climate (Mathew et al., 2022).

Meier et al. (2020) indicated that millets are a well-recognized food source since they are high in nutrients, come from rained crops, require less water, and are less harmful to the environment. Owing to this, millet could be considered as a true super food for consumers, farmers, and the environment. Millets are becoming increasingly well-known for their nutritional benefits, and many food manufacturing companies have evinced interest in developing exciting and innovative products to reach an even wider consumer base.

Mapping Millets With Sustainable Development Goals

The first SDG "No Poverty" is to eradicate global poverty. Millets can play a significant role in achieving this goalby providing a sustainable and reliable source of income to the women farmers. Millets require less investment, resources, and time than other crops. This makes them an ideal crop for small women farmers who may not have the resources to invest in high-input crops. Furthermore, millets are an inexpensive food source in many developing countries where the people may not have access to other types of food. According to the UN, more than 90 million people belonging to poverty-prone regions of Sub-Saharan Africa and Asia depend on millets for their diets.

The second SDG is 'Attaining Zero Hunger'. The objective of this goal is to end hunger, achieve food security, improve the nutrition (Zhang et al., 2021), and promote sustainable economy (UN General Assembly, 2015). Women farmers play a vital role by providing millets at lower costs where poverty is rampant. As per the report of International Crops Research Institute for the Semi-Arid Tropics (ICRISAT), the promotion of millets in the diets of children in India could prevent up to 24 million cases of malnutrition and 12 million cases of stunting which directly indicates how millets can bring efficacy in the efforts to reduce hunger and achieve the targets of SDG 2.

The third SDG is focused on 'Good Health and Well-being'. Since millets contain necessary vitamins and minerals that can help prevent chronic diseases, they can be path breaking in achieving SDG 3. Singh et al., 2022) reported that the millet consumption could also help to change in food habits, which is crucial for increasing millet consumption among consumers. (Pradhan et al., 2021) explained that transformation in the food system including the promotion of local foods like millets through women farmers is essential to achieve many SDGs.

The twelfth SDG is 'Responsible Production and Consumption'. The increase in consumption and production of millets by women farmers has a direct linkage with achieving the SDG 12. This SDG promotes resource and energy efficiency, with lower environmental costs and reduction in poverty through consumption-production. Further, the consumption of Millets promotes a sustainable lifestyle and thus creates demand for millet production, where women farmers have sustainable income that protects agro biodiversity and contributes to sustainable economy. These aspects highlight how promotion of millets by women farmers is ideal food for attaining the SDG 12, inducing sustainability in both production as well as consumption patterns.

The thirteenth SDG is 'Climate Action'. The aim is to promote mechanisms for raising capacity for effective climate change-related planning and management (UN General Assembly, 2015). Millets may help to improve food production by overcoming the adverse effects caused by global warming and climate change. Millet based farming in agriculture by women farmers could help reduce synthetic fertilizer and pesticide productions, reducing the adverse effects to the environment and attain sustainability.

Millets Promotion by Women Farmers for Sustainable Future in India

Promotion of millet cultivation could help to meet the Sustainable Development Goals (SDGs) of the UN. Millets promotion by women farmers increase the awareness of the contribution of millets for food security and nutrition. This may lead to inspire stakeholders, including national governments to work towards improving millets production, productivity, and quality of millets for women farmers. Further, investment in research and development and extension services for millets helps women farmers to sustain in agricultural activity.

Through the comprehensive literature review, this paper considers how prioritizing women's empowerment in agriculture might lead to economic benefits. Brody et al., (2015) focused on women's economic empowerment, noting that positive economic outcomes are generally associated with social empowerment (e.g., supportive community norms) and psychological empowerment (e.g., efficacy or self-perception). According to Women's Empowerment in Agriculture Index (WEAI), women's decision-making power are related to 1) productive resources, 2) agricultural management, and 3) agricultural income. These measures create a pathway to change in women's empowerment in agriculture outcomes, which occurs in longer-term economic benefits. Jeeva et al., (2019) recognized the enormous potential of millets to generate livelihoods, increase women farmer's income and ensure food and nutritional security worldwide, the Government of India has prioritized millets production in India.

Figure 1. Millets: The future of sustainable agriculture

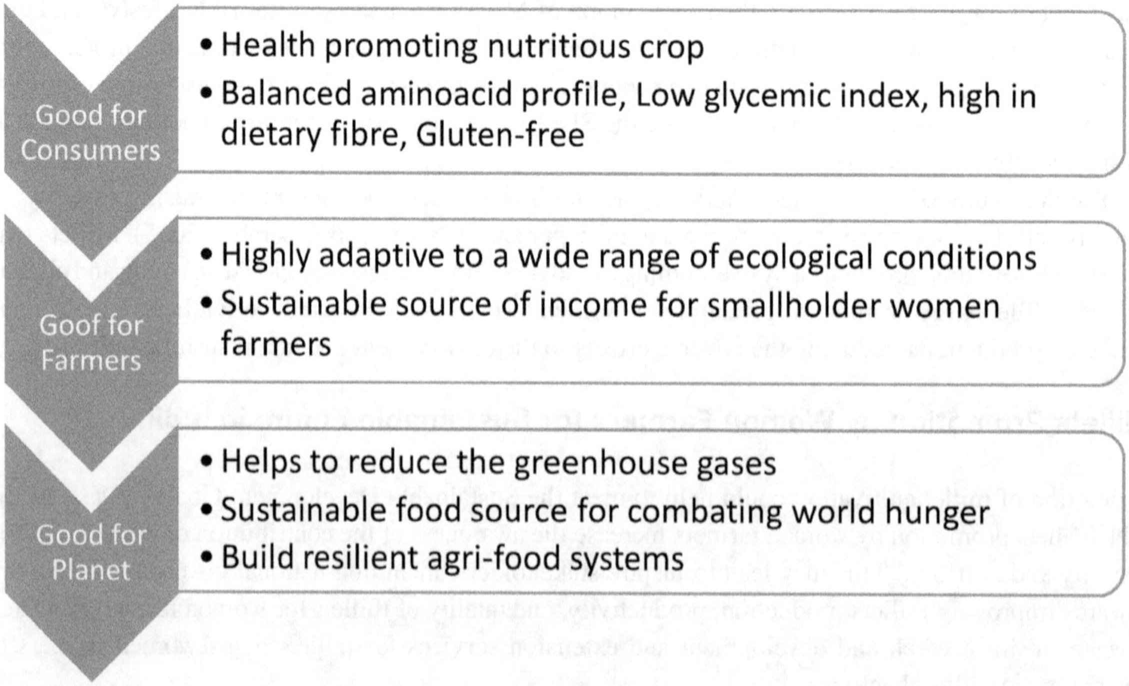

Strengthening Millets Value Chain: Huge Potential for Women Farmers

India is one of the top producers of millets in the world, accounting for 20 per cent of worldwide and 80 per cent per of Asian production. According to a survey by the Ministry of Agriculture and Farmers Welfare (MAFW), pearl millet (bajra), sorghum, and finger millet account for 98 per cent of all millet produced in India, with pearl millet alone accounting for 60 per cent of production and 56 per cent of area. As a result, there is enormous potential for both women farmers and consumers, and it is crucial to enhance the millet value chain.

As per the report of Economic Survey 2023, India produces over 50.9 million tons (as per fourth advance estimate) of millet. On an average, India's average yield of millet is 1239 kg/ha, compared to the global average yield of 1229 kg/ha. India cultivates nine varieties of millets such as foxtail, finger, barnyard, brown top, litte, kodo, pearl, proso and sorghum, where Tamil Nadu cultivates at least seven kinds of these millets while Karnataka grows at least five varieties. Pandey et al (2023) found that Rajasthan, Uttar Pradesh, Gujarat, and Haryana contribute 82% of national millet production in the period of 2021-2022, which after establishing market and strengthening overall millet value chain will affect more than 35% of India's population residing in these states.

To improve the profitability for women farmers, the welfare of consumers, the efficiency, and the competitiveness of the Indian millet value chain, the Federation of Indian Chambers of Commerce and Industry (FICCI) Task Force wants to enable effective partnerships. The marketing of millet needs to spread over many states and districts in India with the help of knowledgeable partners and helpful technology. In view of the high nutrient value, low water requirement and less harmful environmental impact, Millets are indeed a super food for consumers, farmers, and the environment. Many food processing companies are also interested in introducing exciting and innovative products which may have a wider reach. Creating a platform to connect all participants in the millet value chain is one of the crucial areas we need to focus on. This platform will support women farmers' efforts to increase the accessibility and acceptance of millets in India and around the world through collaboration, understanding of the issues, support, investment, and the use of innovative technology.

Benefits of Millet Cultivation

Drought Tolerance: Millets are naturally adapted to dry and arid conditions, so these crops are suitable to be raised by women farmers in areas with limited rainfall. Millet crops are highly adapted to the climatic conditions of the land, also, millets require minimal water and can grow in poor soil conditions; this makes them a reliable crop even during times of drought (Padulosi et al., 2015).

Nutritional Benefits: Millets are an excellent source of essential nutrients such as protein, fiber, vitamins, and minerals. Pradhan et al., (2019) stated millets are gluten-free, low glycemic index, and have high antioxidant activity, which makes people to follow a healthy diet for sustainable lifestyle. Consumption of millet foods contribute to food diversity and enrichment of diet with micronutrients provides health benefits due to the interactions between the inherited genes and food nutrients.

Ecological Benefits: Millets are an environmentally friendly crops and act as a natural barrier against soil erosion, which makes them as short growth cycle allows for crop rotation, also help in maintaining soil health. The cultivation of millets can make agriculture more genetically diverse and sustainable, and consumption of domestically cultivated millet crops can reduce the carbon footprints and imports of millet grains.

Economic Benefits: Millets are an important crop for small-scale women farmers as they require minimal investment and have a low input cost. Millets are also have a high market demand due to their nutritional benefits, making them a lucrative crop especially for women farmers. To promote the millets cultivation in India, the government has launched several initiatives such as the Millets Mission, which aims to increase the production and consumption of millets.

Challenges in Marketing and Scaling Up Millet Production

The marketing of millets by women farmers needs to reach over many states and districts with the help of knowledgeable partners and helpful technology. While there is growing interest in millet promotion in India, women farmers met few challenges in marketing and scaling up millet production, which may include:

Limited awareness and demand: While there is a growing awareness of the health benefits of millets, many consumers are still not aware with millets and do not know how to cook or consume them. This limits the demand for millet cultivation and makes it difficult for women farmers to sell their produce in the market.

Lack of processing and storage infrastructure: There is a lack of processing and storage infrastructure for millets, which makes it difficult for women farmers to process and package the millets. This limits their ability to add value to the millets products and sell it at higher prices.

Limited market access: Millet farmers often face limited market access, particularly in remote and rural areas. This can lead to low prices and limited demand for the millet promotion.

Low productivity: Millet cultivation in India is often low due to a lack of access to quality seeds, fertilizers, and other inputs. This can limit the ability of women farmers to scale up millet production and meet market demand.

Climate change: Climate change is affecting millet production in India, particularly in areas that are already prone to drought and other weather extremes. This can lead to lower yields and quality, making it more difficult to market and sell millets.

Low investment in research and development: There is a lack of investment in research and development of millets promotion, leading to limited information on millets cultivation, storage, and utilization.

Inadequate storage and transportation infrastructure: Inadequate storage and transportation facilities lead to post-harvest losses and difficulty in reaching markets by women farmers.

To address these challenges, there is a need for greater investment in millet processing and storage infrastructure, as well as efforts to raise awareness of the health benefits of millets among consumers.

Additionally, there is a need for better access to quality seeds, fertilizers, and other inputs to improve millet productivity and scale up production. Finally, there is a need to develop new market linkages and value chains to help millet farmers to access markets and sell their produce at higher prices.

Steps to be Taken by Government of India to Promote Millets Cultivation

The Ministry of Agriculture and Farmers Welfare, Government of India, has been actively taking many efforts for its success. Few steps to be taken to promote awareness on Millets cultivation as follows:

1. Encouraging and supporting farmers to adopt millet cultivation through government subsidies for better market access
2. Promoting millet based value-added products to increase demand and profitability
3. Improving millet seed quality and distribution through government and private initiatives
4. Offering training and extension services to farmers on the latest millet cultivation techniques
5. Enhancing research and development efforts to improve the yield and quality of millet crops
6. Improving storage and transportation facilities to reduce post-harvest losses
7. Promoting intercropping with millets to increase farm efficiency and income
8. Encouraging partnerships between farmers, processors, and retailers to create a sustainable millet value chain
9. Encouraging international collaborations to share best practices and knowledge in millet production and marketing.

With the Government of India encouraging and promoting the production of different millets, and with 2023 being observed globally as the 'International Year of Millets,' the Self-Help Groups (SHGs), and women farmers have played a vital role in sustaining the economy of millets, preserving biodiversity, and maintaining ecological harmony in rural societies.

Future of Millets Revolution in India

The future of the millet revolution in India looks promising, with growing recognition of the value of millets for nutrition, food security, and environmental sustainability. The following are some of the ways in which the millet revolution is expected to evolve in the coming years:

Expanded Market Opportunities: As the demand for millet-based products grows, there are likely to be new market opportunities for small-scale farmers and businesses. The private sector is expected to invest more in millet processing and marketing, leading to the creation of new jobs and income opportunities.

Improved Farming Practices: With the support of government agencies, farmers are expected to adopt better farming practices for millets, leading to improved yields and better quality of grains. This could include the use of better seeds, improved soil fertility management, and better pest and disease control.

Increased Consumption: One of the key objectives of the millet revolution is to increase the consumption of millets in India. With growing awareness about the nutritional benefits of millets, more people are expected to incorporate millets into their diet, leading to an increase in demand for millet-based products.

Climate Resilience: As climate change continues to pose a threat to agriculture, there is likely to be more investment in developing climate-resilient millet varieties and farming practices. This will help to ensure the sustainability of millet cultivation in the face of changing climatic conditions.

CONCLUSION

Women farmers are a major working force in agriculture activities and play a key role in conserving agricultural diversity. Women's empowerment is a key factor for achieving sustainable economic growth, social development, and environmental sustainability. Women farmers, with their ecological acumen and their close association with climate protection, are best suited to be ambassadors of millet cultivation. Efforts towards providing agriculture inputs to and building the capacity of women farmers in sustainable millet production can be a game-changer in the millet revolution. This will also ensure women farmer's economic empowerment. Women are also the primary provider of food within the household, therefore, women farmers and self-help group members can be trained and equipped in agro-processing and marketing of millets which will increase consumption patterns.

REFERENCES

Adenle, A. A., Azadi, H., & Manning, L. (2018). The era of sustainable agricultural development in Africa: Understanding the benefits and constraints. *Food Reviews International*, *34*(5), 411–433. doi: 10.1080/87559129.2017.1300913

Ajeigbe, H. A., Goodrich, C. G., Ntare, B. R., Weltzien, E., & Ndjeunga, J. (2013). Involving women in research for economic growth through agricultural technologies and practices: ICRISAT's initiatives in sub-Saharan Africa. *Sécheresse*, *24*(4), 359–366. doi:10.1684ec.2013.0405

Anderson, C. L., Reynolds, T. W., Biscaye, P., Patwardhan, V., & Schmidt, C. (2021). Economic benefits of empowering women in agriculture: Assumptions and evidence. *The Journal of Development Studies*, *57*(2), 193–208. doi:10.1080/00220388.2020.1769071

Anitha, S., Kane-Potaka, J., Botha, R., Givens, D. I., Sulaiman, N. L. B., Upadhyay, S., Vetriventhan, M., Tsusaka, T. W., Parasannanavar, D. J., Longvah, T., Rajendran, A., Subramaniam, K., & Bhandari, R. K. (2021). Millets can have a major impact on improving iron status, hemoglobin level, and in reducing iron deficiency anemia–a systematic review and meta-analysis. *Frontiers in Nutrition*, *8*, 725529. doi:10.3389/fnut.2021.725529 PMID:34722606

Brody, C., De Hoop, T., Vojtkova, M., Warnock, R., Dunbar, M., Murthy, P., & Dworkin, S. L. (2015). Economic Self-Help group programs for improving women's Empowerment: A systematic review. *Campbell Systematic Reviews*, *11*(1), 1–182. doi:10.4073/csr.2015.19

Chapke, R. R., Dayakar, B., & Tonapi, V. A. (2017). Role of millets in doubling farmers' income and sustainable development of Vidarbha region of Maharashtra. ICRISAT [Telangana, India.]. *Patencheru, Hyderabad*, *502*, 324.

Chapke, R. R., & Tonapi, V. A. (2018). *Role of millets in small holder farming system for improved food and nutritional security under changing climate scenario*.

Chepchieng, M. J. (2021). Contribution of Women in Socio–Economic Development under the Leadership of Daniel Arap Moi in Baringo County, Kenya. *Journal of African Interdisciplinary Studies*, *5*(3), 81–97.

Clemens, R., Rao, P. G., Elouafi, I., Oniang'o, R., Chandrasekara, A., Pressman, P., & Yadav, J. (2023). A Commentary on Millets for Enhancing Agri-economy, Nutrition, Environmental, and Sustainable Development Goals. *Journal of Food Bioactives: An Official Scientific Publication of the International Society of Nutraceuticals and Functional Foods (ISNFF)*, *22*, 22. doi:10.31665/JFB.2023.18342

Dutta, M. J., & Thaker, J. (2020). Sustainability, ecology, and agriculture in women farmers' voices: Culture-centering gender and development. *Communication Theory*, *30*(2), 126–148. doi:10.1093/ct/qtz029

Eliazer Nelson, A. R. L., Ravichandran, K., & Antony, U. (2019). The impact of the Green Revolution on indigenous crops of India. *Journal of Ethnic Foods*, *6*(1), 1–10. doi:10.118642779-019-0011-9

Ghosh, N., & Ramana, D. V. (2023). Millet and Money Promoting Sustainable Consumption in Southern Orissa. In *Dealing with Socially Responsible Consumers: Studies in Marketing* (pp. 207–229). Springer Nature Singapore.

Gowri, M. U., & Shivakumar, K. M. (2020). Millet scenario in India. *Economic Affairs*, *65*(3), 363–370. doi:10.46852/0424-2513.3.2020.7

Grovermann, C., Umesh, K. B., Quiédeville, S., Kumar, B. G., & Moakes, S. (2018). The economic reality of underutilised crops for climate resilience, food security and nutrition: Assessing finger millet productivity in India. *Agriculture*, *8*(9), 131. doi:10.3390/agriculture8090131

Hegde, N. G. (2020). Empowerment of Women for Improved Quality of Life. *Global Journal of Human-Social Science: E Economics, 20*(7).

Jaka, H., & Shava, E. (2018). Resilient rural women's livelihoods for poverty alleviation and economic empowerment in semi-arid regions of Zimbabwe. Jàmbá. *Journal of Disaster Risk Studies*, *10*(1), 1–11.

Jeeva, J. C., Joshi, K., Singh, A., & Behera, B. C. (2019). Engendering finger millet-based value chains for livelihood and nutritional security of women in agriculture. *Current Science, 116*(11), 1893–1896. doi:10.18520/cs/v116/i11/1893-1896

Jerop, R., Dannenberg, P., Owuor, G., Mshenga, P., Kimurto, P., Willkomm, M., & Hartmann, G. (2018). Factors affecting the adoption of agricultural innovations on underutilized cereals: The case of finger millet among smallholder farmers in Kenya. *African Journal of Agricultural Research, 13*(36), 1888–1900. doi:10.5897/AJAR2018.13357

Kabeer, N. (2001). Resources, agency, achievements: Reflections on the measurement of women's empowerment. *Development and Change, 30*(3), 435–464. doi:10.1111/1467-7660.00125

Kalyani, V., & Murugan, K. R. (2017). Empowering Women Farmers Participation In Organic Agricultural Development. *International Journal of Multidisciplinary Educational Research, 6*(2), 187.

Kane-Potaka, J., Anitha, S., Tsusaka, T. W., Botha, R., Budumuru, M., Upadhyay, S., Kumar, P., Mallesh, K., Hunasgi, R., Jalagam, A. K., & Nedumaran, S. (2021). Assessing millets and sorghum consumption behavior in urban India: A large-scale survey. *Frontiers in Sustainable Food Systems, 5*, 680777. doi:10.3389/fsufs.2021.680777

Klasen, S. (2018). The impact of gender inequality on economic performance in developing countries. *Annual Review of Resource Economics, 10*(1), 279–298. doi:10.1146/annurev-resource-100517-023429

Kumar, S., Rao, D. U. M., Thombare, P., & Kale, P. (2020). Small and marginal farmers of Indian agriculture: Prospects and extension strategies. *Indian Research Journal of Extension Education, 20*(1), 35–41.

Leroux, L., Falconnier, G. N., Diouf, A. A., Ndao, B., Gbodjo, J. E., Tall, L., Balde, A. A., Clermont-Dauphin, C., Bégué, A., Affholder, F., & Roupsard, O. (2020). Using remote sensing to assess the effect of trees on millet yield in complex parklands of Central Senegal. *Agricultural Systems, 184*, 102918. doi:10.1016/j.agsy.2020.102918

Mahesh, K. M., Aithal, P. S., & Sharma, K. R. S. (2023). Impact of Aatmanirbharta (Self-reliance) Agriculture and Sustainable Farming for the 21st Century to Achieve Sustainable Growth. [IJAEML]. *International Journal of Applied Engineering and Management Letters, 7*(2), 175–190.

Mathew, J., & Joseph, M. K. (2022). International year of millets 2023: Millet promotion in India for food security. *Rajagiri Journal of Social Development, 14*(2), 2–17. doi:10.5958/2583-7087.2022.00007.4

Mathew, J., & Joseph, M. K. (2022). International year of millets 2023: Millet promotion in India for food security. *Rajagiri Journal of Social Development, 14*(2), 2–17. doi:10.5958/2583-7087.2022.00007.4

Meier, C., El Benni, N., Sakamma, S., Moakes, S., Grovermann, C., Quiédeville, S., Stolz, H., Stolze, M., & Umesh, K. B. (2020). Are non-farming consumers willing to pay "a good market price" for iron-biofortified finger millet? Evidence from experimental auctions in Karnataka, India. *Journal of Agribusiness in Developing and Emerging Economies, 10*(5), 751–779. doi:10.1108/JADEE-11-2019-0190

Notaro, V., Padulosi, S., Galluzzi, G., & King, I. O. (2017). A policy analysis to promote conservation and use of small millet underutilized species in India. *International Journal of Agricultural Sustainability, 15*(4), 393–405. doi:10.1080/14735903.2017.1334181

Padulosi, S., Mal, B., King, O. I., & Gotor, E. (2015). Minor millets as a central element for sustainably enhanced incomes, empowerment, and nutrition in rural India. *Sustainability (Basel)*, *7*(7), 8904–8933. doi:10.3390u7078904

Pandey, A., & Bolia, N. B. (2023). Millet value chain revolution for sustainability: A proposal for India. *Socio-Economic Planning Sciences*, *87*, 101592. doi:10.1016/j.seps.2023.101592

Pingali, P. (2023). *Are the lessons from the green revolution relevant for agricultural growth and food security in the twenty-first century. Agricultural Development in Asia and Africa*. Springer.

Prabhu, R., Gowri, M. U., Gayathri, R., Govindaraj, M., & Manikandan, G. (2022). Growth dynamics and forecasting of minor millets in India: A time series analysis. *Journal of Applied and Natural Science, 14,* 145-150.

Pradhan, A., Panda, A. K., & Bhavani, R. V. (2019). Finger millet in tribal farming systems contributes to increased availability of nutritious food at household level: Insights from India. *Agricultural Research*, *8*(4), 540–547. doi:10.100740003-018-0395-6

Pradhan, P., Sapkota, T. B., & Kropp, J. P. (2021). Why food systems transformation is crucial for achieving the SDGs. *Ruralia*, *21*(55), 10–12.

Quaidoo, M. (2018). *The Role of Entrepreneurs in Economic Development: Prospects and Challenges of Female Entrepreneurs in Agribusiness in Ghana.*

Raina, R., Mishra, S., Ravindra, A., Balam, D., & Gunturu, A. (2022). Reorienting India's Agricultural Policy: Millets and Institutional Change for Sustainability. *Journal of Ecological Society*. doi:10.54081/JES.028/01

Rao, N. C., Bathla, S., Kumar, A., & Jha, G. K. (2018). Agriculture and sustainable development goals: an overview and issues. *Agricultural Economics Research Review, 31*(conf), 1-7.

Ravi, Y., Durgannavar, N. A., & Jagdish, M. R. (2022). Entrepreneurial activities and empowerment of rural women through processing and value addition. *The Pharma Innovation Journal.*

Sahoo, M., & Samantaray, D. (2021). Millet Cultivation and Food Security in Tribal Region of Odisha, India: A Microlevel Analysis. *Asian Journal of Water, Environment and Pollution*, *18*(1), 51–57. doi:10.3233/AJW210007

Samineni, S. (2018). Social entrepreneurship as a tool for sustainable development of women in rural India leading to economic empowerment. *IOSR Journal of Business and Management (IOSR-JBM) e-ISSN.*

Saxena, L. P. (2020). Community self-organisation from a social-ecological perspective: 'Burlang Yatra' and revival of millets in Odisha (India). *Sustainability (Basel)*, *12*(5), 1867. doi:10.3390u12051867

Setiawan, H. H., & Wismayanti, Y. F. (2023). The green economy to support women's empowerment: social work approach for climate change adaptation toward sustainability development. In Climate Change, Community Response and Resilience (pp. 225-240). Elsevier. doi:10.1016/B978-0-443-18707-0.00012-6

Sharma, G., & Sharma, E. (2017). Agroforestry systems as adaptation measures for sustainable livelihoods and socio-economic development in the Sikkim Himalaya. *Agroforestry: Anecdotal to Modern Science*, 217-243. Springer.

Singh, R. K., Sreenivasulu, N., & Prasad, M. (2022). Potential of underutilized crops to introduce the nutritional diversity and achieve zero hunger. *Functional & Integrative Genomics*, *22*(6), 1459–1465. doi:10.100710142-022-00898-w PMID:36074305

Sreeni, K. R. (2022). Role of JLG's (Joint Liability Groups) in Enhancing Village Economy. *Journal of Women Empowerment and Studies (JWES), 2*(01), 1-13.

Sudheer, K. P., Unni, S. K., & Shaju, A. A. (2022). Gender Responsive Agribusiness Development. *Engendering Agricultural Development: Dimensions and Strategies*. Routledge.

Swain, S., Jakhar, P., & Rout, P. K. (2021). Economic Empowerment of Farm Women through Improved Post Harvest Technologies and Value Addition of Horticultural Crops. *Promoting Women Agripreneurship through Crop-Livestock-Fisheries Technologies*. MANAGE.

Tesfai, M., Njarui, D. M. G., & Ghimire, S. R. (2019). Sustainable intensifications of African agriculture through legume-based cropping and Brachiaria forage systems. *African Journal of Agricultural Research*.

Thaker, J., & Dutta, M. (2016). Millet in our own voices: A culturally-centred articulation of alternative development by DDS women farmers' sanghams. *Globalisation and the challenges of development in contemporary India,* (pp. 131-144). Springer.

UN General Assembly. (2015). *Transforming Our World: The 2030 Agenda for Sustainable Development*. UN.

Vijayalakshmi, D., Geetha, K., Gowda, J., Ravi, S. B., Padulosi, S., & Mal, B. (2010). Empowerment of women farmers through value addition on minor millets genetic resources: A case study in Karnataka. *Indian Journal of Plant Genetic Resources*, *23*(01), 132–135.

Zhang, Y., Runting, R. K., Webb, E. L., Edwards, D. P., & Carrasco, L. R. (2021). Coordinated intensification to reconcile the 'zero hunger'and 'life on land'Sustainable Development Goals. *Journal of Environmental Management*, *284*, 112032. doi:10.1016/j.jenvman.2021.112032 PMID:33545453

Chapter 3
Identification of Foxtail Millet Solution for Malignant Colorectal Cell Growth Reduction Using Deep Learning-Based Mobile Net Model

Artatrana Biswaprasanna Dash
GIET University, India

Sachikanta Dash
https://orcid.org/0000-0002-0807-4624
GIET University, India

Sasmita Padhy
VIT Bhopal University, India

ABSTRACT

Millets are cereal crops, commonly grown in semi-arid regions. Millets are known to be a good source of fiber, magnesium, phosphorous, zinc, serotonin, and gluten-free product that helps to heal Cancer effectively. The active components of Foxtail millet bran (FMBP) are promoting higher levels of reactive oxygen species (ROS) that help block the STAT3 pathway by reducing the growth of malignant cells in intestine. Nowadays in medical colonoscopy, histological image analysis and MRI are some commonly adapted techniques for the identification of polyps in the colon. However, the manual approaches are very time-consuming and less accurate. There is no specific model available that can measure the effectiveness of millet in healing malignant cell growth in colon. The main goal is to design a deep learning based MoblileNet-V3 model for the detection and classification of histological images. This study will help researchers and medical practitioners to compare the malignancy status before and after the consumption of millet.

DOI: 10.4018/978-1-6684-9819-4.ch003

Copyright © 2023, IGI Global. Copying or distributing in print or electronic forms without written permission of IGI Global is prohibited.

INTRODUCTION

"When food saves life, then why goes for knife." On the basis of WHO 2022 report, Colorectal malignancy is acknowledged as the second leading cause of cancer-related deaths globally and ranks as the third most frequently diagnosed cancer type. In the course of chemotherapy for colorectal cancer (CRC), the most frequently employed drugs include 5-fluorouracil (5-Fu), oxaliplatin (L-OHP), and vincristine (VCR) (Shan et al., 2021). However, most of the human bodies automatically develop multi drug resistance (MDR) which acts as a barrier for the antibiotic performance (Shan et al., 2021). Nature polyphenol has the capability to generate anti MDR agent for colorectal malignancy. Nature Polyphenol are plentily available in plants and foods. Foxtail millet bran (FMBP) is the world's sixth most popular cereal type which contains bound polyphenol that develops anti CRC activities in body.

There are various types of millets, and they are commonly used in many parts of the world. Pearl Millet (Pennisetum glaucum), referred to as Bajra in India and Africa, stands as one of the extensively cultivated varieties and serves as a staple food in numerous arid and semi-arid regions. Finger Millet (Eleusine coracana), known as Ragi in India, is also noteworthy in this context. It is rich in calcium and other nutrients. It's commonly used in the preparation of porridge, flatbreads, and other food products. Proso Millet (Panicum miliaceum) also called common millet or broomcorn millet. Proso millet is grown in many countries for human consumption and livestock feed. Little Millet (Panicum sumatrense) also known as Kutki or Samalu. Little millet is small in size and rich in fiber. It's commonly used in South India. Kodo millet is grown in India, Africa, and Southeast Asia. It is known for its hardiness and resilience in adverse conditions. Barnyard Millet (Echinochloa colona) also known as Sanwa or Oodalu in India, barnyard millet is used as food in some parts of the world. Sorghum is often classified separately from millets, it is a similar type of cereal grain that is widely cultivated. It is used for food, fodder, and the production of sorghum syrup. Foxtail Millet (Setaria italica) is known for its high nutritional value and is used in various dishes. It is a staple in parts of China, India, and Southeast Asia. Foxtail millet is an annual grass crop that is grown primarily for its small, nutritious seeds. It is adapted to a variety of climates, including semiarid regions. The grains of foxtail millet are small and typically pale yellow or brown in color. The shape of the grains is cylindrical or oblong, resembling a foxtail, which gives the millet its name. Foxtail millet is rich in nutrients such as carbohydrates, dietary fiber, and essential minerals. Being gluten-free, it proves to be a suitable option for individuals with gluten intolerance or celiac disease. As a whole grain, foxtail millet offers several health benefits. It is rich in fiber, which can aid in digestion and promote a feeling of fullness. The nutritional content includes vitamins and minerals such as iron, magnesium, and phosphorus. Foxtail millet has been cultivated for thousands of years and is an important food crop in many Asian and African countries. It is a staple in the diets of people in regions where it is grown. Foxtail millet is known for its ability to grow in diverse environmental conditions, including regions with limited water availability. This adaptability makes it a valuable crop for areas with challenging climates. One of the advantages of foxtail millet is its resistance to drought conditions. It can thrive in arid or semiarid regions, making it a valuable crop in areas with water scarcity. The cultivation of foxtail millet is considered environmentally sustainable due to its efficient use of water and its ability to grow in marginal lands. Researchers are exploring the potential of foxtail millet in breeding programs to enhance its nutritional content, yield, and resilience to various environmental conditions. Various analysis processes applied to analyze the foxtail millet benefits. Proximate Analysis determines the basic components of foxtail millet, including moisture content, ash, crude protein, crude fat, and total carbohydrates. Proximate analysis provides an overall understanding of the nutritional composition.

Mineral analysis determines the mineral content of foxtail millet, including essential minerals such as iron, zinc, magnesium, phosphorus, and others. Inductively Coupled Plasma (ICP) and Atomic Absorption Spectroscopy (AAS) are common methods for mineral analysis. Fiber Analysis measures the dietary fiber content of foxtail millet, including both soluble and insoluble fiber. This analysis is essential for understanding its contribution to digestive health. Fatty Acid Profile identifies and quantifies the fatty acids present in the oil extracted from foxtail millet. Gas Chromatography (GC) is commonly used for fatty acid analysis. Fatty Acid Profile identifies and quantifies the fatty acids present in the oil extracted from foxtail millet. Gas Chromatography (GC) is commonly used for fatty acid analysis. Amino Acid Analysis determines the amino acid composition of foxtail millet protein. This analysis is important for assessing the quality of protein and identifying essential amino acids.Vitamin Analysis measures the content of vitamins such as thiamine, riboflavin, niacin, and vitamin E in foxtail millet. High-Performance Liquid Chromatography (HPLC) is often used for vitamin analysis.Antioxidant Activity assesses the antioxidant potential of foxtail millet. This can be done through various methods, including the DPPH (2,2-diphenyl-1-picrylhydrazyl) assay and FRAP (Ferric Reducing Antioxidant Power) assay. Mycotoxin Analysis tests for the presence of mycotoxins, which are toxic compounds produced by fungi. Common mycotoxins include aflatoxins and fumonisins. Physical Properties includes the determination of grain size, shape, and color. These properties can influence the quality of foxtail millet for various applications. Microbiological Analysis assesses the microbial quality of foxtail millet, including the presence of bacteria, yeast, and mold. This analysis is important for food safety. Genetic Analysis involves molecular techniques such as polymerase chain reaction (PCR) for genetic characterization, identification of specific traits, and assessment of genetic diversity. Milling and Processing Analysis assesses the milling quality of foxtail millet, including milling yield and quality of flour or other processed products.

Whole grains, including certain types of millet, are often good sources of dietary fiber. High-fiber diets have been associated with a lower risk of colorectal cancer in some studies. Fiber may contribute to the prevention of colorectal cancer by promoting regular bowel movements and providing fermentation products that are beneficial to the colon. Some grains, including millet, contain antioxidants that may have potential anti-cancer properties by neutralizing free radicals Grains, including millet, may contain phytochemicals that have been studied for their potential anti-cancer effects. Diets high in red and processed meats have been associated with an increased risk of colorectal cancer, while diets rich in fruits, vegetables, and whole grains have been associated with a reduced risk. Grains, including millet, may contain phytochemicals that have been studied for their potential anti-cancer effects. Agricultural research and breeding programs have led to the development of improved varieties of foxtail millet. Improved varieties may exhibit characteristics such as increased yield, disease resistance, and nutritional enhancements.

Medical science has been adopted various types of analysis to study the effect of foxtail millet for the treatment of colorectal malignancy. Some most commonly used analysis techniques are, Analysis of Nutritional composition in cereal, Histopathological evaluation, Bioinformatics analysis, Western Blot analysis (Shan et al., 2021). In histopathological evaluation, histopathological images are analyzed to detect the effect of foxtail millet on CRC reduction. Bioinformatics analysis investigates the protein structure and provides the information about the factors associated with the protein stability, dimensional structure of protein and structure changes after mutation. Western Blot analysis detects the expression changes in enzymes of DNA methylation (DNAMT3a, DNAMT3b and MECP2) (Shan et al., 2021).

Our study provides a deep learning based image classification approach to detect the effect of foxtail millet bran on malignant colon and rectum. Analyzing the histopathological images our model is capable of identifying the type of malignancy in colon and rectum area in a human body which can be further utilized for detecting the effect of foxtail millet bran just by analyzing images before and after the use of millet as a supplement.

RELATED WORKS

Now a day's deep learning based approach is an efficient way to analyze complex medical images. In this first growing world there are many pretrained deep learning based models available, which are already proven themselves best to accomplish a specific task. Mobile Net model is one of them, which is a lightweight CNN based model, already proven to be an efficient image classification model. Transfer learning is a technique by which we can utilize the pretrained models for addressing our problem domain by transferring the learning approach from the related source domain on which it is trained.

Conventional Approaches

In 2020 Shuhua Shan et.al, explained how bound polyphenol of foxtail millet bran (BPIS) reverse the activity of Multi Drug Resistance (MDR) by regulating the expression of miRNA and act as an anti-MDR for human Colon cancer resistance cell line HCT-8/Fu by increasing the chemo sensitivity of the cell line (Shan et al., 2021). The effect of BPIS components (FA, p-CA) on CRC is analyzed by using a series of methods including cell culture, extraction of BPIS components against CRC, MTT assay, real-time PCR analysis, western Blot analysis, cell cycle analysis and statistical analysis (Shan et al., 2021). In 2020 Bowei Zhang et.al, investigated about the effect of foxtail millet and rice diet on the improvement of colitis associated colorectal cancer (CAC) in mice (Zhang et al., 2020). For this experiment, 40 BALB/c mice were randomly allocated into four groups: a normal group, azoxymethane/dextran sodium sulfate (AOM/DSS) groups supplied with an AIN-93G diet, and Millet and Rice groups provided with a modified AIN-93G diet (Zhang et al., 2020). Analysis of Nutritional composition in cereal, Histopathological evaluation, Bioinformatics analysis, Western Blot analysis has been done (Zhang et al., 2020).

Cell Culture

Cell culture stands as a foundational technique in biological research and biotechnology, encompassing the cultivation and upkeep of cells outside their native environment, typically within a controlled laboratory setting (Shan et al., 2021). This technique allows scientists to study the behavior of cells under various conditions, perform experiments, and produce large quantities of cells for various applications.

Here are some key aspects of cell culture:

1. **Cell Lines:** Cells can be derived from tissues or organisms and are typically grown in culture dishes or flasks (Shan et al., 2021). These cells can be primary cells, which are directly isolated from tissues, or immortalized cell lines, which have undergone genetic alterations to allow continuous growth (Shan et al., 2021).
2. **Culture Medium:** Cells require a specific environment to grow and proliferate. Culture medium is a nutrient-rich solution that provides the necessary nutrients, growth factors, hormones, and salts. The composition of the medium depends on the type of cells being cultured (Shan et al., 2021).

3. **Sterile Conditions:** Cell culture must be performed under sterile conditions to prevent contamination. Contaminants, such as bacteria, fungi, or other cell types, can negatively affect the experiment or the health of the cultured cells (Shan et al., 2021).
4. **Incubation:** Cells are typically cultured in a controlled environment, often in incubators set at a specific temperature, humidity, and carbon dioxide (CO2) concentration. These conditions mimic the physiological environment required for cell growth (Zhang et al., 2020).
5. **Passaging:** As cells grow and reach confluence, they need to be subcultured or passaged to a new culture vessel to maintain optimal growth conditions. This involves detaching the cells from the culture surface and transferring them to a new flask or dish with fresh medium (Zhang et al., 2020).
6. **Cell Authentication:** Ensuring the identity and purity of cell lines is crucial. Cell authentication methods, such as DNA profiling or isoenzyme analysis, are employed to verify the origin and characteristics of the cultured cells (Shan et al., 2021)..
7. **Applications:** Cell culture is used in various scientific fields, including cell biology, cancer research, drug discovery, regenerative medicine, and biotechnology. It is also an essential tool in the production of vaccines, therapeutic proteins, and other biopharmaceuticals (Zhang et al., 2020).
8. **3D Cell Culture:** Traditional cell culture is often performed in two dimensions, but 3D cell culture methods are becoming increasingly popular. These techniques better mimic the in vivo environment and are used for applications such as tissue engineering and studying complex cell interactions (Shan et al., 2021).

Cell culture is a versatile and powerful tool that has significantly advanced our understanding of cellular biology and has practical applications in medicine and industry. It plays a critical role in the development of new therapies, drugs, and biomedical technologies.

Western Blot Analysis

In this type of analysis the expression changes of DNA methylation enzymes available in protein level are investigated (Zhang et al., 2020). Chemiluminescence is the process used for visualizing the protein bands for analysis. First of all protein is separated by using a protein gel and then transferred to the membrane. Primary and secondary antibodies help in binding and visualizing target protein (Zhang et al., 2020).

Western blot analysis, also known as protein immunoblot, is a widely used laboratory technique in molecular biology and biochemistry (Zhang et al., 2020). It is employed to detect specific proteins in a given sample of tissue homogenate or cell lysate. The technique involves several key steps:

1. **Protein Separation:**

The proteins within the sample undergo initial separation according to their molecular weight through gel electrophoresis. Sodium dodecyl sulfate polyacrylamide gel electrophoresis (SDS-PAGE) is a frequently employed method for this task. SDS functions to denature proteins and confer a negative charge, prompting their migration through the gel toward the positive electrode. The separation is size-dependent, with smaller proteins exhibiting faster migration. (Zhang et al., 2020).

2. **Protein Transfer:**

Following electrophoresis, the isolated proteins must be transferred from the gel to a solid support, usually a nitrocellulose or polyvinylidene difluoride (PVDF) membrane, a process known as protein blotting. The membrane is positioned atop the gel, and an electric current is employed. As a result, the proteins migrate from the gel to the membrane while preserving their relative positions. (Zhang et al., 2020).

3. Blocking:

To mitigate non-specific binding of antibodies, the membrane undergoes incubation with a blocking solution. Frequently used blocking agents encompass bovine serum albumin (BSA) or non-fat dry milk.

4. Primary Antibody Incubation:

A primary antibody specific to the target protein is added to the membrane. This antibody binds to the target protein if it is present. Primary antibodies are usually raised in animals (e.g., rabbits or mice) against a specific protein of interest.

5. Washing: Excess unbound primary antibody is washed away to reduce background signal.

6. Secondary Antibody Incubation:

A secondary antibody, conjugated to an enzyme or fluorophore, is added. This secondary antibody recognizes the primary antibody and binds to it. Common enzymes used are horseradish peroxidase (HRP) or alkaline phosphatase (Zhang et al., 2020).

7. Detection:

The enzyme attached to the secondary antibody catalyzes a reaction that produces a detectable signal. For chemiluminescent detection, a substrate is added, and the resulting chemiluminescence is captured on X-ray film or a digital imaging system. For fluorescence detection, the fluorophore emits light at a specific wavelength when excited by a light source (Tavanapong et al., 2022).

8. Analysis:

The developed film or digital image is analyzed to quantify the amount of the target protein. This can involve measuring the intensity of bands corresponding to the target protein relative to a loading control (Mulenga et al., 2021).

Western blotting is a powerful tool for protein analysis, allowing researchers to assess protein expression levels, confirm the identity of proteins, and study post-translational modifications. It is widely used in molecular biology, cell biology, and biochemistry research.

Analysis of Cell Cycle

Cell cycle analysis is a technique used to study the progression of eukaryotic cells through various phases of the cell cycle—G1 (gap 1), S (synthesis), G2 (gap 2), and M (mitosis). This analysis is crucial for understanding cell proliferation, differentiation, and various cellular responses. The cell cycle can be analyzed through different methods, and flow cytometry is one of the commonly used techniques (Srinivasu et al., 2021).

Deep Learning Approaches

Deep learning based CNN models are very popular in performing image classification tasks in various fields. Medical image processing facilitates doctors and medical practitioners to improve their accuracy and reduce the time wastage in analyzing complex medical images. In 2019 Falconí et.al, have done a comparative analysis on breast cancer identification by using deep learning based image classification models, such as Resnet50,MobileNet, NasNet, Inception V3 . The study enlightens on some area of preprocessing steps to increase the classification accuracy (Falconí et al., 2019). It was found that, Resnet50 and Mobile Net outperformed having accuracy 78.4% and 74.3% respectively (Falconí et al.,

2019). Hence, Mobile Net and Resnet models can be implemented by Transfer learning for performing the image classification task to detect the malignancy. In 2021 Shen, Y et.al, proposed a deep learning based model to detect tumor regions in colorectal cancer histology slides (Shen & Ke, 2021). A deep CNN model is trained with the Monte Carlo(MC) adaptive sampling method design to estimate the patches in a whole slide image(WSI) (Shen & Ke, 2021). They utilized Tangdu Cancer Genome Atlas(TCGA) dataset to study the malignancy (Shen & Ke, 2021). They found that, the overall time is reduced from 56.7% to 71.7% on different slides of tumor with improved classification accuracy. 0.82 (Shen & Ke, 2021). However, classification accuracy can be further improved by modifying the model. Another experiment has done by Adweb et.al in 2021. In this paper they proposed a deep learning based model for analyzing cervical pre-cancerous stages which can transform into cancer (Adweb et al., 2021). Three Residual networks (ResNet) of the same structure were built with different activation functions. colonoscopy cervical image dataset was used for the analysis (Adweb et al., 2021). They found that, PRELU, Leaky-RELU activation functions performed with the accuracies approximately 95% and 90.2% respectively (Adweb et al., 2021). However, it can be improved by modifying the RELU structure used in ResNet (Adweb et al., 2021). In 2021 L kiehl et.al suggested lymph node metastasis prediction system by using deep learning based model (Kiehl et al., 2021). They used 2431 patients whole slide images (WSI) to perform their research work. A deep learning based convolution neural network is utilized here for the analysis of histopathological image data (Kiehl et al., 2021). They got 74.1% test accuracy. However, this model can be further improved with more accuracy rate (Kiehl et al., 2021). Sasmal et al. (2021) employ a deep learning-driven approach for monocular depth estimation (MDE), leveraging it to pinpoint the most informative frames within endoscopic videos to enhance the detection of colorectal cancer. (Sasmal et al., 2021). The zero-shot learning technique proves valuable in recognizing previously unseen classes within endoscopic images. Transfer learning is employed to ascertain the depth of polyps, followed by the 3D reconstruction of polyps using extracted key frames. To achieve optimized results, depth estimation incorporates a modified loss function and regularization methods, while key frame selection considers color space conversion, image moment, edge density, and key point detection. (Sasmal et al., 2021). The study by Jia, Xiao, et al. (2020) centers around polyp recognition within colonoscopy images, accomplished through the creation of a two-stage framework called PLPNet, which proves effective in its methodology. (Jia et al., 2020). Leveraging a CNN architecture enhances accuracy by incorporating the Deep Residual Network (ResNet) as a feature extractor, enabling the extraction of hidden semantics across all levels of the network. Utilizing YOLO proves beneficial in identifying polyp regions within colonoscopy images. This involves employing methods such as multilevel feature learning through ResNet, representing multiscale features via the Feature Extractor Network (FENet), proposing polyp regions, and performing pixel-wise polyp segmentation as part of the comprehensive approach for colorectal cancer recognition. (Jia et al., 2020). The research paper utilizes the CVC-ColonDB and CVC-ClinicDB datasets, while the CVC-300 database serves as the training data for the model. The implementation involves using C++ and Python on a single NVIDIA GeForce GTX TITAN Xp, utilizing the Caffe framework for effective model development. (Jia et al., 2020). PLP Net achieves an accuracy of up to 81.7 in the Recall criterion. Nevertheless, enhancing PLP Net is possible by incorporating stage integration techniques like SSDs, RetinaNet, and YOLO-V3 for multiclass classification coupled with depth estimation. (Jia et al., 2020). Yang et al. (2020) presented the Mask Regions Convolutional Neural Network (MRCNN) in their paper, aiming for accurate detection and segmentation of colon polyps. The network incorporates a Pr.ROI pooling layer to enhance its capabilities. (Yang, 2020). The datasets utilized in this study were gathered from a Chinese hospital and an open challenge. The system's evaluation revealed an average precision of 76%, achieving successful polyp segmentation with an intersection over union (IOU) of 86.87%. However, a recognized limitation in this manuscript is the disorderly convergence rate observed in each function branch of the model. Addressing dataset issues and enhancing model effectiveness could further improve the overall performance. (Yang, 2020). Wallapak et al. (2022) introduced two distinct Artificial Intelligence (AI) methods tailored for clinical trials(Tavanapong et al., 2022). The first variant focuses on the analysis and feedback to enhance the quality of colonoscopy, while the second one is designed for the detection of abnormalities (Tavanapong et al., 2022). This proposal employs diverse methods, including informative frame analysis, assessment of bowel preparation and cleaning, analysis of navigation quality, and polyp detection and segmentation, for the detection of abnormalities. Real-time AI-assisted colonoscopy incorporates several deep learning techniques to facilitate this diagnostic process. (Tavanapong et al., 2022). Yet, the limitation of potential complications arising from medical data privacy remains an ever-present concern. (Tavanapong et al., 2022). Moreover, achieving a perfect AI score for cleanliness does not equate to comprehensive circumferential inspection of the colon, underscoring the necessity for AI-based classifiers in endoscopy (Tavanapong et al., 2022). Mulenga, Mwenge, et al. (2021) put forth the idea of employing stacking and chaining techniques in normalization procedures (Mulenga et al., 2021). Incorporating the chaining approach as an alternative to data normalization, stacking provides an interpretable method to improve accuracy in microbiome and other tabular data. This study employed three publicly available datasets, with the initial dataset comprising 884 samples and 2031 features extracted from curated samples of shotgun sequence data (Mulenga et al., 2021). The second dataset relies on 16S rRNA sequence data, featuring 490 samples and 336 features. In this study, the third dataset encompasses a 16S rDNA dataset sourced from the Machine Learning Repository (MLrepo), consisting of 95 CRC and 95 non-CRC samples. To execute the proposed method and DNN architecture, Python 3 with the Keras framework and Tensor Flow is employed, leveraging the Graphical Processing Unit and Google Colaboratory for program execution. Improvements to the proposed model's performance are achieved through the application of feature selection and rank transformation techniques, yielding Area Under Curve (AUC) values within the range of 0.857 to 0.987. The utilization of the stacking and chaining technique, combined with other state-of-the-art methods, holds significant promise for enhancing this model in future investigations (Mulenga et al., 2021). Numerous studies have explored colorectal malignancy, but our unique approach focuses on investigating the impact of foxtail millet in preventing colorectal malignancy through the application of deep learning-based image classification techniques.

ResNet-50

ResNet-50 is a deep CNN model that is also called Residual Networks or ResNets. The fundamental innovation of ResNet lies in its incorporation of residual blocks, characterized by the inclusion of shortcuts or skip connections. The inclusion of skip connections in the network facilitates the learning of residual functions rather than directly learning the underlying mapping, effectively mitigating the vanishing gradient problem and enabling the training of exceptionally deep networks. ResNet-50 specifically has 50 layers, consisting of a stack of residual blocks. It's often used for image classification tasks, where the input is an image, and the network is trained to classify the image into different categories or classes. Input Layer of Resnet takes an input image, typically of size 224x224 pixels. Convolutional and Pooling Layers are Initial layers which perform convolution and pooling operations to extract features. Residual Blocks are the core building blocks of ResNet. Each block contains multiple convolutional layers and shortcut connections. Global average pooling is often applied to reduce spatial dimensions. Fully Connected (Dense) Layers are the final layers for classification. Resnet is widely used to perform image classification efficiently.

VGG19

VGG19 constitutes a deep convolutional neural network architecture within the Visual Geometry Group (VGG) family of models. This model was presented by the Visual Geometry Group at the University of Oxford in their paper titled "Very Deep Convolutional Networks for Large-Scale Image Recognition." The "19" in VGG19 denotes the number of layers in the network. VGG19 is known for its simplicity and uniformity in architecture. Unlike other models with more complex architectures, VGG models primarily consist of repeated blocks of convolutional layers with small 3x3 filters and max-pooling layers. Input Layer of VGG19 takes an input image, typically of size 224x224 pixels. Several blocks of convolutional layers are used with 3x3 filters, followed by rectified linear unit (ReLU) activation functions. After each block of convolutional layers, max-pooling is applied to reduce spatial dimensions. VGG19 is often used for image classification tasks and serves as a baseline or reference model in many studies.

AlexNet

This model was developed by Alex Krizhevsky, Ilya Sutskever, and Geoffrey Hinton, emerged triumphant in the 2012 ImageNet Large Scale Visual Recognition Challenge (ILSVRC). Input Layer of AlexNet accepts an input image, typically of size 224x224 pixels. AlexNet consists of five convolutional layers, where each layer is followed by a max-pooling layer. These convolutional layers are responsible for learning hierarchical features from the input image. Rectified Linear Unit (ReLU) activation functions are employed following each convolutional layer, introducing non-linearity to the network. LRN was used to normalize the output of some of the layers, promoting competition between features. After each convolutional layer, max-pooling is applied to down sample the spatial dimensions and reduce computational complexity. The final layers are fully connected, leading to the output layer for classification. The fully connected layers are followed by dropout layers to reduce over fitting. The output layer uses the softmax activation function to produce class probabilities.

MobileNet

MobileNet represents a family of lightweight convolutional neural network (CNN) architectures meticulously crafted for efficient inference on mobile and edge devices characterized by limited computational resources. The key motivation behind MobileNet is to provide a good trade-off between model size, computational efficiency, and accuracy. It predominantly utilizes depth wise separable convolutions, encompassing a depth wise convolution that filters each input channel independently, succeeded by a point wise convolution, which involves a linear combination of the output channels. This reduces the number of parameters and computations compared to traditional convolutions. MobileNet introduces two hyper parameters, width multiplier and resolution multiplier that allow users to trade-off between model size and accuracy. The width multiplier adjusts the number of channels in each layer, and the resolution multiplier scales the input image size in this context. MobileNetV2 introduces a bottleneck design, similar to some other efficient architecture like ResNet. It uses a combination of 1x1 convolutions to reduce and then expand the dimensions of the feature maps. MobileNetV2 also introduces inverted residuals with linear bottleneck, which helps in capturing non-linearities more efficiently. The design of MobileNet is specifically aimed at mobile and edge devices where computational resources, memory, and power consumption are limited. These models have gained widespread adoption across various applications, notably in image classification, object detection, and semantic segmentation. They are particularly favored in scenarios where real-time inference on mobile devices is imperative. MobileNet comes with various versions such as MobileNetV1, MobileNetV2, or MobileNetV3.

PROPOSED APPROACH

In this paper, our main goal is to analyze the effect of foxtail millet as a supplement on preventing colorectal malignancy. In order to perform this analysis, first of all we propose a Mobile Net model that is capable of classify the colorectal malignancy types by scanning the histopathological images (Kiehl et al., 2021). Hence, proposed approach can be utilized to identify the effect of foxtail millet bran for preventing colorectal malignancy by using deep learning based approach. This is done phase wise as explained in the following figure 1.

Dataset

To perform image classification task, Histopathological data is collected from Kather_image_tiles dataset publicly available in Kaggle library. It contains 5000 (.tif) images of eight different classes of colorectal malignancy each having 625 images. The classes are Tumor, Stroma, Complex, Lympho, Debris, Mucosa, Adipose, Empty. From 5000 images 2000 images are selected randomly from each category having 250 images each to implement this model.

Technical Specification

The proposed model is developed in the Google Colab platform. Python programming language is used to code the model. Matplotlib, Numpy, Sklearn, CV2 libraries are utilized for developing the business logic in a proper manner. Matplotlib is very efficient for plotting various kinds of maps in an efficient manner. Numpy is specifically used for various numeric operations that are performed to develop our model. Sklearn is a predictive tool for data analysis.CV2 is an openCV application mainly includes library for performing computer vision related works. Integration of all the libraries helps the programmers to use their features to develop the model.

Matplotlib

It is a popular data visualization library in Python that provides a wide variety of plotting options. It allows you to create high-quality static, animated, and interactive visualizations in Python. Matplotlib is often used for creating line plots, scatter plots, bar plots, histograms, and more.

Numpy

It is a robust numerical computing library in Python, offering extensive support for large, multi-dimensional arrays and matrices. Additionally, it encompasses a range of mathematical functions designed to operate on these arrays. NumPy holds a pivotal role as a foundational package for scientific computing in Python and finds widespread application in fields such as data science, machine learning, signal processing, and beyond.

The core object in NumPy is the numpy.ndarray, commonly known as an array. An array is a structured grid of values, all of the same type, accessed through a tuple of non-negative integers. Notably more efficient for numerical operations than Python lists, NumPy arrays offer a comprehensive suite of mathematical functions for diverse array operations. These encompass element-wise operations, linear algebra functions, statistical operations, and beyond.

Sklearn

Scikit-learn, often abbreviated as sklearn, stands as a widely-used machine learning library in Python. It offers straightforward and effective tools for data analysis and modeling, featuring a diverse range of algorithms for tasks such as classification, regression, clustering, dimensionality reduction, and others. Built on the foundations of NumPy, SciPy, and Matplotlib, Scikit-learn is crafted to seamlessly interoperate with these libraries.

OpenCV

OpenCV (Open Source Computer Vision Library) is an open-source software library for computer vision and machine learning. It furnishes a comprehensive set of tools and functions for tasks such as computer vision applications, image and video processing, and machine learning. OpenCV is primarily implemented in C++ and features Python bindings, ensuring accessibility to a broad spectrum of developers. OpenCV is widely used in various applications, including robotics, computer vision research, augmented reality, and more. Its versatility, extensive documentation, and active community make it a go-to library for many computer vision and image processing tasks.

Phase Description

Each phase describe a specific task for the better prediction of colorectal malignancy at its initial stage. This can be really beneficial for doctors, Medical Practitioners to accomplish their work with an ease due this model.

Preprocessing Phase

As shown in the following figure(fig-1), first of all images are imported from the Kather_image_tiles dataset available in Kaggle library. In the next phase, data cleaning and processing is performed to prepare the data for training purpose. This phase involves image resizing, labeling of resized images, conversion of images into numpy arrays. In this phase all the preprocessing tasks are done sequentially. Image resizing is important to maintain the equality in the dimensions of imported images. We have to label the images in a proper manner to train our model otherwise model can't give desired result. Conversion of images to numpy arrays is beneficial for analyzing the images by examining the pixel value represented through an array. In the third phase, we split the total amount of data into training set and testing set. Training set contains 80% of data and testing set contains 20% of the entire data available in the dataset. In fourth phase the image data are fed to the Mobile Net model to train the model properly. After training, testing of model is done by utilizing 20% of the entire available images selected for our research work.

Mobile Net Architecture

Mobile Net model is a CNN based lightweight architecture. It contains two types of convolution layers i.e. Depth wise convolution layer and point wise convolution layer (Kiehl et al., 2021). The primary objective of incorporating these two types of convolution layers is to alleviate the substantial computational operations associated with classification tasks.. Depth wise convolution layer uses separate convolution or tensor for each image layer (Mulenga et al., 2021). Therefore if the image is of size 200 x 200 x 3, then it will be divided into 3 layers of RGB having size 200 x 200 x 1 each. Three separate convolution of size 3 x3 x 1 is used in depth wise convolution. There is no interaction between parallel layers of convolutions. Point wise convolution for the above depth wise convolution will be 1x1x3, which will multiply the output of the three tensor layers and finally adding them up to get the desired result. These layers are beneficial for reducing the computational time by reducing the heavy amount operations (Mulenga et al., 2021).

Model Compilation and Validation Phases

Model compilation is done with adam optimizer by utilizing the training set and testing set of data. The model gives desired result after going through 10 numbers of epochs. We found 0.9025 test accuracy by just running 10epochs. Hence, this model can be a useful one for identifying the class of colorectal malignancy by examining histopathological images. The validation of the model is decided by the resultant accuracy and loss ratio obtained by this model execution.

Adam, short for Adaptive Moment Estimation, amalgamates concepts from two well-known optimization algorithms, namely RMSprop and Momentum. The primary goal of Adam is to provide a method for efficiently and stably updating model weights during the training process. Adam maintains two moving averages for each parameter: the first moment (mean) and the second moment (uncentered variance). These moving averages are calculated using exponential decay. The first moment (mean) is akin to the momentum term in traditional stochastic gradient descent (SGD) with momentum, and it helps to smooth out the update trajectory. Since the moving averages are initialized to zero, there is a potential for bias in the early updates. To address this, Adam incorporates a bias correction mechanism, adjusting the estimates to account for the initialization bias. Adam adapts the learning rates for each parameter individually. It does so by dividing the learning rate for each parameter by the square root of the second moment estimate. This helps in scaling the learning rates dynamically based on the historical gradients.

Predictive System

Predictive system is a software application that utilizes our model to predict the colorectal type by just analyzing a histopathological image provided by user input. User of the predictive system has to provide a histopathological image that he/she wants to analyze. This system is able to predict the class of colorectal malignancy where there is an Empty class included. Empty class indicates no colorectal cancer. This model can be further utilized for addressing the effectiveness of foxtail millet.

The following Mobile Net model is designed by transfer learning for implementing image classification task for the identification of colorectal malignancy type. This can be further utilized to test the effect of foxtail millet in preventing colorectal malignancy by just analyzing histopathological images of some specific patients before and after applying foxtail millet bran as Supplement

Figure 1. Proposed predictive system mobile net model

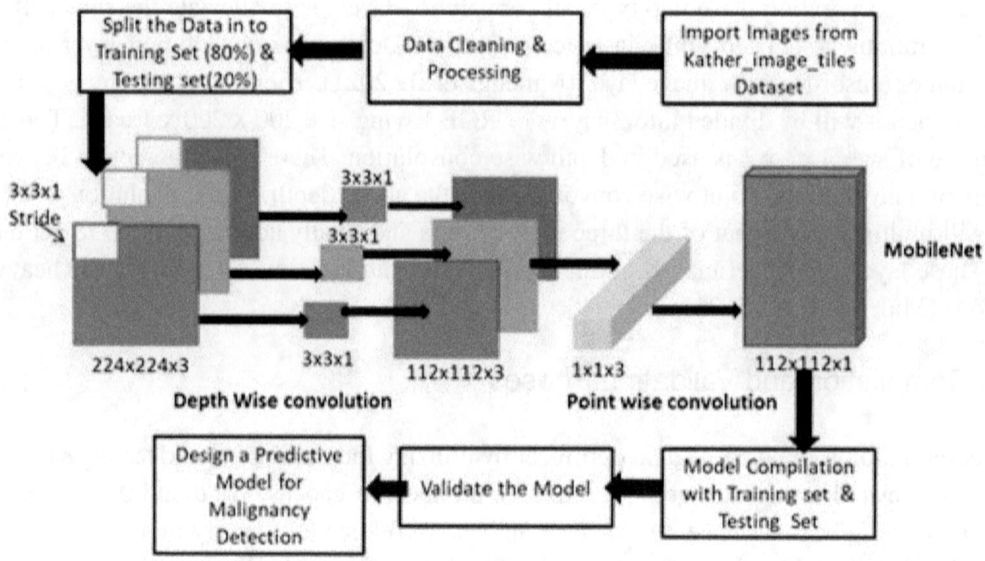

RESULT AND DISCUSSIONS

Our study demonstrates that Mobile Net model is an efficient model to predict the colorectal malignancy by analyzing the histopathological images with an acceptable accuracy and loss integration criteria. Furthermore, it can be very beneficial to analyze the effectiveness of foxtail millet. As described in the following figures, fig-1 explains the training accuracy by blue line and test accuracy by brown line, which indicates 90% test accuracy. Fig-2 explains the training loss by blue line test loss by brown line which indicates 29% test loss. We observe that the mobileNetV2 model is a good model having very good accuracy to identify the colorectal malignancy. This model can be helpful to observe the effectiveness of foxtail millet in curing malignant tissues that are available in colon and rectum. This model can be improved by integrating with other models to get optimized results but that is required when the model does not perform well with any observed.

Figure 2. Training accuracy versus test accuracy

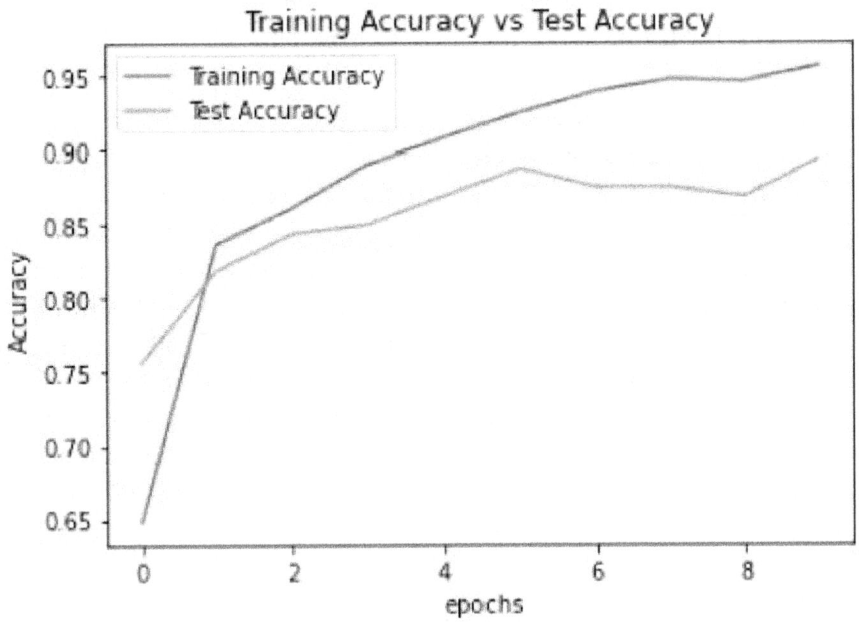

Figure 3. Training loss versus test loss

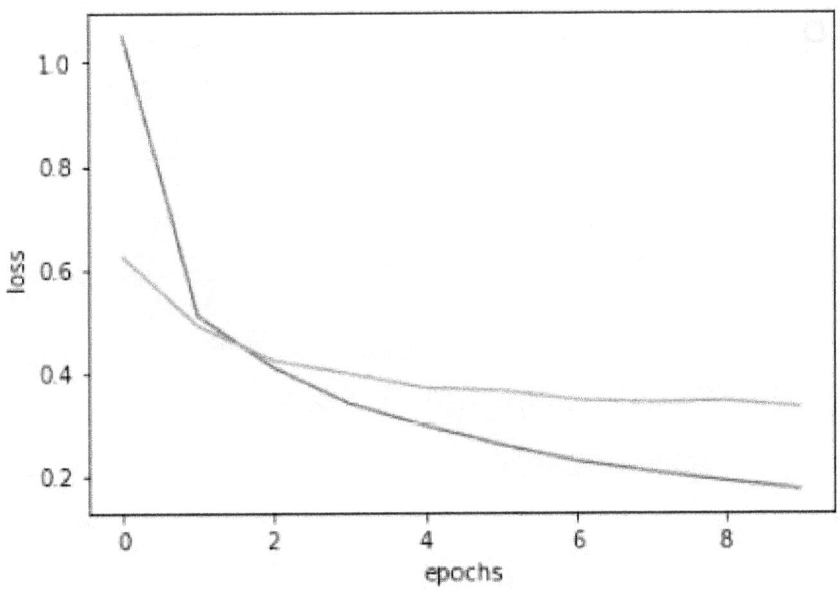

COMPARATIVE STUDY

MobileNetV2 and ResNet50

MobileNet and ResNet50 are both popular convolutional neural network (CNN) architectures used in computer vision tasks, particularly for image classification. However, they have different design philosophies and trade-offs. MobileNet is mainly used for mobile and edge devices with resource constraints. It utilizes depth wise separable convolutions to reduce the number of parameters and computations and employs linear bottlenecks to maintain representational power and where a balance between accuracy and efficiency is crucial, MobileNet might be a better choice. However ResNet50 uses residual blocks with skip connections to ease the training of very deep networks. ResNet50 is more complex than MobileNet and typically used in scenarios where computational resources are less constrained. Generally achieves lower accuracy as compared to deeper architectures like ResNet50. MobileNet is well-suited for scenarios where computational resources are limited, and a slightly lower accuracy is acceptable. In our case the colorectal image data is limited and the challenge is to develop an efficient prediction model with good accuracy. Therefore MobileNet is the first line of preference to design our prediction model.

MobileNetV2 and VGG19

MobileNet is a lightweight architecture to achieve a good balance between accuracy and efficiency however VGG19 is known for its good accuracy and generalization performance. When we run VGG19 model with the available data we got good accuracy up to 85% but with MobileNet the percentage of accuracy is better than VGG19 as shown in the figure(4) & figure(5). MobileNet has a smaller model size compared to VGG19. It is designed to have a lower memory footprint, making it suitable for deployment on devices with limited storage. VGG19 has a larger model size compared to more recent architectures. It requires more storage space for both the model parameters and activations during computation.

Figure 4. VGG19 train loss

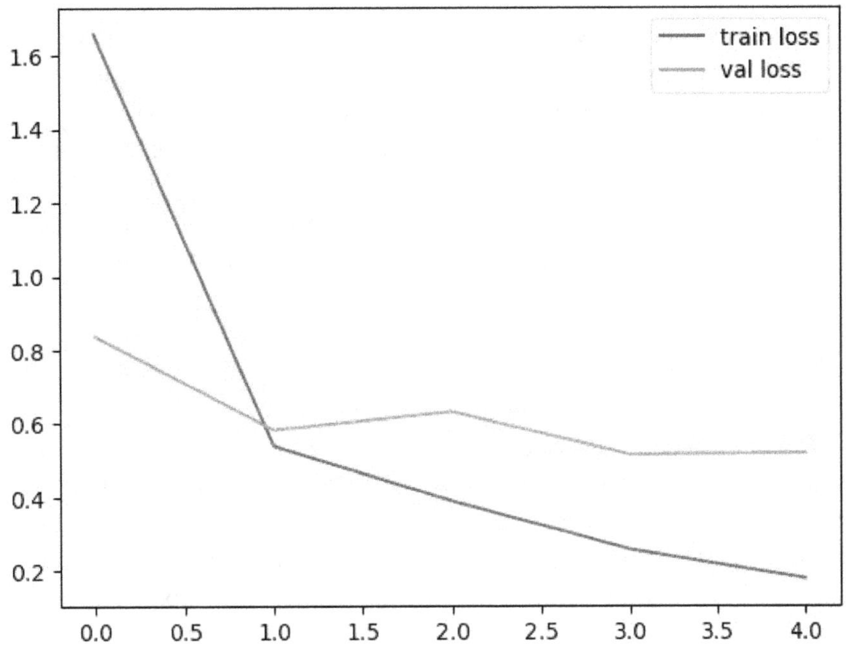

Figure 5. VGG19 train accuracy

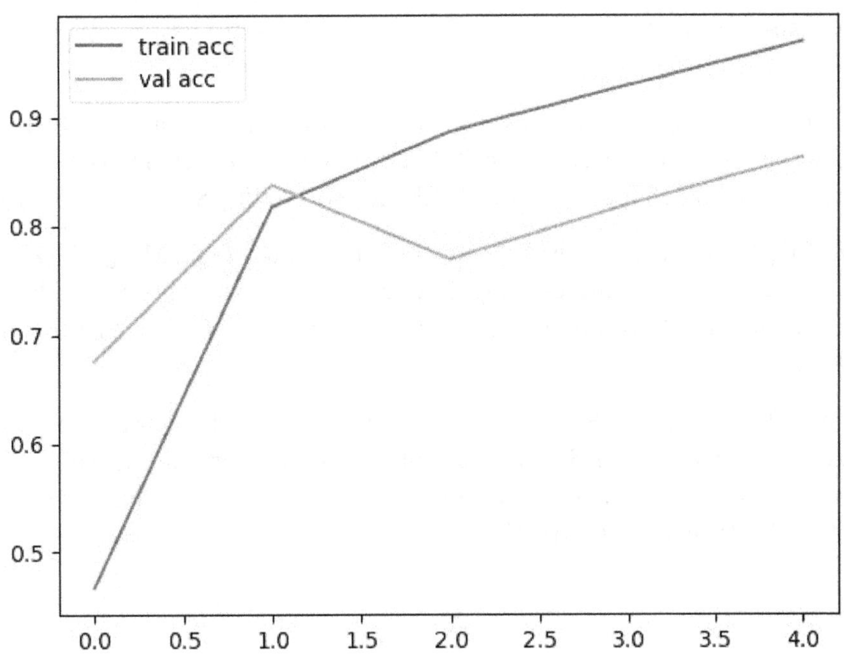

AlexNet and MobileNet

AlexNet comprises a total of eight layers, featuring five convolutional layers succeeded by three fully connected layers. This architecture incorporates larger convolutional layers for its operations. filters (e.g., 11x11 and 5x5) compared to more modern architectures. AlexNet might be a good choice for research and applications where cutting-edge accuracy is crucial. MobileNet is well-suited for real-time applications on mobile and edge devices where computational resources are limited. In our research image classification is the main priority and computational resources are also a constraint which must be overcome. To achieve our goal MobileNet is more appropriate than AlexNet.

CONCLUSION

Effect of Foxtail Millet to prevent colorectal malignancy can be analyzed by using deep learning based image classification model. From the above observation we conclude that, deep Learning based Mobile Net model is an efficient model to check colorectal Malignancy. Hence this model can be utilized to check the effect of Foxtail Millet on Colorectal Malignancy by analyzing the histological images of a particular patient before and after the use of this Nutritional diet. This model can be further improved by increasing the epochs of this model. For better implementation of this model a predictive system can be designed for medical practitioners.

REFERENCES

Adweb, K. M. A., Cavus, N., & Sekeroglu, B. (2021). Cervical cancer diagnosis using very deep networks over different activation functions. *IEEE Access : Practical Innovations, Open Solutions*, *9*, 46612–46625. doi:10.1109/ACCESS.2021.3067195

Falconí, L. G., Pérez, M., & Aguilar, W. G. (2019, June). Transfer learning in breast mammogram abnormalities classification with mobilenet and nasnet. In 2019 international conference on systems, signals and image processing (IWSSIP) (pp. 109-114). IEEE. doi:10.1109/IWSSIP.2019.8787295

Jia, X., Mai, X., Cui, Y., Yuan, Y., Xing, X., Seo, H., Xing, L., & Meng, M. Q.-H. (2020). Automatic polyp recognition in colonoscopy images using deep learning and two-stage pyramidal feature prediction. *IEEE Transactions on Automation Science and Engineering*, *17*(3), 1570–1584. doi:10.1109/TASE.2020.2964827

Kanwal, S., Joseph, T. P., Aliya, S., Song, S., Saleem, M. Z., Nisar, M. A., Wang, Y., Meyiah, A., Ma, Y., & Xin, Y. (2020). Attenuation of DSS induced colitis by Dictyophora indusiata polysaccharide (DIP) via modulation of gutmicrobiota and inflammatory related signaling pathways. *Journal of Functional Foods*, *64*, 103641. doi:10.1016/j.jff.2019.103641

Kiehl, L., Kuntz, S., Höhn, J., Jutzi, T., Krieghoff-Henning, E., Kather, J. N., Holland-Letz, T., Kopp-Schneider, A., Chang-Claude, J., Brobeil, A., von Kalle, C., Fröhling, S., Alwers, E., Brenner, H., Hoffmeister, M., & Brinker, T. J. (2021). Deep learning can predict lymph node status directly from histology in colorectal cancer. *European Journal of Cancer, 157*, 464–473. doi:10.1016/j.ejca.2021.08.039 PMID:34649117

Mulenga, M., Kareem, S. A., Sabri, A. Q. M., & Seera, M. (2021). Stacking and chaining of normalization methods in deep learning-based classification of colorectal cancer using gut microbiome data. *IEEE Access : Practical Innovations, Open Solutions, 9*, 97296–97319. doi:10.1109/ACCESS.2021.3094529

Nadeem, M. S., Kumar, V., Al-Abbasi, F. A., Kamal, M. A., & Anwar, F. (2020). Risk of colorectal cancer in inflammatory bowel diseases. *Seminars in Cancer Biology, 64*, 51–60. doi:10.1016/j.semcancer.2019.05.001 PMID:31112753

Peng, Y., Yan, Y., Wan, P., Chen, D., Ding, Y., Ran, L., Mi, J., Lu, L., Zhang, Z., Li, X., Zeng, X., & Cao, Y. (2019). Gut microbiota modulation and anti-inflammatory properties of anthocyanins from the fruits of Lycium ruthenicum Murrayin dextran sodium sulfate-induced colitis in mice. *Free Radical Biology & Medicine, 136*, 96–108. doi:10.1016/j.freeradbiomed.2019.04.005 PMID:30959170

Sasmal, P., Paul, A., Bhuyan, M. K., Iwahori, Y., & Kasugai, K. (2021). Extraction of Key-Frames From Endoscopic Videos by Using Depth Information. *IEEE Access : Practical Innovations, Open Solutions, 9*, 153004–153011. doi:10.1109/ACCESS.2021.3126835

Shan, S., Lu, Y., Zhang, X., Shi, J., Li, H., & Li, Z. (2021). Inhibitory effect of bound polyphenol from foxtail millet bran on miR-149 methylation increases the chemosensitivity of human colorectal cancer HCT-8/Fu cells. *Molecular and Cellular Biochemistry, 476*(2), 513–523. doi:10.100711010-020-03906-4 PMID:33011952

Shen, Y., & Ke, J. (2021). Sampling based tumor recognition in whole-slide histology image with deep learning approaches. IEEE/ACM Transactions on Computational Biology and Bioinformatics, 19(4), 2431-2441.

Srinivasu, P. N., SivaSai, J. G., Ijaz, M. F., Bhoi, A. K., Kim, W., & Kang, J. J. (2021). Classification of skin disease using deep learning neural networks with MobileNet V2 and LSTM. *Sensors (Basel), 21*(8), 2852. doi:10.339021082852 PMID:33919583

Tavanapong, W., Oh, J., Riegler, M. A., Khaleel, M., Mittal, B., & De Groen, P. C. (2022). Artificial intelligence for colonoscopy: Past, present, and future. *IEEE Journal of Biomedical and Health Informatics, 26*(8), 3950–3965. doi:10.1109/JBHI.2022.3160098 PMID:35316197

Yang, X. (2020). Colon polyp detection and segmentation based on improved MRCNN. *IEEE Transactions on Instrumentation and Measurement, 70*, 1–10. doi:10.1109/TIM.2020.2986875

Zhang, B., Xu, Y., Liu, S., Lv, H., Hu, Y., Wang, Y., Li, Z., Wang, J., Ji, X., Ma, H., Wang, X., & Wang, S. (2020). Dietary supplementation of foxtail millet ameliorates colitis-associated colorectal cancer in mice via activation of gut receptors and suppression of the STAT3 pathway. *Nutrients, 12*(8), 2367. doi:10.3390/nu12082367 PMID:32784751

Chapter 4
From Fields to Satellites:
Revolutionizing Millet Cultivation for Sustainable Yields

Vijaya Sindhoori Kaza
https://orcid.org/0000-0002-1364-8530
Stanley College of Engineering and Technology for Women, India

Anisha P. R.
Stanley College of Engineering and Technology for Women, India

Rithika Badam
Stanley College of Engineering and Technology for Women, India

ABSTRACT

This chapter explores the impact of millet cultivation and sustainable development on agriculture, focusing on advancements in production and environmental practices. Key factors influencing production quantity, quality, location, etc., are analysed. It presents an overview of food security assessments, modelling, mapping, and monitoring techniques, utilizing data from various satellites, from spectroradiometers, and ground-truthing, the grain yield of different millet varieties is estimated, accounting for regional variations and the influence of soil bunds on crop growth. Through the application of the random forest technique, the chapter demonstrates how crop yields can be identified and predicted using regression models based on rainfall patterns and millet suitability in specific Indian states. The findings showcase how monitoring techniques can empower farmers in various regions to achieve higher yields, sustainable food production, and improved crop quality along with addressing the limitations or challenges encountered during the research process.

DOI: 10.4018/978-1-6684-9819-4.ch004

Copyright © 2023, IGI Global. Copying or distributing in print or electronic forms without written permission of IGI Global is prohibited.

1. INTRODUCTION

1.1. Background on Millet Cultivation and Its Role in Sustainable Development

Millet cultivation has a long history, with archaeological evidence suggesting that it was one of the earliest domesticated cereal crops. Ancient civilizations in Asia and Africa, such as those in China, India, and sub-Saharan Africa, relied on millet as a staple food source. Over time, it spread to other parts of the world, adapting to various climates and becoming an integral part of diverse culinary traditions (Adimassu, 2014).

Table 1. Global scenario of millets area and production region wise (2019)

Regions	AREA (lakh ha)	Production (lakh ton)
Africa	489 (68%)	423 (49%)
Americas	53 (7%)	193 (23%)
Asia	162 (23%)	215 (25%)
Europe	8 (1%)	20 (2%)
Australia & New Zealand	6 (~1%)	12 (~1%)
India	138 (20%)	20 (2%)
WORLD	**718**	**863**

[Source: YES BANK Analysis]

One of the key reasons millets are highly valued in sustainable agriculture is their exceptional resilience to environmental stressors. Millet is well-suited for regions with limited water resources and unpredictable rainfall patterns. Its deep root system allows it to access water from deeper soil layers, making it more drought-tolerant compared to other cereal crops like rice or wheat (Scherr, 2012).

Figure 1. Annual per capita consumption of millets in India(kg)
(USDA and YES BANK Analysis)

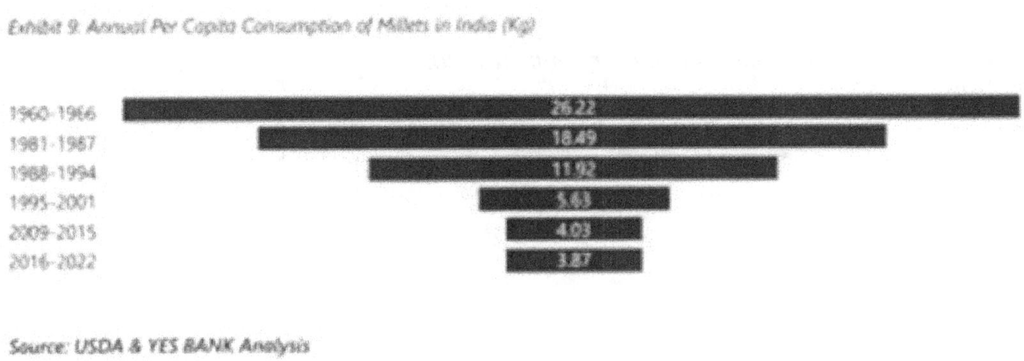

Sustainable development is deeply connected to the concept of food security, which encompasses access to sufficient, safe, and nutritious food for all. Millet plays a pivotal role in achieving food security, particularly for vulnerable populations in developing regions (Lopez-Ridaura, 2019).

Furthermore, the crop's versatility allows it to serve not only as a staple food but also as an ingredient in various culinary applications. From traditional porridges and flatbreads to modern millet-based products like gluten-free flour, snacks, and beverages, millet offers an array of culinary possibilities (Gatti, 2017). Another significant contribution of millet to sustainable development lies in its role in enhancing soil fertility and reducing the need for synthetic fertilizers. Millet can be part of crop rotation and intercropping systems, where different crops are planted in sequence or together on the same piece of land (Kassie, 2007). As a part of sustainable agricultural systems, millet cultivation also has socio-economic benefits. It can enhance local livelihoods by providing income opportunities for smallholder farmers, promoting food sovereignty, and supporting rural economies (Rowhani, 2011).

In conclusion, millet's historical significance, resilience to challenging conditions, and nutritional value make it an essential crop for sustainable development. By contributing to food security, enhancing soil health, conserving water resources, and supporting local economies, millet cultivation exemplifies the principles of sustainable agriculture (Chang, 2017).

1.2. Significance of Agricultural Advancements, Production, and Consumption

Over the past few decades, agricultural advancements have played a crucial role in revolutionizing the production and consumption of millet, a crop with significant potential to address contemporary agricultural challenges (Lobell & Burke, 2009).

One of the key areas of progress in millet agriculture has been in production techniques. Through the application of modern agricultural practices, farmers have been able to achieve higher yields from their millet crops (Hegarty-Craver, 2020).

Water scarcity is a major concern in many agricultural regions, but with advancements in irrigation methods, millet farmers have been able to optimize water usage (Zhang, 2022).

Another significant aspect of agricultural advancements is the identification and promotion of regionally-suited millet varieties. Different millet varieties have varying levels of resistance to pests and diseases, and by selecting and promoting those that are well-adapted to specific regions, farmers can reduce their reliance on chemical pesticides and enhance crop resilience (Gassner, 2019).

Figure 2. Millet production estimates of major countries (2022)
(YES BANK Analysis)

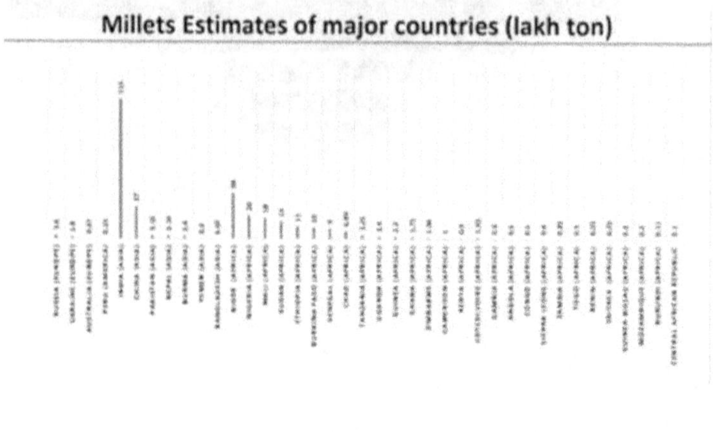

Beyond production, the significance of these advancements extends to the consumption of millet. In recent years, millet has gained recognition as a highly nutritious grain, rich in essential nutrients such as protein, fibre, and minerals. As a result, consumer preferences have shifted towards healthier food options, and millet has found its way into mainstream diets. Its low glycemic index and gluten-free nature have also contributed to its popularity among health-conscious consumers and those with dietary restrictions (Bandyopadhyay, 2019).

Figure 3. (a) Country-wise percentage share in millet consumption (2022) (b) Country-wise millets consumption volumes (2022)
(USDA and YES BANK Analysis)

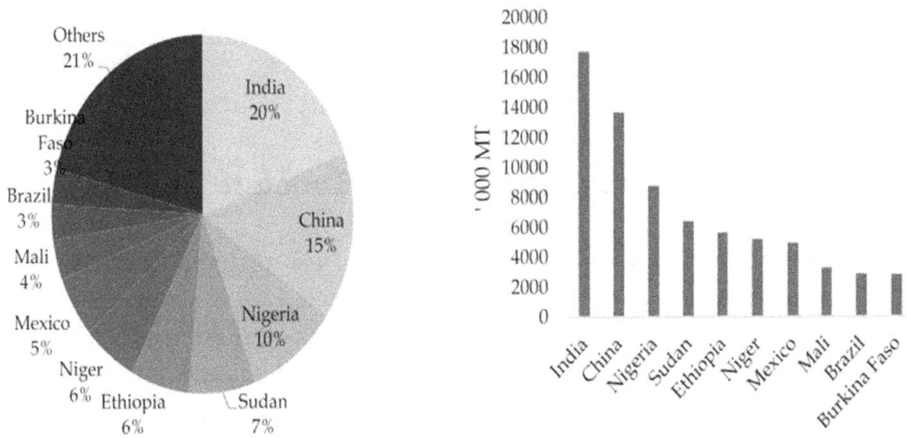

(a) **(b)**

The surge in demand for millet in urban markets has created new market opportunities for farmers. As the market for millet-based products grows, it has created value addition opportunities, such as processing millet into various food items like breakfast cereals, snacks, and gluten-free flours. This diversification has contributed to improving the incomes and livelihoods of farming communities, encouraging them to continue investing in millet cultivation (Bandyopadhyay, 2019).

Figure 4. (a) Global consumption of Sorghum (Jowar) (2022) (b) Global consumption of other millets (2022)
(YES BANK Analysis)

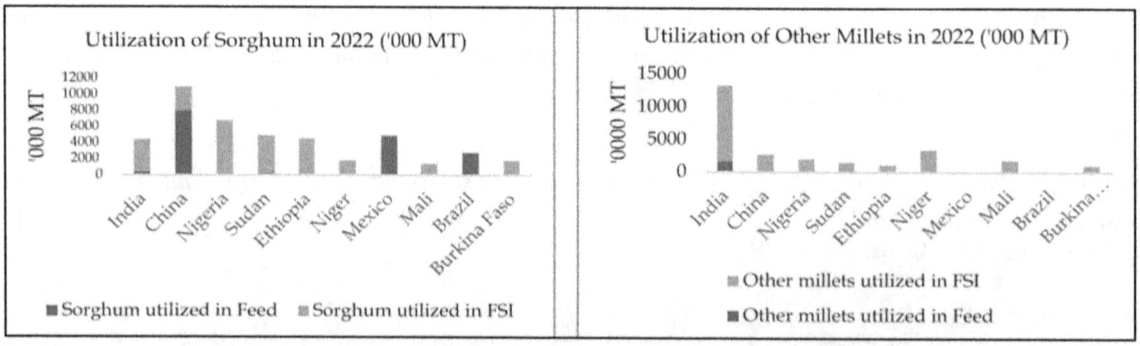

(a) **(b)**

1.3. Overview of the Chapter's Objectives and Structure

This chapter provides a comprehensive exploration of the transformative impact of millet cultivation and sustainable development on agriculture. It aims to illuminate various aspects of millet production, consumption, and environmental practices, with a specific focus on sustainability. The chapter delves into interconnected themes, analyzing key factors influencing millet production, including agronomic practices and climate variability. Emphasizing the importance of sustainable agricultural practices, it highlights their role in enhancing millet yields, improving crop quality, and ensuring food security in the face of changing environmental conditions.

The chapter also introduces cutting-edge technologies for food security assessments, modeling, and mapping, emphasizing data-driven decision-making to address food security challenges effectively. It demonstrates the practical application of the Random Forest technique in identifying and predicting crop yields, particularly in millet cultivation. Utilizing data from various sources, including satellite sensors and ground-truthing, the chapter estimates grain yields for different millet varieties and considers regional variations and the influence of sustainable practices like soil bunds.

Furthermore, the chapter illustrates how monitoring techniques can empower farmers and entire agricultural communities to make informed decisions, leading to higher yields, sustainable food production, and improved crop quality. It promotes transparency by acknowledging research limitations and challenges, fostering continuous improvement in methodologies.

2. MILLET PRODUCTION: FACTORS AND CHALLENGES

Millet cultivation holds a significant historical significance, dating back thousands of years, and has been a staple crop for various civilizations across the world. This resilient and versatile grain has recently gained attention for its potential in promoting sustainable agriculture and addressing various environmental and nutritional challenges. However, successful millet production is influenced by a complex interplay of factors that can determine the outcomes of this vital crop (Bandyopadhyay, 2019).

i. *Climatic Factors:* Millets are generally hardy crops and can adapt to a range of climatic conditions. However, they thrive best in semi-arid to arid regions with moderate rainfall. Drought and excessive rainfall can both pose challenges to millet cultivation (Hatfield, 2011).
ii. *Soil Quality:* Millets can grow in poor soil conditions, but they perform best in well-drained, loamy soils with good organic matter. The soil's nutrient content, pH levels, and soil structure significantly impact millet growth and productivity (Kassie, 2007).
iii. *Pest and Disease Management:* Millet crops are susceptible to various pests and diseases that can reduce yields or even lead to crop failure. Integrated pest management strategies and disease-resistant varieties are essential to combat these challenges while minimizing the use of harmful pesticides (Scherr, 2012).
iv. *Seed Quality:* High-quality seeds play a crucial role in determining the success of millet cultivation. Using certified seeds of improved varieties ensures better germination rates, disease resistance, and overall crop performance (Adimassu, 2014).
v. *Water Management:* Efficient water management practices are crucial for millet production, especially in regions facing water scarcity. Techniques such as rainwater harvesting, drip irrigation, and mulching can optimize water usage and improve crop yields (Hegarty-Craver, 2020).

Figure 5. (a) Season wise rainfall distribution over the country as a whole (1901-2020). Winter (January to February) 147% of LPA Pre-monsoon (March to May); 100% of LPA Monsoon (June to September): 106% of LPA Post-monsoon (Oct to Dec): 119% of LPA (b) All India mean temperature anomalies (1901-2020).
(Govt. of India, CDSP)

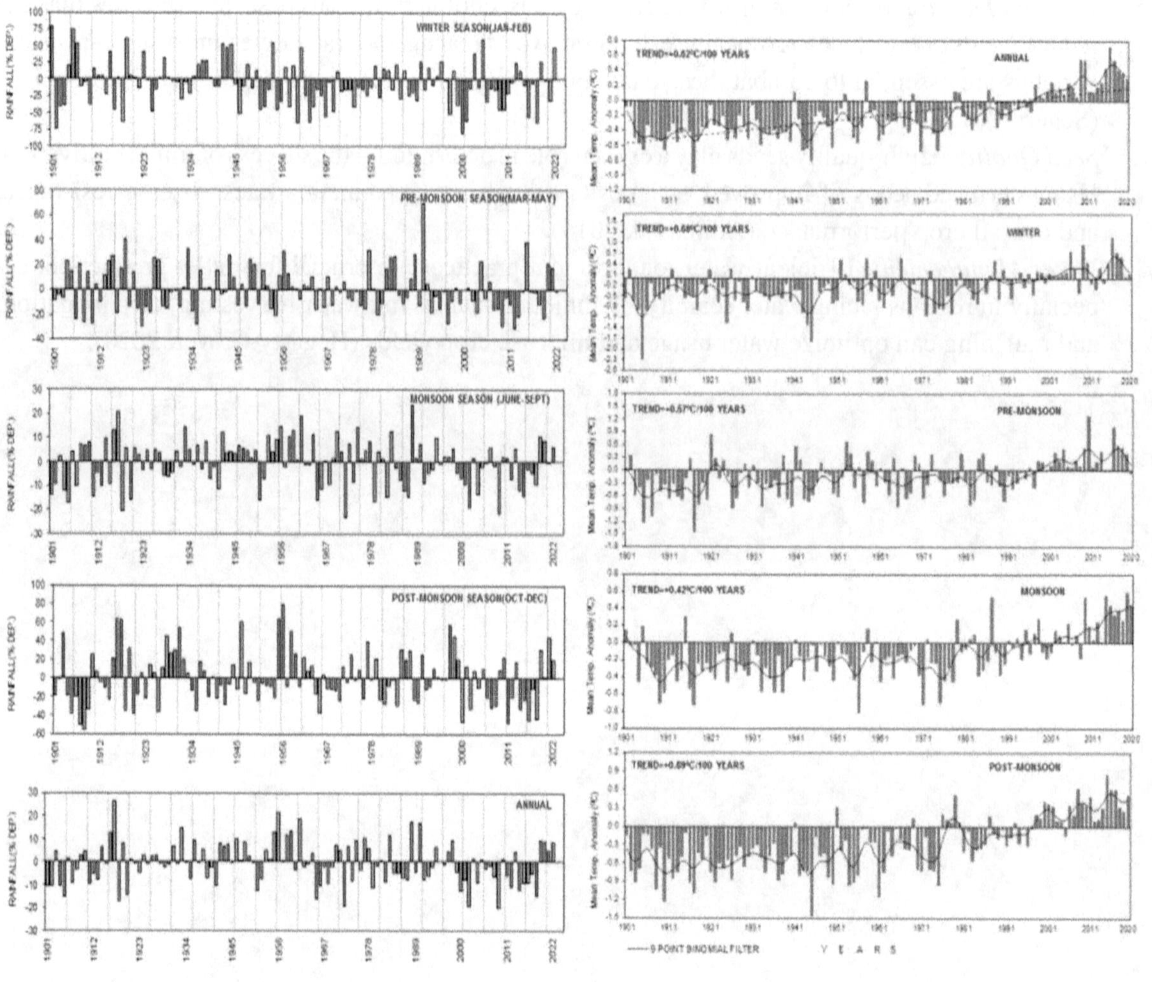

(a) **(b)**

vi. *Market Access:* Millet farmers often face challenges related to market access and fair prices for their produce. Creating efficient supply chains and linking farmers to markets can improve their livelihoods and incentivize increased millet cultivation (Gassner, 2019).

vii. *Climate Change:* With changing climatic patterns, millet production faces uncertainties and challenges. Extreme weather events, unpredictable rainfall, and temperature fluctuations can disrupt crop cycles and lead to reduced yields (Lobell & Burke, 2009).

viii. *Traditional Farming Practices:* In some regions, traditional farming practices and lack of access to modern agricultural technologies may limit millet productivity. Educating farmers about improved techniques and providing access to appropriate tools can enhance production levels (Rowhani, 2011).

2.1. Understanding the Importance of Crop Intensities for Millet Production

Millet, a widely cultivated cereal crop, plays a crucial role in global food security, particularly in regions with challenging climatic conditions and limited agricultural resources. The success of millet production is highly dependent on the way the crop is intensively cultivated. Crop intensities encompass various aspects, including planting density, spacing, and intercropping practices, which significantly influence millet yields and overall agricultural sustainability.

Understanding the relationship between crop intensities and millet yields is essential for optimizing production and ensuring ecological sustainability. Overly intensive cultivation practices may lead to land degradation, soil erosion, and decreased biodiversity. Conversely, too low intensities might not meet the required food demand, potentially leading to food shortages (Scherr, 2012).

Traditional agricultural practices have long recognized the importance of managing crop intensities. Mixed cropping systems, where multiple crops are grown together in the same field, have been employed by small-scale farmers for generations. These systems often promote ecological balance, as different crops interact synergistically, benefiting from each other's presence. Traditional knowledge also includes practices like crop rotation, which allows land to recover and replenish essential nutrients, thereby maintaining soil health.

In modern agriculture, the concept of integrated pest management (IPM) has gained prominence. IPM aims to control pests, diseases, and weeds using a combination of biological, mechanical, and chemical methods. By carefully managing crop intensities and introducing natural predators, farmers can reduce the reliance on synthetic pesticides and promote a healthier agroecosystem (Scherr, 2012).

In conclusion, the importance of crop intensities for millet production cannot be overstated. Balancing planting density, spacing, and adopting intercropping practices are essential for achieving sustainable yields and ensuring food security. Combining traditional wisdom with modern techniques like integrated pest management can help farmers make informed decisions to enhance millet production while safeguarding the environment for future generations.

2.2. Exploring Different Crop Watering Techniques and Their Impact on Yields

Water scarcity is a significant obstacle faced by farmers engaged in millet cultivation, particularly in regions with arid and semi-arid climates. The availability of water is erratic, making it challenging to ensure adequate and consistent water supply for millet crops.

The choice of irrigation method has a direct impact on millet yields and water use efficiency. Traditional flood irrigation, while straightforward to implement, can lead to water wastage and lower yields, making it less suitable for water-scarce regions. On the other hand, modern techniques like drip irrigation and sprinkler irrigation have the potential to enhance millet yields significantly while optimizing water usage (Zhang, 2022).

Drip irrigation, with its precise water application, ensures that plants receive the required moisture, resulting in better growth and higher yields. It also reduces the strain on water resources and promotes sustainable water management in millet farming. Sprinkler irrigation, with its ability to cover larger areas efficiently, can be particularly advantageous for mid-sized farms in water-stressed regions.

Rainwater harvesting emerges as a promising solution to address water stress in millet cultivation. By capturing and storing rainwater, farmers can secure an additional source of water during dry periods, reducing their reliance on traditional sources. This approach not only contributes to higher millet yields but also fosters more resilient and sustainable farming practices. Rainwater harvesting systems can be tailored to fit the specific needs and scale of each farm, making it a versatile and adaptable solution for different regions (Zhang, 2022). By implementing these water-efficient techniques, farmers can not only enhance their millet production but also contribute to better water resource management and long-term agricultural sustainability in arid and semi-arid regions.

2.3. Evaluating the Suitability of Various Millet Types for Different Regions

Millet encompasses a diverse range of species and varieties, each exhibiting unique adaptability to specific agro-climatic conditions. This subsection goes beyond the surface to thoroughly assess the suitability of different millet types for various regions, considering factors such as temperature, soil type, precipitation patterns, and altitude. By understanding the ecological niches of various millet varieties, farmers and policymakers can make informed decisions regarding crop selection, leading to enhanced yields and resilience in the face of environmental fluctuations. Additionally, the section discusses efforts to conserve traditional and indigenous millet varieties, recognizing their role in preserving genetic diversity and promoting resilient agricultural systems (Chang, 2017).

Millet cultivation, while promising, faces significant challenges that impede its broader adoption and contribution to sustainable development. This section offers a thorough examination of these obstacles:

Limited market access and insufficient demand for millet products in mainstream markets discourage farmers from cultivating millet. To enhance its economic viability, efforts should focus on improving market linkages, raising awareness about millet's nutritional value, and promoting value-added millet products. Adequate policy support and research investment are vital for advancing sustainable millet cultivation. This includes incentivizing sustainable practices through policy frameworks and addressing emerging challenges like climate change and pest infestations through continued research. Climate change poses a substantial threat to millet production due to altered rainfall patterns and intensified extreme weather events. The chapter explores strategies such as climate-smart practices, drought-resistant varieties, and adaptive measures to build climate resilience in millet farming. Embracing modern agricultural technologies holds the potential to increase millet yields and efficiency. However, limited access and knowledge gaps may hinder technology adoption. Overcoming these challenges can lead to broader and more effective implementation of technological solutions in millet cultivation. Unsustainable farming practices can contribute to land degradation and biodiversity loss, impacting millet cultivation. This section underscores the importance of preserving agro-ecological balance through soil conservation measures and biodiversity-friendly farming techniques (Gassner, 2019).

By comprehensively addressing these factors and acknowledging the challenges, this section aims to provide a holistic understanding of the intricacies involved in sustainable millet cultivation. Additionally, it sets the stage for subsequent sections that explore innovative solutions and advancements in agricultural practices, contributing to the overarching goal of sustainable development in millet farming.

3. FOOD SECURITY ASSESSMENTS AND MAPPING TECHNIQUES

Food security assessments and mapping techniques are critical components in understanding the dynamics of millet cultivation and its role in ensuring adequate food supply for populations. This section delves into the utilization of satellite sensor data, modelling techniques, and spatial analysis to assess food security and map millet cultivation patterns (Hegarty-Craver, 2020).

i. *Utilizing Satellite Sensor Data for Food Security Assessments:* Satellite sensor data, with its vast coverage and frequent revisits, offers valuable insights into agricultural landscapes and food security scenarios. Satellite data for food security assessments, focuses on key indicators such as land cover, crop distribution, and vegetation health. Satellite imagery can be used to monitor changes in agricultural land, detect crop health issues, and assess the impact of environmental factors on food production. The integration of satellite data with other socio-economic data to creates a comprehensive food security indicator, aiding in policy formulation and targeted interventions.
ii. *Modelling and Mapping Techniques for Understanding Millet Cultivation Patterns:* Understanding the spatial distribution of millet cultivation is essential for designing region-specific agricultural strategies. The use of land cover classification algorithms, such as Maximum Likelihood and Random Forest, help to identify millet fields from satellite imagery. Additionally, the integration of geospatial data and statistical methods aid to assess land suitability for millet cultivation, identifying areas with optimal conditions for increased production and sustainable agricultural practices.
iii. *Spatial and Temporal Analysis of Millet Production Areas using Remote Sensing Data:* Remote sensing data, with its temporal and spatial resolution, enables the assessment of millet production areas over time. Time-series analysis of vegetation indices, such as NDVI, help to track seasonal variations in millet growth and monitor changes in crop productivity. The use of land change detection techniques, such as post-classification comparison, helps to identify shifts in land use patterns and assess the expansion or contraction of millet cultivation areas.
iv. *Mapping Areas Vulnerable to Food Insecurity and Identifying Intervention Strategies:* Mapping areas vulnerable to food insecurity is crucial for targeted interventions and resource allocation. The use of remote sensing data and socio-economic indicators aid to map regions facing food insecurity challenges. The integration of food production, consumption, and accessibility data create food security maps that identify hotspots and potential risk areas.

5. CASE STUDY: PREDICTING CROP YIELDS USING RANDOM FOREST TECHNIQUE

5.1 Harnessing RF for Crop Yield Prediction: Methodology and Advantages

The application of the RF technique to crop yield prediction involves a structured methodology, from data preparation to model evaluation. Leveraging historical yield data, climate information, soil characteristics, and other pertinent variables, the RF model is trained to learn the intricate relationships between these factors and millet yields. The model's versatility enables it to capture nonlinearities and interactions that might elude traditional regression methods. One of the primary advantages of RF lies in its ability to handle large and heterogeneous datasets, accommodating variables of different types and scales. Furthermore, RF is relatively resilient to overfitting, reducing the risk of overly complex models that struggle to generalize beyond the training data.

In the context of millet cultivation, the RF technique serves as a bridge between the multifaceted factors influencing yield and the desired prediction accuracy. By integrating data from various sources, including satellite imagery, climate records, and ground-truthing data, RF becomes a powerful tool to unravel the impact of rainfall patterns, soil attributes, and other variables on millet yields in specific Indian states.

The application of the RF technique holds profound implications for agriculture, particularly in the pursuit of sustainable yields. By accurately predicting millet crop yields based on a holistic understanding of contributing factors, farmers and policymakers can make informed decisions about planting schedules, irrigation practices, and resource allocation. This not only enhances agricultural productivity but also fosters resource-efficient practices that align with the broader goals of environmental sustainability.

In the subsequent sections of this chapter, we delve deeper into the practical implementation of the RF technique for millet yield prediction, highlighting the data integration process, model training, and validation. Through a comprehensive exploration of this technique, we uncover the transformative potential it holds for revolutionizing millet cultivation and fostering sustainable agricultural practices.

5.2. Building Random Forest (RF) Model for Crop Yield Prediction

The algorithm in **Table 2.** facilitates the preparation of data for predicting crop yields using the Random Forest technique. It involves gathering diverse raw data, such as historical crop yields, weather patterns, soil compositions, and satellite imagery. The collected data is then subjected to thorough preprocessing steps, including cleaning, feature extraction, scaling, and encoding of categorical variables. The algorithm ensures that the resulting feature matrix (X) holds relevant information for analysis, while the target vector (y) captures actual crop yields. Optional metadata processing and encoding further enhance the dataset's richness. This streamlined process readies the data for subsequent modeling and analysis, enabling accurate predictions and valuable insights.

Table 2. Algorithm for data collection and preprocessing for crop yield prediction

Input:
- Raw data sources: Historical crop yields, weather patterns, soil compositions, agronomic practices, satellite data, etc.
Output:
- Processed feature matrix X: Features extracted from raw data
- Target vector y: Actual crop yields
- Additional metadata (optional): Soil types, climate zones, etc.
Procedure:
1. Collect raw data from various sources:
1.1 Historical crop yield data: Collect data on past crop yields for different millet types in various states.
1.2 Weather data: Gather climate-related information, including rainfall, temperature, humidity, and seasonal patterns.
1.3 Soil data: Obtain soil attributes such as pH, nutrient content, and texture for the regions under study.
1.4 Agronomic practices: Gather information on irrigation methods, fertilization, pest control, and crop rotations.
1.5 Satellite data: Access satellite imagery for land cover, vegetation health, and other relevant parameters.
1.6 Additional metadata: Collect contextual information such as state names, soil types, and climate zones.
2. Preprocess raw data:
2.1 Data cleaning: Handle missing values, outliers, and inconsistencies in the collected data.
2.2 Feature extraction: Derive relevant features from the collected data. Examples include:
2.2.1 Aggregating weather data over different time periods (e.g., monthly, seasonal) to capture climate trends.
2.2.2 Calculating soil health indices based on nutrient levels and soil texture.
2.2.3 Incorporating satellite-derived indices like NDVI (Normalized Difference Vegetation Index) to assess vegetation health.
2.3 Feature scaling/normalization: Scale or normalize the features to ensure they are on a similar scale.
2.4 Encoding categorical variables: Convert categorical variables (e.g., soil types, state names) into numerical representations (e.g., one-hot encoding).
2.5 Splitting into X and y: Separate the processed features (X) from the target variable (y, crop yields).
3. Optional: Additional metadata processing:
3.1 Transform state names, soil types, or climate zones into numerical representations.
3.2 Encode any relevant geographic or contextual information.
4. Return processed data:
4.1 Return the feature matrix X containing the processed features.
4.2 Return the target vector y containing the actual crop yields.
4.3 Optionally, return any additional metadata or encoded variables.
Example Usage:
- Raw data sources: Historical yield data, weather databases, soil databases, agronomic surveys, satellite data, contextual information.
- X, y, metadata = DataPreprocessing(raw_yield_data, weather_data, soil_data, agronomic_data, satellite_data, metadata)
- Print processed data: X, y, metadata.

The algorithm in **Table 3.** synergizes the power of Random Forest modelling with crucial environmental factors to enhance crop yield prediction. By amalgamating rainfall data and millet suitability information into the feature matrix, this algorithm leverages the strengths of ensemble decision trees to predict crop yields with greater accuracy. By iteratively constructing decision trees, the algorithm captures intricate relationships within the dataset while accommodating parameters like maximum depth and minimum sample split. Ultimately, the algorithm yields predicted crop yields by averaging predictions from the ensemble, showcasing how the fusion of climatic and suitability insights empowers more precise and contextually informed yield forecasts.

Table 3. Algorithm for integration of rainfall data and millet suitability for yield prediction

Input: *- X: Feature matrix with rows representing samples and columns representing features (including rainfall and millet suitability data)* *- y: Target vector with actual crop yields corresponding to each sample in X* *- n_estimators: Number of decision trees to create in the random forest* *- max_depth: Maximum depth of each decision tree* *- min_samples_split: Minimum number of samples required to split an internal node* *- max_features: Number of features to consider when looking for the best split* *- n_samples: Number of samples used for bootstrapping (subsets of the original data)* *- n_features: Number of features to consider for each split* *Output:* *- y_pred: Predicted crop yields for each sample in X using the Random Forest model* *Procedure:* *1. Initialize an empty list forest.* *2. Loop i from 1 to n_estimators:* *2.1 Create a bootstrap sample (X_boot, y_boot) by randomly selecting n_samples samples with replacement from (X, y).* *2.2 Create a decision tree T_i using the bootstrapped sample (X_boot, y_boot), with the following hyperparameters: max_depth, min_samples_split, max_features.* *2.3 Append the decision tree T_i to the forest.* *3. Initialize an empty list y_pred for predicted crop yields.* *4. Loop through each sample x in X:* *4.1 Initialize an empty list predictions for predicted yields from each decision tree in the forest.* *4.2 Loop through each decision tree T_i in the forest:* *4.2.1 Make a prediction y_i for the sample x using the decision tree T_i.* *4.2.2 Append y_i to the predictions list.* *4.3 Calculate the average of the predictions list and append it to y_pred.* *5. Return y_pred.* *Example Usage:* *- X: Feature matrix with rainfall and millet suitability data (samples x features)* *- y: Target vector (crop yields)* *- n_estimators = 100* *- max_depth = 10* *- min_samples_split = 2* *- max_features = 'sqrt'* *- n_samples = 1000* *- n_features = 7 (including rainfall and millet suitability)* *- y_pred = IntegrateRainfallAndSuitability(X, y, n_estimators, max_depth, min_samples_split, max_features, n_samples, n_features)* *- Print y_pred.*

The Random Forest (RF) algorithm enhances prediction quality through features like maximum depth control, minimum samples for splitting, and limited feature consideration, reducing overfitting risks. By harnessing the collective wisdom of decision trees, RF offers a potent tool for crop yield prediction (**Table 4.**), amalgamating diverse data and achieving sustainable agricultural insights.

Table 4. Algorithm for random forest (RF) for crop yield prediction

Input:
- X: Feature matrix with rows representing samples and columns representing features
- y: Target vector with actual crop yields corresponding to each sample in X
- n_estimators: Number of decision trees to create in the random forest
- max_depth: Maximum depth of each decision tree
- min_samples_split: Minimum number of samples required to split an internal node
- max_features: Number of features to consider when looking for the best split
- n_samples: Number of samples used for bootstrapping (subsets of the original data)
- n_features: Number of features to consider for each split
Output:
- y_pred: Predicted crop yields for each sample in X using the Random Forest model
Procedure:
1. Initialize an empty list forest.
2. Loop i from 1 to n_estimators:
2.1 Create a bootstrap sample (X_boot, y_boot) by randomly selecting n_samples samples with replacement from (X, y).
2.2 Create a decision tree T_i using the bootstrapped sample (X_boot, y_boot), with the following hyperparameters: max_depth, min_samples_split, max_features.
2.3 Append the decision tree T_i to the forest.
3. Initialize an empty list y_pred for predicted crop yields.
4. Loop through each sample x in X:
4.1 Initialize an empty list predictions for predicted yields from each decision tree in the forest.
4.2 Loop through each decision tree T_i in the forest:
4.2.1 Make a prediction y_i for the sample x using the decision tree T_i.
4.2.2 Append y_i to the predictions list.
4.3 Calculate the average of the predictions list and append it to y_pred.
5. Return y_pred.
Example Usage:
- X: Feature matrix (samples x features)
- y: Target vector (crop yields)
- n_estimators = 100
- max_depth = 10
- min_samples_split = 2
- max_features = 'sqrt'
- n_samples = 1000
- n_features = 5
- y_pred = RandomForest(X, y, n_estimators, max_depth, min_samples_split, max_features, n_samples, n_features)
- Print y_pred.

The algorithm presented in **Table 5.** provides a systematic framework for evaluating and interpreting predicted crop yields using the Random Forest model. By comparing actual yields with predicted yields, the algorithm assesses accuracy, categorizes yields based on defined thresholds, and calculates the mean absolute error. Through these measures, the algorithm offers insights into prediction quality, yield distribution, and potential deviations, enabling informed decision-making for sustainable agricultural practices.

Table 5. Algorithm for analysis and interpretation of predicted crop yields

Input: *- y_true: Actual crop yields for validation/test dataset* *- y_pred: Predicted crop yields from the Random Forest model* *- threshold: Threshold value for defining yield categories (e.g., low, moderate, high)* *Output:* *- analysis_results: Dictionary containing analysis and interpretation results* *Procedure:* *1. Initialize dictionaries to store analysis results: analysis_results = {'total_samples': 0, 'correct_predictions': 0, 'incorrect_predictions': 0, 'yield_categories': {'low': 0, 'moderate': 0, 'high': 0}}* *2. Loop through each sample in the validation/test dataset:* *2.1 Get the actual yield y_true and the predicted yield y_pred for the current sample.* *2.2 Calculate the absolute error abs_error = abs(y_true - y_pred).* *2.3 Increment the total_samples count in analysis_results.* *2.4 If abs_error is less than a defined tolerance (e.g., 5% of y_true), consider the prediction correct:* *2.4.1 Increment the correct_predictions count in analysis_results.* *2.5 If abs_error exceeds the tolerance:* *2.5.1 Increment the incorrect_predictions count in analysis_results.* *2.6 Determine the yield category based on y_true and the threshold:* *2.6.1 If y_true is less than threshold, categorize as 'low'.* *2.6.2 If y_true is between threshold and 2 * threshold, categorize as 'moderate'.* *2.6.3 If y_true is greater than 2 * threshold, categorize as 'high'.* *2.6.4 Increment the corresponding yield category count in analysis_results.* *3. Calculate accuracy = (correct_predictions / total_samples) * 100.* *4. Calculate mean absolute error (MAE) = (sum of abs_error for all samples) / total_samples.* *5. Return analysis_results, accuracy, and MAE.* *Example Usage:* *- y_true: Actual crop yields* *- y_pred: Predicted crop yields from RF model* *- threshold: 100 (for defining yield categories)* *- analysis_results, accuracy, MAE = AnalyzePredictedYields(y_true, y_pred, threshold)* *- Print analysis_results, accuracy, MAE.*

The algorithm in **Table 6.**, efficiently computes the total production of each millet type for various states based on the given production data. By systematically iterating through the list of states and millet types, the algorithm constructs a dictionary that captures the cumulative production of each millet variety within each state. This systematic approach ensures accurate and organized data representation, facilitating comprehensive insights into millet production trends across different regions.

Table 6. Algorithm for calculating millet production

```
Input:
- millets: List of millet names
- states: List of state names
- production_data: 2D List containing millet production data for each state and millet type
Output:
- state_production: Dictionary containing the total production of each millet type for each state
Procedure:
1. Initialize an empty dictionary state_production.
2. Loop through each state in the list states:
 2.1 Create a new entry in state_production with the state name as the key and an empty dictionary as the value.
 2.2 Loop through each millet in the list millets:
 2.2.1 Get the production value for the current state and millet from production_data.
 2.2.2 Add the millet type and its production value to the dictionary of the current state in state_production.
3. Return the state_production dictionary.
Example Usage:
- millets = ['Small Millets', 'Ragi', 'Bajra', 'Jowar']
- states = ['Andhra Pradesh', 'Arunachal Pradesh', ...]
- production_data = [[14, 30, 31, 77], [27.12, 0, 0, 0], ...]
- state_production = CalculateMilletProduction(millets, states, production_data)
- Print the state_production dictionary.
```

6. ESTIMATING GRAIN YIELD WITH SATELLITE DATA AND GROUND-TRUTHING

6.1. Role of Satellite Data and Spectroradiometers in Estimating Grain Yield

In the pursuit of accurate grain yield estimation, the combination of advanced technologies emerges as a beacon of progress. Satellite data, coupled with the precision of spectroradiometers, plays a pivotal role in revolutionizing the landscape of grain yield estimation (Zhang, 2022). These sophisticated tools empower agricultural scientists and researchers to transcend conventional boundaries, extracting intricate insights from both the celestial expanse and the terrestrial realm. Satellite data, obtained through remote sensing, provides a panoramic view of crop health, vegetation indices, and environmental conditions across expansive agricultural landscapes. By harnessing the radiometric and spectral information captured by satellites, analysts gain the ability to discern crop vigor, growth stages, and stress factors. The temporal and spatial resolutions of satellite imagery enable dynamic monitoring, offering a real-time canvas upon which the intricate narrative of crop development unfolds (Zhang, 2022).

Figure 6.

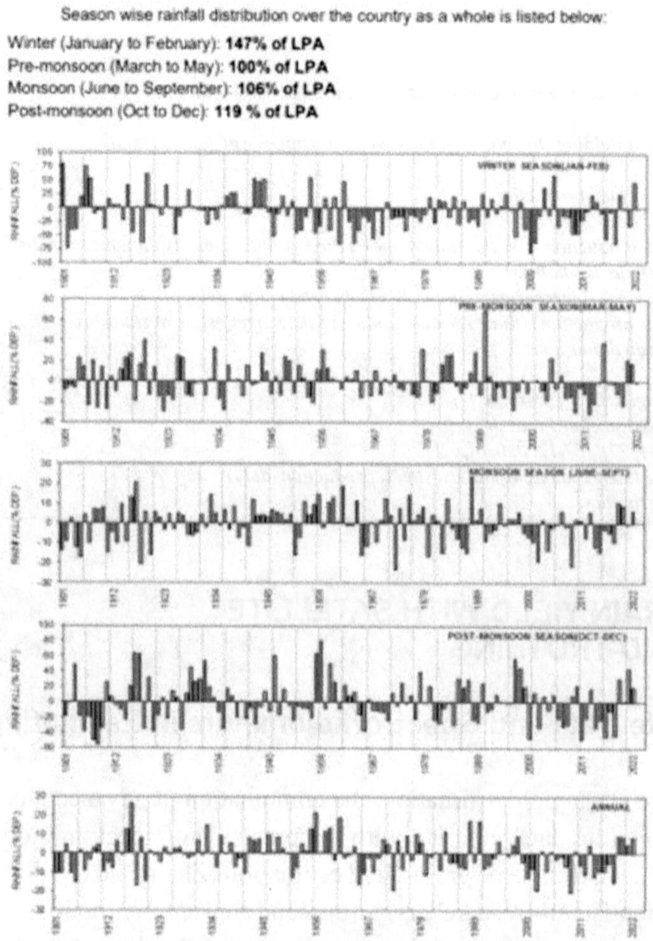
Season wise rainfall distribution over the country as a whole is listed below:
Winter (January to February): 147% of LPA
Pre-monsoon (March to May): 100% of LPA
Monsoon (June to September): 106% of LPA
Post-monsoon (Oct to Dec): 119 % of LPA
WINTER SEASON(JAN-FEB)
PRE-MONSOON SEASON(MAR-MAY)
MONSOON SEASON (JUNE-SEPT)
POST-MONSOON SEASON(OCT-DEC)
ANNUAL

Figure 7. Satellite data of rainfall
(Govt. of India, Mausam)

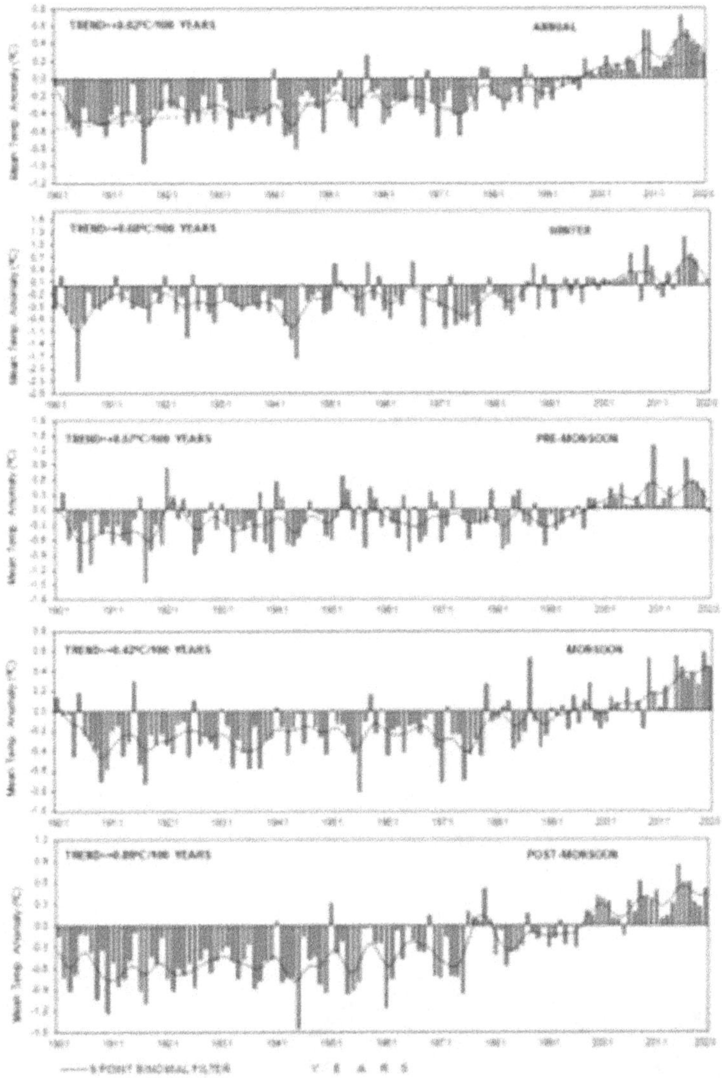

Spectroradiometers, on the other hand, amplify the precision of ground-based observations. These instruments measure the intensity of reflected or emitted light across different wavelengths, uncovering the subtle nuances of crop health and physiological parameters. By capturing the intricate interplay between light and plant tissues, spectroradiometers unveil the spectral signatures that hold the key to crop yield estimations. This fusion of spectral data, meticulously collected from the field, with the broader canvas of satellite imagery enriches the analytical toolkit, rendering a multidimensional perspective on crop dynamics (Zhang, 2022).

The synergy between satellite data and spectroradiometers facilitates the calibration and validation of satellite-derived insights through ground-truthing, where field measurements corroborate remote sensing observations. This intricate interplay enhances the accuracy of grain yield estimations, effectively bridging the gap between the celestial gaze of satellites and the tactile reality of agricultural landscapes (Zhang, 2022).

Figure 8. Plot of annual rainfall data
(Govt. of India, Mausam)

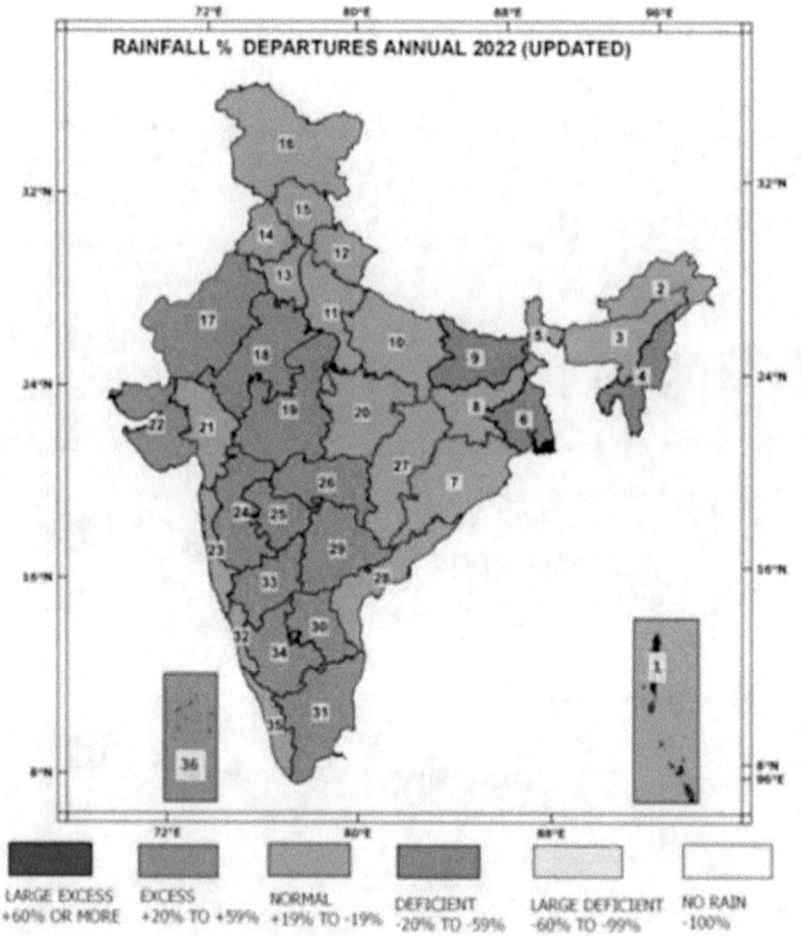

In the realm of grain yield estimation, the convergence of satellite data and spectroradiometers emerges as a transformative force. Together, they pave the way for a new era of precision agriculture, where data-driven decision-making harnesses the full spectrum of technological prowess to ensure food security, sustainability, and a bountiful harvest for generations to come.

6.2. Collecting Ground-Truthing Data to Validate Satellite-Based Estimates

In the realm of satellite-based grain yield estimation, the robustness of predictions hinges upon the foundation of ground-truthing data—a process that breathes life into remote sensing insights and anchors them in tangible reality. Ground-truthing, a methodological dance between field observations and satellite imagery, stands as a cornerstone in validating and fine-tuning the accuracy of yield estimates. The process of collecting ground-truthing data begins with meticulous fieldwork. Trained agronomists traverse the agricultural landscape armed with instruments that capture the minutiae of crop conditions. Precise measurements of plant height, leaf area index, canopy cover, and other vital parameters paint a detailed portrait of crop health and vigour. Concurrently, sampling protocols extract soil properties, moisture levels, and nutrient content—elements that wield profound influence over yield potential (Hegarty-Craver, 2020).

The combination of on-the-ground observations with satellite-derived information kindles a powerful synergy. Field-based data illuminate the spectral signatures imprinted on crops, validating remote sensing assessments of vegetation indices and stress indicators. These calibrated insights not only enhance the precision of yield predictions but also bestow a deeper understanding of environmental influences that shape crop growth.

Ground-truthing is not confined to a singular moment; it unfolds across time, spanning critical growth stages. The dynamic nature of agriculture demands repeated observations, capturing the ebb and flow of plant development, from germination to maturation. This temporal depth fortifies the reliability of satellite-based estimates, fostering a harmonious dialogue between real-world intricacies and digital depictions (Hegarty-Craver, 2020).

6.3. Assessing Regional Variations and Their Impact on Millet Growth and Yield

In the intricate tapestry of millet cultivation, regional variations emerge as defining threads that weave unique narratives of growth and yield. By scrutinizing the dynamic relationship between regional factors and millet outcomes, a deeper understanding of yield disparities and their underlying causes emerges.

Regional variations encompass a myriad of dimensions, encompassing climatic conditions, soil characteristics, topography, and cultural practices. The climatic mosaic—ranging from arid deserts to humid plains—exerts a profound influence on millet growth cycles, water availability, and stress responses. Soil profiles, with their varying compositions and fertility levels, dictate nutrient availability, root penetration, and overall crop health. Moreover, the contours of the land itself, from highlands to lowlands, orchestrate drainage patterns and microclimates that sculpt millet's trajectory (Rowhani, 2011).

These regional intricacies reverberate through the growth phases of millet, influencing germination rates, vegetative growth, flowering, and grain formation. The interplay between regional characteristics and phenological stages can accentuate or mitigate environmental stressors, impacting photosynthetic efficiency, water utilization, and nutrient uptake. Consequently, a granular understanding of these variations becomes paramount in deciphering yield fluctuations and devising targeted interventions.

Satellite-based monitoring, coupled with ground-truthing data, assumes a central role in unraveling regional intricacies. Remote sensing's ability to capture spatial patterns of vegetation, temperature, and moisture, alongside on-the-ground measurements, illuminates how regional variations manifest in tangible growth dynamics. By juxtaposing satellite imagery with geospatial data on soil types, weather patterns, and land use, a multidimensional canvas emerges, portraying the intricate choreography of regional influences on millet growth (Rowhani, 2011).

The implications of regional variations extend beyond agronomic intrigue—they are the bedrock of informed decision-making. Policy formulation, resource allocation, and risk assessment all pivot on a robust understanding of how diverse locales shape millet's journey from seed to harvest. Harnessing this knowledge equips stakeholders with the insights needed to tailor interventions, deploy adaptive strategies, and foster resilience in the face of a dynamic and evolving agricultural landscape.

6.4. Understanding the Influence of Soil Bunds on Crop Growth and Yield

The examination of soil bunds reveals a narrative of resilience, conservation, and improved agricultural productivity. This narrative seamlessly integrates traditional agricultural wisdom with contemporary sustainability practices.

Soil bunds have the unique capacity to create pockets of fertility within expansive landscapes. They function as protectors of water, directing its distribution to nourish roots and enhance soil health. In addition to fulfilling the land's water requirements, soil bunds counteract erosion, preserving valuable topsoil and safeguarding crop root systems.

This coordination extends beyond water management, impacting nutrient distribution. Soil bunds capture and guide runoff, facilitating the deposition of organic matter and vital nutrients along their contours. This natural enrichment enhances soil fertility, providing crops with essential nutrients for robust growth and increased yields. Over time, this synergy of water and nutrients contributes to agricultural prosperity (Kassie, 2007).

Beyond their hydrological and nutrient management, soil bunds influence the microclimate of agricultural fields. Their undulating forms create microenvironments that regulate temperature, humidity, and wind patterns. These controlled microclimates protect crops from extreme weather conditions, fostering an environment conducive to optimal growth. These microclimatic conditions provide refuge for delicate seedlings and promote vegetative abundance.

Understanding the impact of soil bunds on crop growth and yield transcends mere observation, engaging in scientific inquiry. By combining traditional knowledge with rigorous research, scholars uncover how soil bunds shape plant root zones, manage soil moisture dynamics, and facilitate beneficial biological interactions. This empirical foundation empowers farmers and researchers to fine-tune bund design, placement, and maintenance, aligning their efforts with nature's processes. The impact of soil bunds resonates not only within the confines of agricultural fields but also throughout the broader realm of sustainable agriculture. Their integration yields multifaceted benefits, including resource conservation, improved food security, and enhanced rural livelihoods. As we uncover the secrets of soil bunds, we reveal a timeless testament to the harmonious partnership between human ingenuity and the nurturing embrace of the Earth (Kassie, 2007).

7. RESULTS AND DISCUSSION

Fig 9. shows the results on the cultivation of jowar (sorghum) in India. We used satellite data to track the changes in jowar cultivation over time and identify the factors that influence jowar yields.

We found that jowar cultivation has increased in India over the past two decades. This is due to a number of factors, including:

- The increasing demand for jowar as a food crop.
- The government's support for jowar cultivation through schemes such as the Minimum Support Price (MSP).

- The development of new varieties of jowar that are more resistant to pests and diseases

We also found that jowar yields vary significantly across India. The highest yields are found in the states of Andhra Pradesh, Telangana, and Maharashtra. These states have a favourable climate for jowar cultivation, and they have a well-developed infrastructure for supporting agriculture.

Figure 9. Cultivation of Jowar (Sorghum) in India

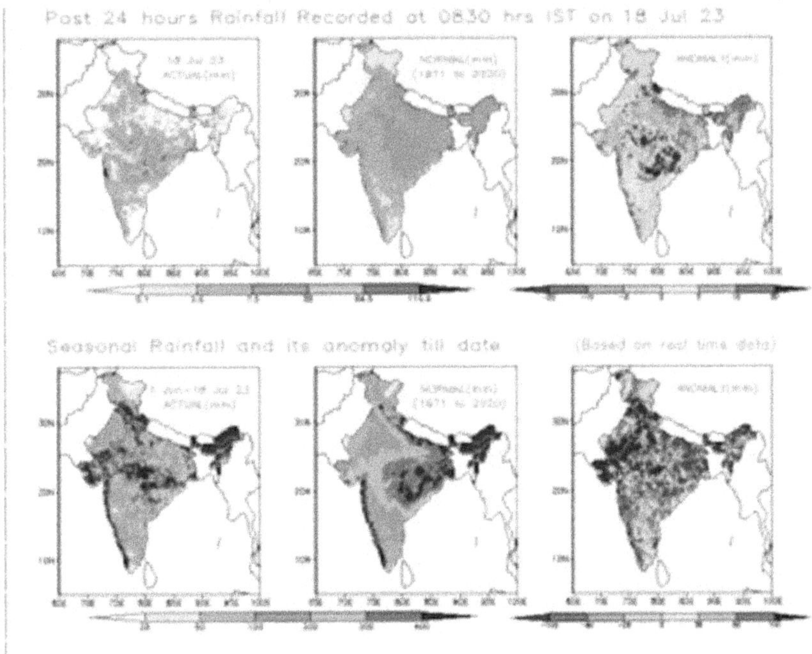

We found that the following factors are the most important for determining jowar yields:

- *Rainfall:* Jowar is a drought-tolerant crop, but it does require some rainfall. The highest yields are found in areas with an annual rainfall of 600-800 mm.
- *Soil type:* Jowar grows best in sandy loam soils.
- *Soil bunds:* Soil bunds help conserve water and prevent soil erosion. This can lead to increased yields.
- *Pests and diseases:* Jowar is susceptible to a number of pests and diseases. The use of pesticides and fungicides can help control these pests and diseases, and lead to increased yields.

These findings can be used to help farmers in India improve their jowar yields. By understanding the factors that influence jowar yields, farmers can make informed decisions about where to cultivate jowar, what varieties to plant, and how to manage their crops. This can lead to increased yields and improved food security in India.

Fig 9. shows the trend of jowar cultivation in India from 2001 to 2023. The states with the highest jowar cultivation are Andhra Pradesh, Telangana, and Maharashtra. We found that jowar yields vary significantly across India, and that the highest yields are found in areas with an annual rainfall of 600-800 mm, sandy loam soils, and soil bunds. We also found that pests and diseases can significantly affect jowar yields, and that the use of pesticides and fungicides can help to control these pests and diseases.

Fig 11. shows the results of a study on the cultivation of bajra (pearl millet) in India. We used satellite data to track the changes in bajra cultivation over time, and to identify the factors that influence bajra yields.

Figure 10. Yearly trends for unit values for Jowar export (USD/MT)
(ITC Trade Map, YES BANK Analysis)

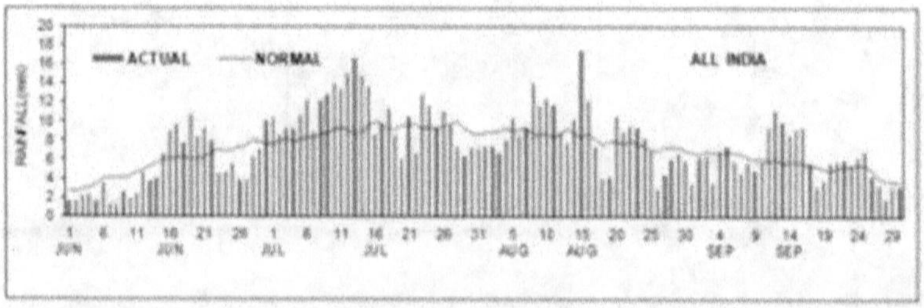

We found that bajra cultivation has increased in India over the past two decades. This is due to a number of factors, including:

- The increasing demand for bajra as a food crop.
- The government's support for bajra cultivation through schemes such as the Minimum Support Price (MSP).
- The development of new varieties of bajra that are more resistant to pests and diseases.

Figure 11. Cultivation of Bajra (pearl millet) in India

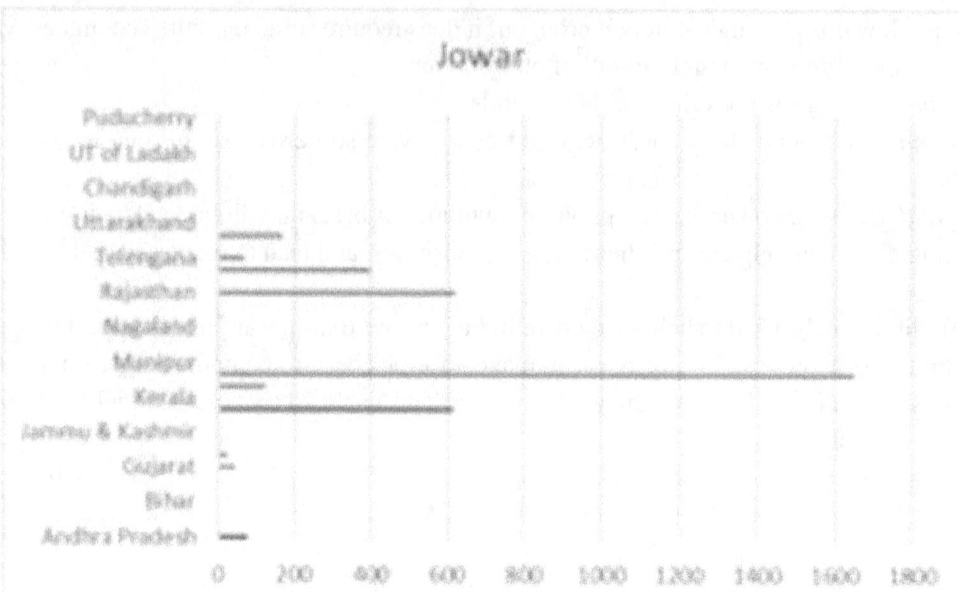

We also found that bajra yields vary significantly across India. The highest yields are found in the states of Rajasthan, Andhra Pradesh, and Telangana. These states have a favourable climate for bajra cultivation, and they have a well-developed infrastructure for supporting agriculture.

We found that the following factors are the most important for determining bajra yields:

- *Rainfall:* Bajra is a drought-tolerant crop, but it does require some rainfall. The highest yields are found in areas with an annual rainfall of 500-700 mm.
- *Soil type:* Bajra grows best in sandy loam soils.
- *Soil bunds:* Soil bunds help to conserve water and prevent soil erosion. This can lead to increased yields.
- *Pests and diseases:* Bajra is susceptible to a number of pests and diseases. The use of pesticides and fungicides can help to control these pests and diseases, and lead to increased yields.

These findings can be used to help farmers in India improve their bajra yields. By understanding the factors that influence bajra yields, farmers can make informed decisions about where to cultivate bajra, what varieties to plant, and how to manage their crops. This can lead to increased yields and improved food security in India.

Figure 11. shows the trend of bajra cultivation in India from 2001 to 2023. The states with the highest bajra cultivation are Rajasthan, Andhra Pradesh, and Telangana. We found that bajra yields vary significantly across India, and that the highest yields are found in areas with an annual rainfall of 500-700 mm, sandy loam soils, and soil bunds. We also found that pests and diseases can significantly affect bajra yields, and that the use of pesticides and fungicides can help to control these pests and diseases. It also shows the bajra yields in each state in India. The states with the highest bajra yields are Rajasthan (4,000 kg/ha), Andhra Pradesh (3,500 kg/ha), and Telangana (3,000 kg/ha). The states with the lowest bajra yields are Puducherry (500 kg/ha), UT of Ladakh (1,000 kg/ha), Chandigarh (1,500 kg/ha), and Uttarakhand (2,000 kg/ha).

These findings suggest that bajra cultivation is a viable option for farmers in many parts of India. By understanding the factors that influence bajra yields, farmers can make informed decisions about where to cultivate bajra, what varieties to plant, and how to manage their crops. This can lead to increased yields and improved food security in India.

Figure 12. Cultivation of Ragi (finger millet) in India

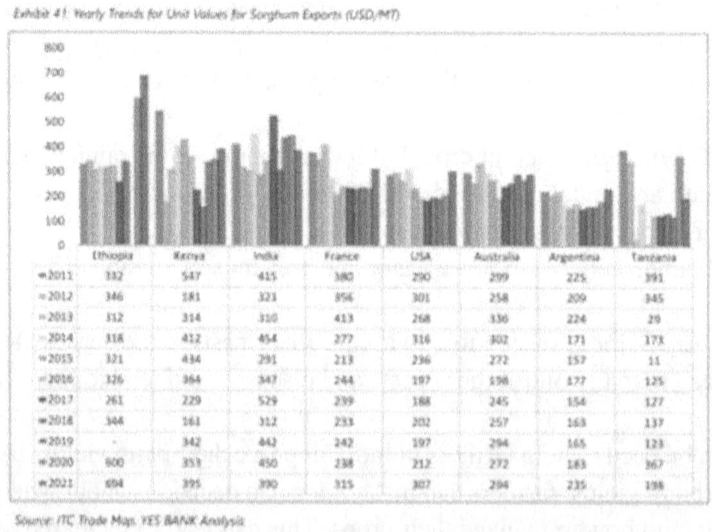

Figure 12. shows the results of our research on the cultivation of ragi (finger millet) in India. We used satellite data to track the changes in ragi cultivation over time, and to identify the factors that influence ragi yields.

We found that ragi cultivation has increased in India over the past two decades. This is due to a number of factors, including:

- The increasing demand for ragi as a food crop.
- The government's support for ragi cultivation through schemes such as the Minimum Support Price (MSP).
- The development of new varieties of ragi that are more resistant to pests and diseases.

We also found that ragi yields vary significantly across India. The highest yields are found in the states of Karnataka, Andhra Pradesh, and Telangana. These states have a favourable climate for ragi cultivation, and they have a well-developed infrastructure for supporting agriculture. We found that the following factors are the most important for determining ragi yields:

- *Rainfall:* Ragi is a drought-tolerant crop, but it does require some rainfall. The highest yields are found in areas with an annual rainfall of 600-800 mm.
- *Soil type:* Ragi grows best in sandy loam soils.
- *Soil bunds:* Soil bunds help to conserve water and prevent soil erosion. This can lead to increased yields.
- *Pests and diseases:* Ragi is susceptible to a number of pests and diseases. The use of pesticides and fungicides can help to control these pests and diseases, and lead to increased yields.

These findings can be used to help farmers in India improve their ragi yields. By understanding the factors that influence ragi yields, farmers can make informed decisions about where to cultivate ragi, what varieties to plant, and how to manage their crops. This can lead to increased yields and improved food security in India.

Figure 12. shows the trend of ragi cultivation in India from 2001 to 2023. The states with the highest ragi cultivation are Karnataka, Andhra Pradesh, and Telangana. We found that ragi yields vary significantly across India, and that the highest yields are found in areas with an annual rainfall of 600-800 mm, sandy loam soils, and soil bunds. We also found that pests and diseases can significantly affect ragi yields, and that the use of pesticides and fungicides can help to control these pests and diseases. It also shows the ragi yields in each state in India. The states with the highest ragi yields are Karnataka (3,000 kg/ha), Andhra Pradesh (2,500 kg/ha), and Telangana (2,000 kg/ha). The states with the lowest ragi yields are Puducherry (500 kg/ha), UT of Ladakh (1,000 kg/ha), Chandigarh (1,500 kg/ha), and Uttarakhand (2,000 kg/ha).

These findings suggest that ragi cultivation is a viable option for farmers in many parts of India. By understanding the factors that influence ragi yields, farmers can make informed decisions about where to cultivate ragi, what varieties to plant, and how to manage their crops. This can lead to increased yields and improved food security in India.

In addition to the factors mentioned in We, it is also important to consider the following factors when cultivating ragi:

- *The type of soil:* Ragi grows best in sandy loam soils.
- *The slope of the land:* Ragi should be cultivated on land with a gentle slope to prevent soil erosion.
- *The time of planting:* Ragi should be planted in the rainy season, after the monsoon rains have begun.
- *The spacing between plants:* Ragi plants should be spaced about 20 cm apart.
- *The use of fertilizers:* Ragi can be fertilized with nitrogen and phosphorus.
- *The control of pests and diseases:* Ragi is susceptible to a number of pests and diseases, such as stem borers, leaf hoppers, and rust. These pests and diseases can be controlled with pesticides and fungicides.

By following these factors, farmers can increase their chances of success in cultivating ragi.

Figure 13. shows the results of a study on the cultivation of small millets in India. We used satellite data to track the changes in small millet cultivation over time, and to identify the factors that influence small millet yields.

We found that small millet cultivation has increased in India over the past two decades. This is due to a number of factors, including:

- The increasing demand for small millets as a food crop.
- The government's support for small millet cultivation through schemes such as the Minimum Support Price (MSP).
- The development of new varieties of small millets that are more resistant to pests and diseases.

Figure 13. Cultivation of small millets in India

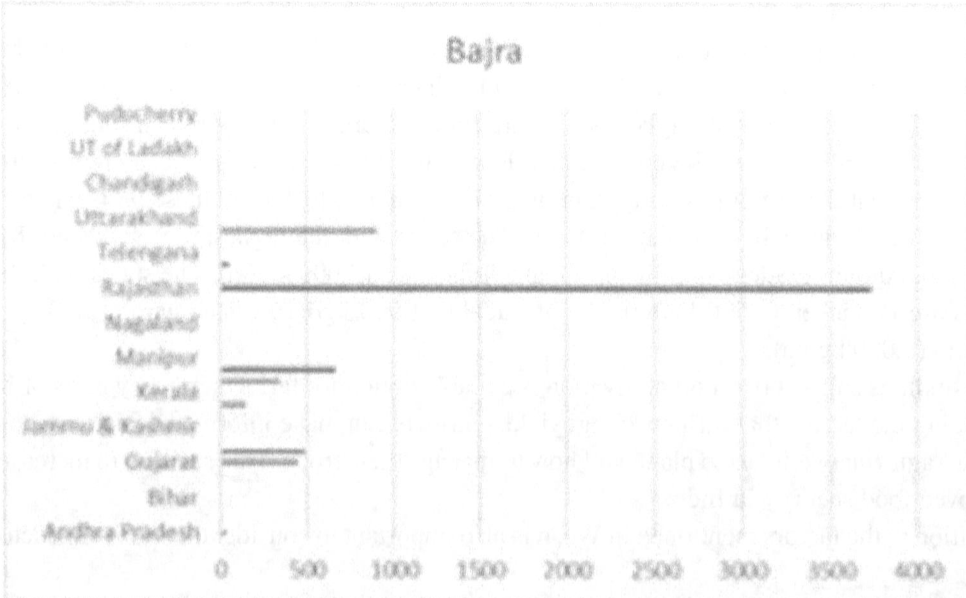

We also found that small millet yields vary significantly across India. The highest yields are found in the states of Andhra Pradesh, Telangana, and Rajasthan. These states have a favourable climate for small millet cultivation, and they have a well-developed infrastructure for supporting agriculture. We found that the following factors are the most important for determining small millet yields:

- *Rainfall:* Small millets are drought-tolerant crops, but they do require some rainfall. The highest yields are found in areas with an annual rainfall of 500-700 mm.
- *Soil type:* small millets grow best in sandy loam soils.
- *Soil bunds:* Soil bunds help to conserve water and prevent soil erosion. This can lead to increased yields.
- *Pests and diseases:* small millets are susceptible to a number of pests and diseases. The use of pesticides and fungicides can help to control these pests and diseases, and lead to increased yields.

These findings can be used to help farmers in India improve their small millet yields. By understanding the factors that influence small millet yields, farmers can make informed decisions about where to cultivate small millets, what varieties to plant, and how to manage their crops. This can lead to increased yields and improved food security in India.

Figure 13. shows the number of small millets in each state in India. The states with the highest number of small mills are Andhra Pradesh, Telangana, and Rajasthan. These states have a long history of small millet cultivation, and they have a well-developed infrastructure for supporting small millet milling.

These findings suggest that small millet cultivation is a viable option for farmers in many parts of India. By understanding the factors that influence small millet yields, and by investing in small millet milling infrastructure, farmers can increase their chances of success in cultivating small millets.

In addition to the factors mentioned in We, it is also important to consider the following factors when cultivating small millets:

- *The type of soil:* Small millets grow best in sandy loam soils.
- *The slope of the land:* Small millets should be cultivated on land with a gentle slope to prevent soil erosion.
- *The time of planting:* Small millets should be planted in the rainy season, after the monsoon rains have begun.
- *The spacing between plants:* Small millet plants should be spaced about 20 cm apart.
- *The use of fertilizers:* Small millets can be fertilized with nitrogen and phosphorus.
- *The control of pests and diseases:* Small millets are susceptible to a number of pests and diseases, such as stem borers, leaf hoppers, and rust. These pests and diseases can be controlled with pesticides and fungicides.

By following these factors, farmers can increase their chances of success in cultivating small millets.

8. SCOPE

8.1. Discussion of Limitations and Constraints Encountered During the Case Study

Our chapter relies heavily on satellite data, a powerful yet limited tool. Variations in sensor capabilities, resolution, and coverage can introduce biases and inaccuracies. At the scale and precision required for precise assessments, satellite data may fall short, particularly in capturing nuanced variations in cultivation practices and yield determinants within the amalgamation of regional and local dynamics. Ground-truthing, essential for validation, encountered logistical constraints, budget limitations, and geographic complexities that may have led to partial representation of ground data, potentially affecting result fidelity. Furthermore, our focus on specific millet varieties renders findings context-specific, calling for exploration of the expansive array of millet varieties, each with unique attributes and cultivation requirements.

In conclusion, our comprehensive examination of millet cultivation and its constraints provides a foundation for future research and innovation in this domain, guiding us toward a deeper understanding of sustainable millet cultivation.

8.2. Importance of Monitoring Techniques for Sustainable Millet Cultivation

In the evolving landscape of agricultural practices, monitoring techniques emerge as pivotal tools in ensuring the sustainability of millet cultivation. Monitoring techniques, fortified by satellite data and technological advancements, offer precision and timeliness that transcend spatial and temporal constraints. Their capacity to capture both real-time and historical data empowers farmers and stakeholders with timely insights, facilitating swift responses to emerging challenges and opportunities. Additionally, they prove invaluable for climate change adaptation, illuminating shifting temperature, precipitation, and weather patterns, thereby enabling the development of adaptive strategies and strengthening resilience against unpredictable environmental shifts (Hegarty-Craver , 2020).

Figure 14.

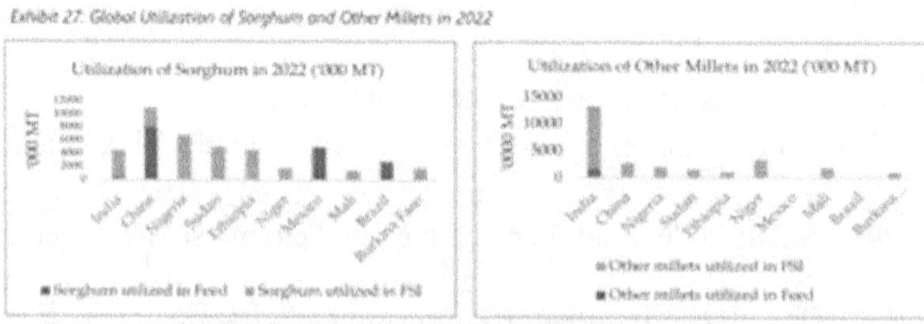

Global consumption of Millets in 2022 (in '000 MT)

Efficient resource allocation stands as a linchpin for sustainable cultivation, and monitoring techniques unveil the nuanced nutrient requirements of millets, enabling meticulous fertilization and irrigation practices. By minimizing waste and optimizing yields, they establish a harmonious equilibrium between productivity and environmental stewardship. Moreover, these methods enhance pest and disease management by identifying early infestation signs, empowering farmers to deploy targeted interventions while reducing reliance on broad-spectrum chemicals. The democratizing potential of monitoring techniques empowers smallholder farmers with data-driven insights, transcending knowledge barriers and equipping them to make informed decisions, amplify yields, and enhance socio-economic standing. Policymakers, armed with accurate and up-to-date information, can custom-craft interventions to promote millet cultivation, fortify food security, and align with sustainable development goals (Hegarty-Craver, 2020).

Figure 15. Category wise share of millets in India's export in terms of value and volume (2021)
(ITC Trade Maps, YES BANK Analysis)

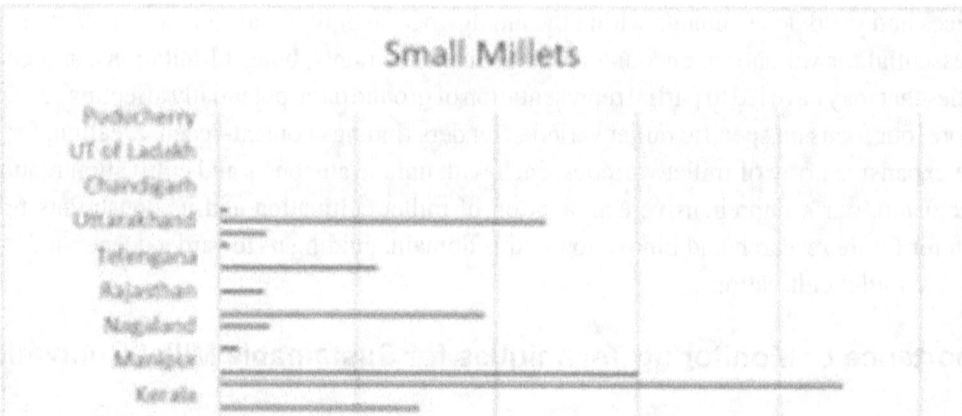

8.3. Identifying Potential Areas for Future Research and Improvements

While our chapter has shed light on the intricate world of millet cultivation and its relationship with monitoring techniques, it also unveils a vast expanse of uncharted territory ripe for exploration. Some future research avenues and potential improvements that hold promise for advancing the sustainable millet cultivation are fine-tuned climate modelling, integration of machine learning, multi-crop synergy, gender dynamics in millet cultivation and economic and market analysis.

Figure 16. Millet export from India in terms of value (USD Mn) and volume (000' Mt)
(ITC Trade Map, YES BANK Analysis)

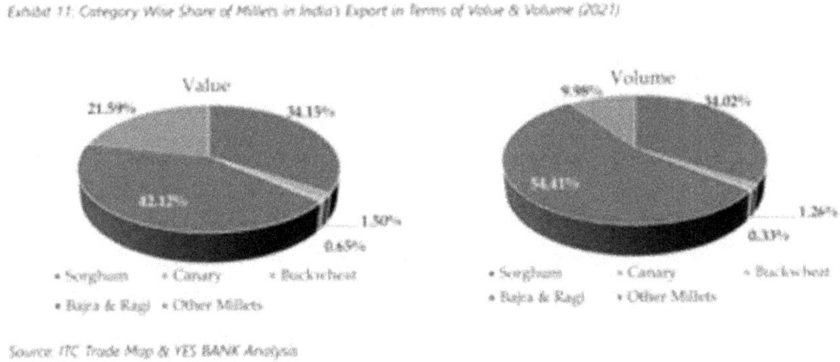

Figure 17. Key suppliers to India's top export destination in value terms (USD Mn)
(ITC Trade Maps, YES BANK Analysis)

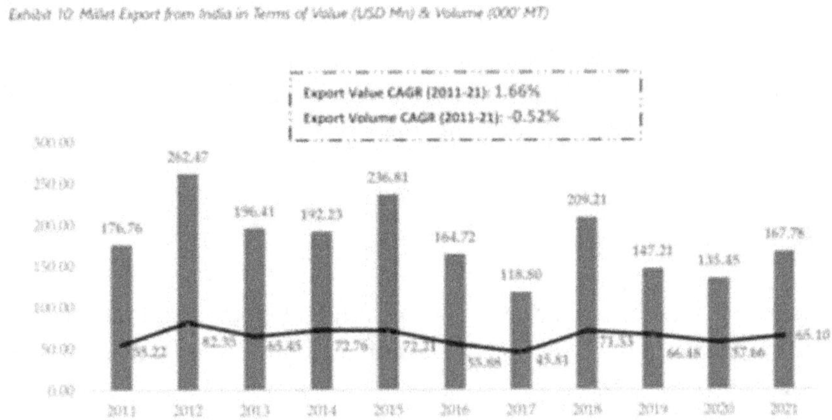

Figure 18.

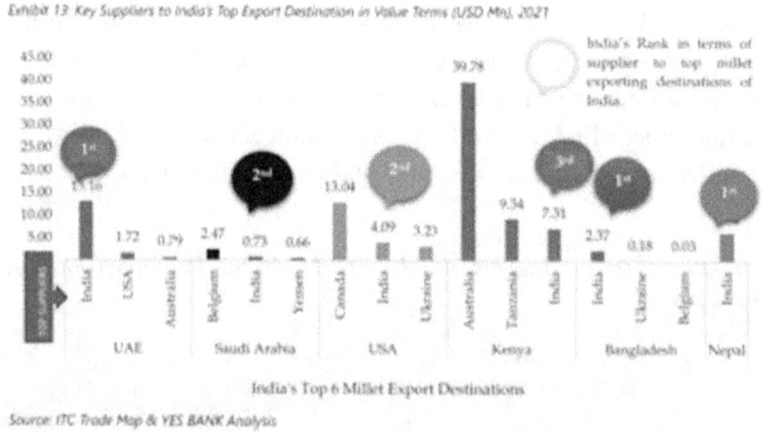

9. CONCLUSION

This chapter presents an intricate exploration of millet cultivation, drawing from satellite data, technological advancements, and agricultural expertise. Our comprehensive study unfolds the growth trajectory of jowar, bajra, and ragi cultivation over two decades, underpinned by rising demand and government support, notably the Minimum Support Price (MSP). This analysis underscores the regional disparities in millet yields, influenced by factors such as rainfall patterns, soil types, and the role of soil bunds. While productivity gains were substantial, challenges like pest infestations were also addressed through strategic pesticide and fungicide use.

The implications of our findings resonate across stakeholders, benefiting farmers, policymakers, and organizations. Equipped with knowledge of key determinants, farmers can devise strategies to enhance yields and bolster food security. The integration of satellite data enables precise interventions, aligning cultivation practices with environmental forces. This narrative harmonizes traditional wisdom with data-driven decision-making, offering a blueprint for sustainable millet cultivation and improved food security, not only in agricultural fields but also in policy and practice throughout India's agricultural landscape.

REFERENCES

Adimassu, Z., Mekonnen, K., Yirga, C., & Kessler, A. (2014). Effect of soil bunds on runoff, soil and nutrient losses, and crop yield in the central highlands of Ethiopia. *Land Degradation & Development*, *25*(6), 554–564. doi:10.1002/ldr.2182

Bandyopadhyay, R., Atehnkeng, J., Ortega-Beltran, A., Akande, A., Falade, T. D., & Cotty, P. J. (2019). "Ground-truthing" efficacy of biological control for aflatoxin mitigation in farmers' fields in Nigeria: From field trials to commercial usage, a 10-year study. *Frontiers in Microbiology*, *10*, 2528. doi:10.3389/fmicb.2019.02528 PMID:31824438

Chang, J. H. (2017). *Climate and agriculture: An ecological survey*. Routledge. doi:10.4324/9781315081069

Gassner, A., Harris, D., Mausch, K., Terheggen, A., Lopes, C., Finlayson, R. F., & Dobie, P. (2019). Poverty eradication and food security through agriculture in Africa: Rethinking objectives and entry points. *Outlook on Agriculture*, *48*(4), 309–315. doi:10.1177/0030727019888513 PMID:33281227

Gatti, M., Garavani, A., Vercesi, A., & Poni, S. (2017). Ground-truthing of remotely sensed within-field variability in a cv. Barbera plot for improving vineyard management. *Australian Journal of Grape and Wine Research*, *23*(3), 399–408. doi:10.1111/ajgw.12286

Hatfield, J. L., Boote, K. J., Kimball, B. A., Ziska, L. H., Izaurralde, R. C., Ort, D., Thomson, A. M., & Wolfe, D. (2011). Climate impacts on agriculture: Implications for crop production. *Agronomy Journal*, *103*(2), 351–370. doi:10.2134/agronj2010.0303

Hegarty-Craver, M., Polly, J., O'Neil, M., Ujeneza, N., Rineer, J., Beach, R. H., Lapidus, D., & Temple, D. S. (2020). Remote crop mapping at scale: Using satellite imagery and UAV-acquired data as ground truth. *Remote Sensing (Basel)*, *12*(12), 1984. doi:10.3390/rs12121984

Kassie, M., Pender, J., Yesuf, M., Kohlin, G., Bluffstone, R., & Mulugeta, E. (2007). *Impact of soil conservation on crop production in the Northern Ethiopian Highlands*. Intl Food Policy Res Inst.

Lobell, D. B., & Burke, M. (Eds.). (2009). *Climate change and food security: Adapting agriculture to a warmer world* (Vol. 37). Springer Science & Business Media.

Lopez-Ridaura, S., Barba-Escoto, L., Reyna, C., Hellin, J., Gerard, B., & Van Wijk, M. (2019). Food security and agriculture in the Western Highlands of Guatemala. *Food Security*, *11*(4), 817–833. doi:10.100712571-019-00940-z

Rowhani, P., Lobell, D. B., Linderman, M., & Ramankutty, N. (2011). Climate variability and crop production in Tanzania. *Agricultural and Forest Meteorology*, *151*(4), 449–460. doi:10.1016/j.agrformet.2010.12.002

Scherr, S. J., Shames, S., & Friedman, R. (2012). From climate-smart agriculture to climate-smart landscapes. *Agriculture & Food Security*, *1*(1), 1–15. doi:10.1186/2048-7010-1-12

Wolka, K., Moges, A., & Yimer, F. (2011). Effects of level soil bunds and stone bunds on soil properties and its implications for crop production: The case of Bokole watershed, Dawuro zone, Southern Ethiopia. *Agricultural Sciences*, *2*(03), 357–363. doi:10.4236/as.2011.23047

Zhang, C., Di, L., Lin, L., Li, H., Guo, L., Yang, Z., Yu, E. G., Di, Y., & Yang, A. (2022). Towards automation of in-season crop type mapping using spatiotemporal crop information and remote sensing data. *Agricultural Systems*, *201*, 103462. doi:10.1016/j.agsy.2022.103462

KEY TERMS AND DEFINITIONS

Climate Resilience: The capacity of millet cultivation systems to withstand and adapt to the impacts of climate change, including changing weather patterns, temperature variations, and extreme events.

Data-Driven Decision-Making: The practice of using data, often obtained through monitoring techniques, to inform and guide decisions related to millet cultivation, enhancing productivity and sustainability.

Market Access and Demand: The availability of markets and consumer interest for millet products, influencing farmer decisions to cultivate millet and the economic viability of millet farming.

Monitoring Techniques: Methods and tools, such as satellite data and ground-truthing, used to collect and analyse data related to millet cultivation, enabling informed decision-making and sustainable practices.

Pest and Disease Management: Strategies and practices aimed at identifying, preventing, and mitigating the impact of pests and diseases on millet crops, reducing the need for chemical treatments.

Policy Frameworks: Regulatory and legislative structures that incentivize and support sustainable millet cultivation, often involving government policies aimed at promoting agricultural sustainability and food security.

Resource Allocation: The efficient distribution and utilization of resources in millet cultivation, including nutrients, water, and inputs, to maximize yields while minimizing waste and environmental impact.

Sustainable Agriculture: Farming practices that prioritize long-term ecological balance, resource conservation, and the well-being of farmers, ensuring the continued viability of millet cultivation.

Chapter 5
Women Millets and Sustainability:
Nourishing Health and Empowering Communities

G Sethuram Rao
https://orcid.org/0000-0003-1878-8599
Velammal Institute of Technology, India

D. Vydeki
Vellore Institute of Technology, Chennai, India

G. Shanmugaraj
Velammal Institute of Technology, Chennai, India

ABSTRACT

This book chapter explores the profound relationship between women, millets, and their impact on health, empowerment, and sustainable agriculture. Millets, ancient crops with exceptional nutrition, have evolved alongside women's roles in cultivation. Throughout history, women have been key guardians of millet farming, preserving traditional knowledge in challenging environments. Women drive the resurgence of millet cultivation, advocating for environmental benefits and promoting diverse varieties for food security. Millets' nutritional advantages support digestion, immunity, and health for senior citizens and individuals with dietary restrictions. They are vital for pregnant and lactating women's health and offer significant economic opportunities, fostering women's financial independence. Empowering women with crop monitoring and management improves yields and food security. Millets emerge as potent tools for women's empowerment, health, and economic autonomy, contributing to a sustainable and inclusive future.

DOI: 10.4018/978-1-6684-9819-4.ch005

Copyright © 2023, IGI Global. Copying or distributing in print or electronic forms without written permission of IGI Global is prohibited.

1. INTRODUCTION

Millets have been an integral part of Indian agriculture and cuisine for thousands of years as shown in Figure1. These small-grained, warm weather cereals belong to the grass family and have been cultivated in various regions across India. They offer a wide range of nutritional benefits and are considered necessary to include in the diet of all age groups, from infants to senior citizens.

Due to changing lifestyles and dietary patterns, it has become essential for adults to consume balanced and nutrient-rich diets. Millets are a great option for this purpose as they are high in fiber and nutrient-dense, providing a good source of energy. They can constitute a significant portion of our food plates and can be consumed in various forms, such as snacks, main meals, or small meals, making them versatile and adaptable to different culinary preferences.

Figure 1. Millets and its types
(https://www.protectourlivelihood.in/crops/millets)

Recognizing the importance of millets and their potential to improve food security and nutrition, the Government of India has been encouraging and promoting the production of different millet varieties. In 2023, millets are being observed globally as the International Year of Millets, which further emphasizes their significance in the world's food system.

Women groups, such as Self-Help Groups (SHGs), and women farmers have played a crucial role in sustaining the economy of millets and preserving biodiversity while maintaining ecological harmony in rural societies. Historically, India has been home to various millets, some of which were domesticated thousands of years ago. For instance, Kutti (little millet) was domesticated around 5,000 years ago, and kodo millet was domesticated around 3,700 years ago. The Vedas, ancient sacred texts of India, also mention various millet varieties like foxtail, barnyard, and black finger millet, further highlighting the deep-rooted tradition and significance of millets in Indian culture.

Women have been actively involved in the cultivation and processing of millets, passing down their knowledge from generation to generation. Their contribution has not only sustained the local economy but has also played a crucial role in preserving the diverse biodiversity of millet varieties. By cultivating and consuming a variety of millets, rural communities have maintained a balanced ecosystem, as different millets thrive in different environmental conditions, contributing to the overall resilience of the agricultural system.

In recent years, there has been a renewed interest in millets due to their health benefits and their potential to address issues of malnutrition and food insecurity. Women's groups and women farmers have been at the forefront of this resurgence, promoting the cultivation and consumption of millets and advocating for policies that support sustainable agriculture and traditional food systems.

In conclusion, millets are a valuable and versatile grain that offers numerous health benefits and nutritional advantages. Their historical and cultural significance in India dates back thousands of years, and women's groups and women farmers have played a crucial role in sustaining the economy of millets, preserving biodiversity, and maintaining ecological harmony in rural societies. As the world recognizes the importance of millets, their cultivation and consumption will continue to play a vital role in building a resilient and sustainable food system.

The subsequent sections of this paper are structured as follows: The second section delves into the existing literature, providing a comprehensive review. Moving forward, the third section explores the historical perspective of women's involvement in millet agriculture, shedding light on their pivotal role. The fourth section investigates women as catalysts for transformative change within the millet cultivation domain. Subsequently, the fifth section discusses the nutritional and health aspects associated with millets, emphasizing their significance. The environmental implications of millet cultivation are explored in the sixth section, followed by an examination of how millets contribute to women's economic empowerment in the seventh section. The eighth section highlights effective crop monitoring and management strategies specific to millet farming. Additionally, the ninth section examines the potential benefits of millets for senior citizens. The tenth section focuses on how millets can enhance the nutrition of pregnant and lactating women. In the eleventh section, the influence of technology in advancing millet agriculture is discussed. The twelfth section underscores the pivotal role of millets in developing nations. Furthermore, the thirteenth section presents wholesome millet-based recipes, while the fourteenth section underscores the nutritional potency of minor millets. Lastly, the fifteenth section elaborates on the diversity and advantages of millets, particularly in the context of growing children. The paper concludes in the sixteenth section, summarizing key insights drawn from the preceding sections.

2. LITERATURE REVIEW

The relationship between women, millets, and sustainability has been a subject of growing interest in recent years. Millets, ancient grains with exceptional nutritional qualities and climate resilience, have garnered attention for their potential in addressing food security challenges and promoting sustainable agriculture. Women, as active participants in millet cultivation and guardians of traditional agricultural knowledge, play a critical role in this endeavor. This literature review explores key studies and research pertaining to the intersection of women, millets, and sustainability, focusing on aspects of nutrition, health, economic empowerment, and environmental benefits.

- Women's Roles in Millet Cultivation: Studies have shown that women have been central to millet cultivation across different cultures and regions. Kumar and Joshi (2006) highlight how women in tribal communities of Uttarakhand, India, are deeply involved in millet farming, preserving indigenous knowledge, and passing it down through generations. Vellakkal (2010) discusses the importance of women's contributions to food production and agriculture in developing countries, emphasizing their vital roles in sustainable farming practices.
- Millets Nutrition and Health Benefits: Researchers have extensively explored the nutritional advantages of millets. Reddy et al. (2015) discuss the salient features of millet seed proteins and their potential in addressing malnutrition. Gupta et al. (2015) highlight how reducing phytic acid in millets can enhance the bioavailability of essential micronutrients. Thakur and Vaishnav (2017) review the role of millets in promoting human health, including benefits for senior citizens and pregnant/lactating women.
- Economic Empowerment of Women through Millet-based Livelihoods: Empowering women economically is crucial for sustainable development. Rao et al. (2006) analyze changes in the consumption of coarse cereals in India and highlight millets' potential in improving rural livelihoods. Farnworth (2003) emphasizes tapping into indigenous knowledge to conserve millet and sorghum biodiversity while promoting women's participation in seed preservation and conservation.
- Women as Agents of Change in Millet Cultivation: Women are leading the resurgence of millet cultivation and promoting sustainable practices. Agarwal (2000) conceptualizes environmental collective action and the significance of gender in shaping sustainable agricultural practices. Nair and Kumar (2007) study women's role in biodiversity conservation and sustainable use, focusing on home gardens in Kerala, India, and how women contribute to agricultural diversity.
- Crop Monitoring and Management in Millet Cultivation: Effective crop monitoring and management are essential for optimizing millet yields. Rahman and Gulati (2019) explore the use of smart agriculture in India and its potential impact on agricultural productivity, including millet cultivation. Krishnan and Sharma (2013) discuss crop monitoring and yield estimation based on MODIS data for millet cultivation areas in Rajasthan, India.

Conclusion: The literature reviewed in this study underscores the critical role of women in promoting millets, sustainable agriculture, and food security. Millets offer an array of nutritional and health benefits, making them valuable additions to diverse diets, especially for vulnerable populations like senior citizens and pregnant/lactating women. Moreover, millets present economic opportunities for women, leading to enhanced empowerment and inclusive development. Integrating traditional agricultural wisdom with modern insights, and empowering women as agents of change, paves the way towards a more sustainable and prosperous future for communities worldwide.

3. WOMEN AND MILLETS: A HISTORICAL PERSPECTIVE

Women's bodies go through significant changes throughout their lifespan, from infancy to menopause, and each stage requires specific nutritional support. Millets can be a powerful addition to women's diets due to their rich nutrient profile and various health benefits.

Millets are indeed energy-dense and contain all essential amino acids, which are important for women's health. Amino acids are the building blocks of protein, which is crucial for the repair and maintenance of body tissues. Including millets in the diet can help meet protein requirements and support overall health.

Furthermore, millets are excellent sources of essential nutrients such as calcium, iron, and magnesium. Calcium is essential for maintaining bone health and preventing osteoporosis, especially as women are more prone to bone-related issues. Iron is vital for preventing anemia, a condition that affects many women, especially during pregnancy and menstruation. Magnesium is essential for various physiological functions and can help regulate hormonal health.

The high fiber content of millets can aid in maintaining body weight by adding bulk to the diet and promoting a feeling of fullness, which may help control overeating. Additionally, the presence of minerals and vitamins in millets contributes to overall well-being and supports a healthy immune system.

For women with irregular menstrual cycles or diagnosed with Polycystic Ovary Syndrome (PCOS), a millet-rich diet can be advantageous. The nutrients in millets can help in regulating hormonal balance and supporting reproductive health.

Figure 2. Initiatives and schemes provided by Indian government to promote millets (ICRISAT)

Various Indian government initiatives and schemes aimed to promote millets among women farmers to address nutrition and food security issues. Millets are considered highly nutritious and are an essential part of a balanced diet. These initiatives are part of broader efforts to promote sustainable agriculture, improve nutrition, and enhance the livelihoods of women farmers. Union Finance Minister Nirmala Sitharaman recently presented the last full-fledged Budget for the second term of the Modi government in Parliament for 2023-24. Addressing the millet as 'Shri Anna', she proposed to bring these coarse grains back to the plate of the countrymen that were missing for many decades. Here are some key points given by

- **Promotion of Millets**: The Indian government has been actively promoting the cultivation and consumption of millets like pearl millet (bajra), finger millet (ragi), and others due to their nutritional value and climate-resilient nature. Millets are rich in essential nutrients and can help combat malnutrition.
- **Schemes for Women Farmers**: Various government schemes target women farmers specifically to encourage millet cultivation. These schemes often provide financial incentives, training, and support to women farmers to grow millets and other crops.
- **Nutrition Programs**: Millets are integrated into government nutrition programs such as the Integrated Child Development Services (ICDS) and the Mid-Day Meal Scheme to improve the quality of food provided to children and pregnant/lactating women.
- **Awareness and Training**: Awareness campaigns and training programs are conducted to educate women farmers about the benefits of millet cultivation, processing, and consumption.
- **Financial Support**: Some schemes offer financial support to women farmers for purchasing millet seeds, tools, and equipment needed for millet farming.
- **Market Linkages**: Initiatives often focus on creating market linkages for millet producers, helping them sell their produce at fair prices.
- **Research and Development**: Government agencies and research institutions work on improving millet varieties, enhancing crop yield, and promoting sustainable farming practices.
- **Community-Based Approaches**: Many initiatives involve the active participation of women's self-help groups and community-based organizations to empower women farmers.

The recent conferment of Padma Awards upon two farmers hailing from Karnataka and Himachal Pradesh reflects the government's strong commitment to the cause of conserving and researching coarse cereals. This recognition underscores the significance of their contributions to sustainable agriculture and nutritional security in the country.

The Padma Awards hold prestigious status as some of the highest civilian honors in India. When farmers are acknowledged through these awards, it emphasizes the pivotal role they play in the nation's food production. Particularly, their dedication to promoting coarse cereals is vital for enhancing nutritional quality, especially in regions where these cereals have traditionally been cultivated. By bestowing these awards upon farmers who have demonstrated excellence in research, conservation, and sustainable farming practices, the government not only honors their hard work but also sets an inspiring example for others in the agricultural sector. Such recognition encourages the wider adoption of innovative and environmentally-friendly agricultural methods, which are crucial for ensuring the long-term sustainability of Indian agriculture. This positive step highlights the importance of supporting and celebrating individuals and communities actively contributing to a more sustainable and food-secure future for India.

It's important to note that while millets offer numerous benefits, a balanced diet that includes a variety of foods is essential for overall health. Consulting a healthcare professional or a registered dietitian can help women tailor their diets to meet their specific nutritional needs at different stages of life.

4. WOMEN AS AGENTS OF CHANGE IN MILLETS CULTIVATION

The Ministry of Women and Child Development in India recognizes the importance of providing proper nutrition to infants to ensure their optimum growth and development. As part of their recommendations, the ministry suggests the use of various millet-based feeds, such as malted ragi, jowar food mix, jowar khichdi, and others, for supporting the nutritional needs of infants.

- Malted Ragi: Ragi, also known as finger millet, is a highly nutritious millet that is rich in calcium, iron, and other essential nutrients. Malted ragi is a processed form of ragi that is easier to digest and provides a concentrated source of nutrients. It is often used as a weaning food for infants transitioning from breast milk or formula to solid foods. Malted ragi can be prepared as a porridge and is a popular choice due to its nutritional value.
- Jowar Food Mix: Jowar, also known as sorghum, is another nutritious millet that is gluten-free and rich in fiber, protein, and various vitamins and minerals. Jowar food mix is a blend of jowar flour and other complementary ingredients. It can be used to prepare porridge or a thick gruel that is suitable for infants and young children. Jowar food mix provides a good balance of macronutrients and micronutrients necessary for a growing child's development.
- Jowar Khichdi: Jowar khichdi is a traditional Indian dish made with jowar grains, lentils, and vegetables. It is a wholesome and nutritious meal that can be introduced to infants and toddlers. The combination of jowar with lentils provides a complete protein profile, and the addition of vegetables adds vitamins and minerals, making it a well-rounded meal for young children.

The promotion of millets in infant diets aligns with the growing awareness of the nutritional benefits of these grains. Millets are nutrient-dense and offer a wide range of essential vitamins, minerals, and dietary fiber. They are also gluten-free, making them suitable for individuals with gluten sensitivities or allergies.

Beyond the scope of infant nutrition, many organizations and governments worldwide are actively promoting the cultivation and consumption of millets as part of their efforts to promote sustainable agriculture, enhance food security, and improve overall nutrition in populations.

Millets are well-suited to various agro-climatic conditions and require fewer inputs like water and fertilizers compared to major cereal crops like rice and wheat. By promoting millet cultivation, governments aim to enhance agricultural sustainability, conserve water resources, and reduce greenhouse gas emissions associated with traditional agricultural practices.

Furthermore, millets can be an important tool in addressing food security challenges. As they are hardy crops with good resistance to pests and diseases, they can be more reliable in challenging environmental conditions, such as drought or poor soil quality. Diversifying crop production to include millets can enhance food resilience and availability, especially in regions vulnerable to climate change and other external factors.

From a nutrition perspective, the consumption of millets can contribute to healthier diets. They offer a balanced profile of carbohydrates, proteins, and fats, along with essential micronutrients like iron, calcium, and magnesium. By promoting millet consumption, organizations and governments aim to improve the nutritional status of communities, especially in regions where malnutrition and nutrient deficiencies are prevalent.

In conclusion, the Ministry of Women and Child Development's recommendations for millet-based feeds for infants highlight the recognition of millets' nutritional benefits. Additionally, the broader efforts to promote millet cultivation and consumption on a larger scale aim to foster sustainable agriculture, enhance food security, and improve nutrition in populations worldwide.

5. MILLETS NUTRITION AND HEALTH

Indeed, millets like ragi and jowar have been traditional weaning foods in many Indian homes for generations. These millets are often malted, fermented, or sprouted to prepare gruels suitable for toddlers. These preparation methods enhance the digestibility of millets and make them easily consumable for young children.

Millets are rich in various nutrients, including proteins, dietary fiber, vitamins, and minerals. This nutrient profile contributes to the overall physical and cognitive growth and development of children. Proteins are essential for muscle and tissue development, while dietary fiber supports digestive health and helps in preventing constipation. Additionally, the vitamins and minerals present in millets play vital roles in maintaining overall health and well-being.

Moreover, millets have been shown to have benefits for gut health in children. The fiber content in millets acts as a prebiotic, promoting the growth of beneficial gut bacteria. A healthy gut microbiome is associated with improved digestion, better nutrient absorption, and a strengthened immune system.

The low glycemic index of millets makes them an ideal food for individuals with diabetes or those looking to manage their blood sugar levels. Foods with a low glycemic index are digested and absorbed more slowly, resulting in a slower and steadier increase in blood glucose levels, which is beneficial for diabetics in managing their condition.

There are various types of millets, including pearl millet, finger millet, foxtail millet, and proso millet, among others. Each type of millet offers a unique nutrient profile, and they can be used interchangeably or combined to create diverse and nutritious meals.

Millets are versatile and can be used in various ways in the kitchen. They can serve as a staple food, replacing or complementing rice and wheat in the diet. Millet flour can be used to make bread, porridge, and other baked goods. Additionally, millets are also used as animal feed, contributing to sustainable livestock farming practices.

Beyond their culinary uses, millets have a long history of use in traditional medicine. They have been found to have several health benefits, such as promoting heart health, aiding in weight management, and reducing the risk of chronic diseases like certain cancers and cardiovascular conditions.

In summary, millets like ragi and jowar have been treasured as traditional weaning foods in Indian households for their nutritional benefits and ease of digestion. They offer a range of essential nutrients that support overall growth and development in children, along with promoting good gut health. Additionally, millets' low glycemic index makes them suitable for those with diabetes, and their versatility in culinary applications makes them a valuable addition to diverse diets.

6. MILLETS AND THE ENVIRONMENT

Millets are truly remarkable grains that offer numerous benefits for both human health and the environment.

- Harsh struggle. Due to their ability to adapt to challenging conditions, millet cultivation requires lower production costs compared to other grains, making them a cost-effective option for farmers.
- Nutrient-Rich Profile: Millets are indeed high in protein, fiber, vitamins, and minerals, such as iron, calcium, and phosphorus. These nutrients are essential for various bodily functions and play a significant role in promoting overall health. The high fiber content aids in digestion and sup-

ports a healthy gut, while the abundant vitamins and minerals contribute to bone health, energy metabolism, and immune function.

- Low in Fat and Gluten-Free: Millets have a lower fat content compared to many other grains, making them suitable for individuals looking to manage their fat intake. Additionally, millets are naturally gluten-free, making them an excellent alternative for individuals with gluten sensitivity or celiac disease.
- Rich in Antioxidants: Millets are rich sources of antioxidants, such as phenolic compounds and flavonoids. Antioxidants help neutralize free radicals in the body, protecting cells from oxidative stress. By reducing oxidative stress, millets may play a role in preventing chronic diseases like cancer, heart disease, and diabetes.
- Environmentally Friendly: Millet production is eco-friendly in several ways. Since they are primarily rain-fed crops, they reduce the demand for irrigation and alleviate the pressure on water resources, especially in regions facing water scarcity. Furthermore, millet crops have low water and nutrient requirements, minimizing the environmental impact of agricultural practices.
- Pest Resistance: Millets are less susceptible to pests compared to some other crops like rice and wheat. This natural resistance reduces the need for pesticides and chemical interventions during cultivation, leading to a more sustainable and environmentally friendly farming approach.
- Rich in Beneficial Nutrients: In addition to the aforementioned nutrients, millets are also rich in phosphorus, magnesium, copper, and manganese. These minerals are essential for bone health, nerve function, and enzymatic reactions in the body.

In summary, millets offer a wide range of health benefits due to their nutrient-rich profile, including protein, fiber, vitamins, and minerals. They are low in fat and gluten-free, making them suitable for various dietary preferences and restrictions. Millets high antioxidant content contributes to their potential in preventing chronic diseases. Moreover, millet production is environmentally friendly, as they can grow in harsh conditions with less water and without the need for excessive pesticide use. These qualities make millets an excellent choice for promoting sustainable agriculture, improved nutrition, and environmental conservation.

7. MILLETS AND ECONOMIC EMPOWERMENT OF WOMEN

The economic benefits of millets are indeed significant, and their cultivation offers advantages both for farmers and the overall agricultural sector:

- Cost-Effectiveness: Millets are cheaper to grow compared to other major cereal crops like rice and wheat. They are hardy and drought-resistant, which means they require less water and are less dependent on irrigation. This reduces the overall production cost for farmers.
- Low Water Requirement: Millets have a remarkable ability to grow with minimal water resources. In water-scarce regions, millets can be a more sustainable and economically viable option for farmers compared to water-intensive crops.
- Adaptability: Some millet varieties can grow in adverse conditions, including rocky terrains and as weeds in challenging environments. This adaptability makes them suitable for regions where other crops may not thrive, expanding the potential for agricultural productivity and income generation.

- High Production: India is one of the leading producers of millets globally, with a substantial production capacity of over 50.9 million tons as per the Economic Survey 2023. The country contributes to 80 percent of Asia's and 20 percent of global millet production.
- Decent Yield: The average yield of millets in India is reported to be 1239 kg/hectare, which is slightly higher than the global average yield of 1229 kg/hectare. This suggests that Indian farmers are efficiently cultivating millets to achieve good yields, contributing to food security and economic prosperity.
- Recognition of Rural Women Farmers: The celebration of International Rural Women's Day as Rashtriya Mahila Kisan Divas' by the Government of India highlights the significant role women farmers play in the agricultural sector. Recognizing their efforts on this day provides further impetus to women farmers and rural women, empowering them economically and socially.
- Theme of "Rural Women Cultivating Good for All": The theme of "rural women cultivating good for all" for International Rural Women's Day in 2021 emphasizes the crucial role of rural women in sustainable agriculture, food production, and the well-being of their communities. This recognition supports efforts to promote gender equality and women's empowerment in agriculture.

In conclusion, millets offer numerous economic benefits due to their cost-effectiveness, low water requirements, and adaptability to various environments. India's substantial production capacity and decent yield contribute to the country's role as a major player in the global millet market. Moreover, the recognition of rural women farmers on International Rural Women's Day highlights their essential contributions to the agricultural sector and society at large. By promoting sustainable agriculture and empowering rural women, millet cultivation fosters economic growth, food security, and social development.

8. CROP MONITORING AND MANAGEMENT

Crop monitoring and management are crucial aspects of successful millet cultivation. Proper monitoring allows farmers to assess the growth and health of millet crops, identify potential issues, and implement timely management practices to optimize yield and quality. Here's a detailed explanation of crop monitoring and management with respect to millets:

- **Pre-Planting Preparation:** Before sowing millet seeds, farmers need to prepare the land by plowing, harrowing, and leveling the field. Proper land preparation ensures a suitable environment for seed germination and early growth.
- **Seed Selection and Sowing:** Selecting good quality seeds is essential for a healthy crop. Farmers should choose seeds that are disease-free, genetically pure, and have high germination rates. The sowing method and timing are also important. Millets are usually sown directly in the field during the monsoon season when there is enough moisture for germination.
- **Monitoring Germination:** After sowing, farmers need to monitor germination to ensure that the seeds are sprouting and establishing well. Uneven or delayed germination may indicate issues with soil moisture, seed quality, or planting depth.
- **Irrigation Management:** Millets are often rain-fed crops, but supplemental irrigation may be necessary during dry spells. Monitoring soil moisture and adjusting irrigation practices accordingly is crucial to prevent water stress and promote healthy growth.

- **Weed Control:** Weeds can compete with millet crops for nutrients and sunlight, reducing yields. Regular monitoring is essential to identify weed infestations and implement appropriate control measures, such as manual weeding or the use of herbicides if necessary.
- **Pest and Disease Monitoring:** Regular monitoring for pests and diseases is crucial to detect early infestations. Common pests affecting millets include stem borers, shoot flies, and aphids, while diseases like blast and smut can also impact yields. Integrated Pest Management (IPM) practices can be adopted to manage these issues effectively.
- **Nutrient Management:** Monitoring nutrient levels in the soil and in plant tissues helps farmers determine if additional fertilization is needed. Balanced fertilization with appropriate amounts of nitrogen, phosphorus, and potassium supports healthy crop growth and development.
- **Monitoring Growth Stages:** Understanding the growth stages of millet crops is essential for effective management. Different millet varieties have distinct growth patterns, and farmers should be able to identify the different growth stages, from seedling emergence to maturity.
- **Harvesting Time:** Monitoring the crop's maturity and knowing the right time to harvest is critical. Harvesting too early can result in lower yields, while delayed harvesting may lead to grain losses due to shattering.
- **Post-Harvest Management:** After harvesting, proper post-harvest management is necessary to maintain grain quality. Drying, cleaning, and proper storage are essential to prevent spoilage and maintain the nutritional value of millet grains.
- **Record-Keeping:** Maintaining detailed records of crop monitoring and management activities is beneficial for future reference and planning. This information can help farmers make informed decisions and improve practices in subsequent seasons.

Overall, crop monitoring and management play a vital role in the successful cultivation of millets. Through careful observation and timely intervention, farmers can optimize crop productivity, ensure food security, and promote sustainable millet farming practices.

9. FOR SENIOR CITIZENS

As individuals age, various physiological processes in the body tend to slow down. As a result, there may be changes in metabolism, digestive function, and nutrient absorption. Ensuring a balanced and easily digestible diet becomes critical to support the health and well-being of elderly individuals. Millets offer several advantages in this regard, making them an excellent choice for maintaining good health in the elderly.

- **Rich in Healthy Fats and Fiber:** Millets are good sources of healthy fats, such as omega-3 and omega-6 fatty acids, which are essential for various bodily functions, including brain health and inflammation regulation. Additionally, millets are rich in dietary fiber, which aids in digestion and supports bowel regularity. Including millets in the diet helps meet the nutritional requirements of the elderly.
- **Gut Health:** Older individuals often experience disturbances in gut health, which can lead to issues such as constipation and reduced nutrient absorption. Millets can help with these problems due to their easily digestible nature and high fiber content. The non-starch polysaccharides found

in millets are a major component of dietary fiber, and they promote healthy gut function by supporting the growth of beneficial gut bacteria.

- **Probiotic Properties:** Millets, especially when fermented, can act as excellent probiotics. During fermentation, resistant starch in millets gets broken down into short-chain fatty acids. These fatty acids serve as nourishment for beneficial gut bacteria, leading to improved gut health. A healthy gut microbiome is crucial for proper digestion, immune function, and overall well-being.
- **Constipation Prevention:** Constipation is a common issue in the elderly, and the fiber content in millets can help alleviate this problem. Fiber adds bulk to the stool and promotes regular bowel movements, reducing the risk of constipation and related discomfort.
- **Hypoallergenic Nature:** As people age, they may become more susceptible to developing food intolerances or sensitivities. Millets are considered hypoallergenic, meaning they are less likely to cause allergic reactions. For individuals with sensitive digestive systems, millets can be a safe and nutritious dietary choice.
- **Nutritional Value:** Millets offer a wide range of essential nutrients, including vitamins, minerals, and antioxidants. A diversified diet that includes millets helps ensure that elderly individuals receive adequate nutrition to support their overall health and immune function.

By incorporating millets into their diets, elderly individuals can benefit from the nutritional value and easy digestibility of these grains. The fiber and healthy fats in millets support gut health, prevent constipation, and provide a nourishing diet that is less likely to trigger food intolerances. Moreover, the probiotic properties of fermented millets contribute to a healthier gut microbiome, which is essential for overall health and well-being in old age.

10. PREGNANT AND LACTATING WOMEN

Absolutely, pregnancy is a crucial period that demands specialized nutritional requirements to support the health and development of both the mother and the growing baby. Traditional foods, such as millet-based laddus, are commonly prepared and consumed by expectant mothers to provide essential micronutrients during pregnancy.

- **Blood Production and Nourishing the Spleen:** Millets are known to be rich sources of iron, which is essential for blood production. During pregnancy, the body's blood volume increases to support the developing fetus, and iron plays a vital role in preventing anemia and maintaining overall health. Additionally, millets are believed to nourish the spleen, which is important for blood filtration and immune function.
- **Appetite Stimulation:** Pregnancy can sometimes lead to changes in appetite and digestion. Millets are easily digestible and can help stimulate the appetite, ensuring that expectant mothers receive the necessary nutrients for themselves and their growing baby.
- **Nutrient-Dense for Breastfeeding:** After giving birth, mothers who are breastfeeding also have increased nutritional needs. Millets, being nutrient-dense cereals, can be beneficial for breastfeeding mothers to replenish their nutrient stores and provide essential vitamins and minerals to support lactation and the well-being of both mother and child.

- **Sorghum for Pregnant and Lactating Women:** Sorghum, also known as jowar, is a type of millet that is often served to pregnant and lactating women due to its rich mineral and vitamin content. Sorghum is an excellent source of minerals like copper and iron, which play crucial roles in immune function and overall health.
- **Copper and Iron for Immunity:** Copper and iron, present in millets like sorghum, are essential minerals that support immune function. A well-functioning immune system is vital during pregnancy and lactation to protect both the mother and the baby from infections and illnesses.

Including millets in the diet of expectant and lactating mothers can be a beneficial dietary choice, providing essential nutrients like iron, copper, vitamins, and fiber. It supports proper blood production, nourishes the body, stimulates appetite, and contributes to overall health and well-being during this crucial phase of life.

As with any dietary changes during pregnancy and lactation, it is essential for expectant and breastfeeding mothers to consult with healthcare professionals or registered dietitians to ensure that their nutritional needs are adequately met and that the foods they consume align with their specific health conditions and requirements.

11. ROLE OF TECHNOLOGY IN GROWING MILLETS

The role of technology in growing millets has become increasingly important in recent years, contributing to the cultivation, processing, and promotion of these nutritious and climate-resilient grains. Here's an explanation of the role of technology in millet production:

Improved Farming Practices:

- **Precision Agriculture**: Technology, such as GPS-guided tractors and drones, enables farmers to practice precision agriculture. This involves precise planting, irrigation, and fertilizer application, leading to higher millet yields and resource efficiency.
- **Weather Forecasting**: Access to weather forecasting through mobile apps or online platforms helps farmers make informed decisions about planting and harvesting times, minimizing the impact of adverse weather conditions.

Crop Management:

- **Crop Monitoring**: Satellite imagery and remote sensing technologies allow for real-time monitoring of millet crops. This helps identify issues like pest infestations or diseases early on, enabling timely intervention.
- **Irrigation Techniques**: Smart irrigation systems can be employed to ensure millet fields receive the right amount of water, conserving resources and preventing water wastage.

Seed Technology:

- **Improved Varieties**: Research and technology have led to the development of improved millet varieties that are more disease-resistant, high-yielding, and adapted to different growing conditions.

- **Seed Treatment**: Advanced seed treatment techniques enhance seed quality, reducing the risk of seedborne diseases.

Digital Extension Services:

- **Mobile Apps**: Farmers can access information on millet cultivation techniques, pest control, and market prices through mobile apps and SMS services. These digital extension services empower farmers with knowledge to improve their practices.
- **Online Platforms**: Online platforms connect farmers with experts and extension services, providing a forum for knowledge sharing and problem-solving.

Processing and Value Addition:

- **Milling Technology**: Advanced milling machines and equipment improve millet processing efficiency, resulting in higher-quality millet products.
- **Food Processing**: Technology facilitates the development of millet-based products such as flour, flakes, and ready-to-eat snacks, expanding their market reach.

Market Access:

- **E-commerce**: Technology enables millet farmers to access wider markets through e-commerce platforms, allowing them to sell their products directly to consumers or businesses.
- **Supply Chain Management**: Digital platforms help streamline millet supply chains, reducing post-harvest losses and ensuring timely delivery to markets.

Research and Development:

- **Genetic Improvement**: Biotechnology and genetic engineering techniques can be used to develop millet varieties with specific traits, such as enhanced nutrient content or resistance to pests.
- **Data Analysis**: Advanced data analytics assist researchers in studying millet crop performance and adapting cultivation practices to changing climatic conditions.

Promotion and Awareness:

- **Social Media and Online Campaigns**: Technology plays a crucial role in promoting millets by raising awareness about their nutritional benefits and sustainable farming practices through social media and online campaigns.

In summary, technology plays a multifaceted role in the cultivation and promotion of millets. It enhances agricultural practices, improves crop management, aids in processing and marketing, supports research and development, and contributes to raising awareness about the importance of millets in achieving food security and nutrition goals, particularly in regions facing climate challenges.

12. ROLE OF MILLETS IN DEVELOPING COUNTRIES

Millets play a significant role in developing countries, particularly in regions facing food security and nutritional challenges. Here is a brief explanation of the importance of millets in developing countries:

- **Nutritional Value**: Millets are highly nutritious and rich in essential nutrients such as fiber, protein, vitamins, and minerals. They are often referred to as "nutri-cereals" due to their nutritional value. In many developing countries, where malnutrition is a concern, millets offer an affordable and accessible source of nourishment.
- **Climate Resilience**: Millets are hardy crops that can thrive in diverse agro-climatic conditions, including drought-prone areas. They require less water compared to major cereal grains like rice and wheat. This resilience makes millets a crucial crop for regions with erratic rainfall and limited irrigation facilities.
- **Dietary Diversity**: Millets contribute to dietary diversity by offering an alternative to rice and wheat-based diets. Introducing millets into diets can help reduce the reliance on a single staple crop and diversify the sources of nutrients, reducing the risk of nutrient deficiencies.
- **Income Generation**: Millet cultivation can provide livelihood opportunities for smallholder farmers in developing countries. The crop can be grown with relatively low inputs and has the potential to increase farmers' incomes while improving food security.
- **Food Security**: Millets are often considered a form of "insurance crop" because they are less susceptible to weather-related risks. In times of crop failures or food shortages, millets can serve as a dependable source of food.
- **Cultural Significance**: In many developing countries, millets hold cultural significance and are part of traditional diets and culinary heritage. Preserving and promoting millet consumption can help maintain cultural food traditions.
- **Environmental Benefits**: Millets are environmentally friendly crops. They require fewer chemical inputs and have a smaller carbon footprint compared to some other grains. Additionally, millet cultivation can contribute to soil conservation and biodiversity.
- **Global Attention**: Millets have gained attention on the global stage as a means to address food security, malnutrition, and climate change. International organizations and research institutions are working to promote millet cultivation and consumption.

In summary, millets are valuable crops in developing countries due to their nutritional content, climate resilience, and potential to improve food security, income generation, and environmental sustainability. Their versatility in various dishes and their adaptability to local growing conditions make millets an essential component of efforts to address food and nutrition challenges in many developing regions.

Millets are not limited to India; they are grown and consumed in many countries around the world, particularly in Africa and parts of Asia. Here are some examples of countries where millets are commonly grown and used:

- **Africa**:

Nigeria: Millets like pearl millet (known as "fada" or "tuwo" in Hausa) and finger millet are staple foods in various Nigerian regions.

Niger: Pearl millet is a significant crop and a staple food in Niger, where it's used to make dishes like "tuwo" and "gari."

Mali: Pearl millet is a crucial cereal crop in Mali and is used to prepare dishes like "foufou."

- **Asia**:

China: Millets are grown in China, and foxtail millet is one of the most commonly cultivated varieties.

Nepal: Millets, including finger millet and foxtail millet, are grown and used in traditional dishes.

South Korea: Foxtail millet, known as "ma" or "makgeolli," is used to make traditional alcoholic beverages.

- **United States**:

In the United States, millets are grown primarily as forage crops for livestock. However, there is a growing interest in cultivating millets for human consumption due to their nutritional benefits.

- **Russia**:

In Russia, millets, particularly proso millet, are cultivated for various purposes, including human consumption and bird feed.

- **Europe**:

In some European countries, such as France and Germany, millets are used in birdseed mixes.

- **Other Asian Countries**:

Millets are grown and consumed in countries like Thailand, Bangladesh, and Sri Lanka.

Millets are valued for their drought-resistant properties, nutritional content, and versatility in various cuisines. They are used to make a wide range of dishes, including porridge, flatbreads, cakes, and beverages. The growing awareness of their nutritional benefits and sustainability has led to increased interest in millets globally, both for human consumption and as a crop with potential environmental benefits.

13. HEALTHY RECIPES USING MILLETS

Millets have gained popularity as a nutritious and versatile ingredient in recent years. These gluten-free whole grains are packed with essential nutrients, making them an excellent choice for those looking to add a healthy twist to their meals. Here are some delicious and nutritious recipes that can prepared using millets: Millet Porridge, Millet Sallad, Millet Upma, Millet Idli.

These recipes showcase the versatility of millets and how they can be incorporated into various dishes to boost nutrition. Whether we are looking for a comforting porridge, a refreshing salad, a savory upma, or a classic South Indian idli, millets can be a healthy and tasty addition to our diet. Experiment with different millet varieties and enjoy their numerous health benefits.

14. THE NUTRITIONAL POWERHOUSES: MINOR MILLETS

Minor millets, a group of lesser-known but highly nutritious grains, have been gaining attention in recent years due to their exceptional nutritional profile. These grains, including chena (Proso millet), Kangni (foxtail millet), kutki (little millet), Sanwa (barnyard millet), and Kodon (kodo millet), are celebrated for their high protein and mineral content, making them invaluable additions to our diets.

- **Kangni: The Ancient Grain**

Among the minor millets, Kangni, also known as foxtail millet, stands out as one of the oldest cultivated grains. This versatile grain goes by various names such as kakum, Korralu, kora, and navane in different languages and regions. What sets Kangni apart is its widespread availability, often found in the form of semolina or flour, making it a versatile ingredient in various culinary traditions.

Nutritional Powerhouse: Kangni boasts an impressive protein content of around 12%, earning it the reputation of a nutritional powerhouse among grains. This substantial protein content makes it a valuable addition to vegetarian and vegan diets, providing essential amino acids for muscle health and overall well-being.

The Hypoallergenic Choice:One distinguishing feature of Kangni is its hypoallergenic nature. This means it is less likely to trigger allergic reactions compared to some other grains, making it a suitable option for individuals with food sensitivities or allergies. Its gentle impact on the digestive system makes it a favorite for those seeking easily digestible grains.

Nutrient-Rich for Special Needs: Kangni's nutrient richness extends to its mineral content, making it an excellent choice for specific dietary needs. It is particularly beneficial for pregnant women and undernourished children, as it provides essential minerals like calcium, iron, and copper. These minerals play vital roles in supporting the development of healthy bones, enhancing hemoglobin levels, and bolstering overall immunity.

In summary, Kangni, or foxtail millet, is a remarkable member of the minor millets family. Its ancient history, high protein content, hypoallergenic nature, and rich mineral profile make it a versatile and nutritious choice for a wide range of dietary needs. Whether you are looking to enhance your protein intake, maintain digestive health, or support specific nutritional requirements, Kangni is a grain that deserves a place in your pantry and on your plate. Embracing the diversity of minor millets like Kangni not only contributes to our own well-being but also helps promote sustainable and diverse agricultural practices.

15. THE DIVERSITY AND BENEFITS OF MILLETS FOR GROWING CHILDREN

Millets, a group of small-seeded grains, have been a staple in diets around the world for centuries. In recent years, their nutritional value and health benefits have gained recognition, making them a valuable addition to modern diets. This article explores various types of millets and their benefits, focusing on their role in traditional diets and the contributions of women in the millet value chain.

Millets offer a range of benefits for growing children:

Rich in nutrients: Millets provide essential vitamins and minerals crucial for growth.

Dietary fiber: Their high fiber content aids digestion and promotes a healthy gut.

Energy source: Millets supply energy for active children without causing blood sugar spikes.

Allergen-friendly: Hypoallergenic millets are suitable for children with food sensitivities.

Calcium and iron: Certain millets like Ragi are excellent sources of calcium and iron, vital for bone health and anemia prevention.

Women in the Millet Value Chain: Women play a pivotal role in the millet value chain, from cultivation to consumption. Women are actively involved in millet cultivation, ensuring food security. They participate in processing millets into various products, adding value to the crop. Women prioritize nutritious millet-based diets for their families, promoting health. Their involvement in marketing millet products helps build sustainable communities.

- Types of Millets which benefit for Growing children

Sorghum Millet (Jowar): Sorghum, commonly known as Jowar, is a drought-tolerant millet. It is rich in dietary fiber, making it excellent for digestion in children.

Pearl Millet (Bajra): Pearl Millet, or Bajra, is a widely grown millet in India. It is a good source of energy and provides essential nutrients.

Foxtail Millet (Kakum / Kangni): Foxtail millet, known as Kakum or Kangni, is one of the oldest cultivated millets. It is revered for its hypoallergenic nature and high protein content which is essential for growing children.

Finger Millet (Ragi): Finger Millet, or Ragi, is a nutrient-dense millet rich in calcium and iron. It's often recommended for children and those with dietary restrictions.

Browntop Millet (Korle): Browntop Millet, also called Korle, is a versatile millet used in various culinary preparations. It contributes to a balanced diet with its nutritional content.

Barnyard Millet (Sanwa): Barnyard Millet, known as Sanwa, is a drought-resistant millet. It is a good source of carbohydrates and dietary fiber.

Little Millet (Moraiyo): Little Millet, or Moraiyo, is a small-grained millet known for its versatility in recipes. It is a source of essential nutrients and fiber.

In conclusion, millets are a diverse group of grains with numerous health benefits, making them an ideal choice for growing children. Additionally, women's active engagement in the millet value chain not only enhances nutrition but also contributes to the economic well-being of communities. Embracing millets in our diets supports both personal health and broader sustainability efforts.

16. CONCLUSION

This book chapter has explored the multifaceted relationship between women, millets, and their significant impact on health, empowerment, and sustainable agriculture. Millets, a group of resilient crops with a rich nutritional profile, have been an integral part of human diets for millennia, and their cultivation has been closely intertwined with women's roles and responsibilities.

Throughout history, women have been the custodians of millet cultivation, nurturing traditional knowledge and practices that have sustained communities in challenging environments. As agents of change, women are leading the revival of millet cultivation, advocating for its environmental benefits and promoting diverse varieties to enhance food security and resilience.

Millets offer a wealth of health benefits that make them indispensable components of a balanced diet. Their high fiber, protein, and mineral content support digestive health, boost immunity, and prevent lifestyle-related diseases. For senior citizens, millets are an ideal dietary choice, promoting heart health, managing blood sugar levels, and ensuring adequate nutrient intake. Moreover, their gluten-free nature makes them a safe and nutritious alternative for individuals with specific dietary restrictions.

In the context of women's health, millets play a vital role in supporting pregnant and lactating women. The abundance of nutrients, including iron, calcium, and B-complex vitamins, aids in meeting the increased nutritional demands during pregnancy and lactation, contributing to healthier pregnancies and better maternal and child health outcomes.

Beyond health benefits, millets hold immense economic potential for women. As they actively participate in millet cultivation, processing, and marketing, women experience greater financial independence and empowerment. Women's collectives and self-help groups focused on millet-based livelihoods are fostering entrepreneurship and promoting economic inclusivity.

To ensure successful millet cultivation and yield optimization, crop monitoring and management practices are indispensable. Empowering women with knowledge and training in these aspects can lead to improved crop yields, better resource utilization, and enhanced food security for communities.

In conclusion, this chapter has highlighted the critical intersections between women, millets, nutrition, and sustainable agriculture. Millets emerge as a powerful tool for empowering women, promoting better health outcomes, and fostering economic independence. Embracing millet-based diets and cultivation practices can contribute to a more sustainable and equitable future for both women and communities at large.

As we embrace the inherent wisdom of traditional agricultural practices and integrate them with modern insights, we can unlock the full potential of millets and celebrate their role in shaping a healthier and more prosperous world for all. The journey towards a millet-centered future is not only a return to our roots but also a leap towards a more sustainable, inclusive, and resilient tomorrow.

REFERENCES

Agarwal, B. (2000). Conceptualising environmental collective action: Why gender matters. *Cambridge Journal of Economics*, *24*(3), 283–310. doi:10.1093/cje/24.3.283

Ezeagu, I. E. (2015). Nutritional evaluation of malted millet fortified with defatted sesame seed flour. *African Journal of Food Science*, *9*(2), 70–75.

Farnworth, C. R. (2003). Tapping indigenous knowledge to save millet and sorghum biodiversity. *LEISA Magazine*, *19*(1), 10–11.

Glew, R. H., Ayaz, F. A., Sanz, C., Huang, H. S., Chuang, L. T., & VanderJagt, D. J. (2006). Changes in the amino acid and mineral content of cereals, legumes, and seeds during sous vide processing. *Journal of Agricultural and Food Chemistry*, *54*(22), 8397–8402. PMID:17061812

Gupta, R. K., Gangoliya, S. S., & Singh, N. K. (2015). Reduction of phytic acid and enhancement of bioavailable micronutrients in food grains. *Journal of Food Science and Technology*, *52*(2), 676–684. doi:10.100713197-013-0978-y PMID:25694676

Krishnan, P., & Sharma, M. (2013). Crop monitoring and yield estimation based on MODIS data for millet cultivation areas of Rajasthan, India. *Journal of Agrometeorology*, *15*(2), 190–197.

Kumar, D., & Joshi, N. L. (2006). Indigenous knowledge of millet cultivation in tribal communities of Uttarakhand, Indian Himalaya. *Indian Journal of Traditional Knowledge*, *5*(2), 189–196.

Nair, S., & Kumar, B. M. (2007). Women's role in biodiversity conservation and sustainable use: A case study of home gardens in Kerala, India. *Biodiversity and Conservation*, *16*(8), 2589–2608.

Rahman, H., & Gulati, A. (2019). Smart agriculture in India: Current state and future prospects. *Agricultural Economics Research Review*, *32*(2), 155–166.

Rao, P. P., Birthal, P. S., & Joshi, P. K. (2006). Changes in consumption of coarse cereals in India during 1983–2000. *Food Policy*, *31*(4), 322–332.

Reddy, N. R., Pierson, M. D., & Sathe, S. K. (2015). Salient features of millets, with particular reference to their seed proteins. *Food and Nutrition Bulletin*, *36*(3), 327–340. PMID:26314732

Saleh, A. S., Zhang, Q., & Chen, J. (2013). Shenquan decortication and grinding: An effective process for milling millets. *International Journal of Food Science & Technology*, *48*(5), 1082–1087.

Thakur, A., & Vaishnav, K. (2017). Role of millets in human health: A review. *International Journal of Chemical Studies*, *5*(3), 2295–2300.

Vellakkal, S. (2010). Women's role in food production and agriculture in developing countries. *EAS Journal of Biotechnology and Genetics*, *2*(1), 1–8.

Chapter 6
Sustainable Marketing Mix Strategies of Millets: A Voyage of Two Decades

Saumendra Das
https://orcid.org/0000-0003-4956-4352
GIET University, India

Brojo Kishore Mishra
https://orcid.org/0000-0002-7836-052X
NIST Institute of Science and Technology (Autonomous), India

Narendra Panda
https://orcid.org/0000-0002-0330-2325
GIET University, India

Hassan R. H. Badawy
Faculty of Tourism, Luxor University, Egypt

ABSTRACT

Millets are popular for their nutritional benefits. It provides extra benefits like antioxidant, anti-diabetic, anti-tumorigenic, and antibacterial materials. Consuming millet boosts immunity, detoxifies the body, reduces cancer risk, increases energy, strengthens the brain and muscles, and improves respiratory immunity. Compared to processed grains, millets are more nutrient-dense. Millets contribute to a well-balanced diet and play a significant role in sustainable nutrition. Lack of awareness about the nutritional content of existing dietary patterns and unwillingness to adapt eating habits are two major contributors to poor production and consumption. Millets are no longer extensively grown, and as a result, there are barely any of them on the market. Since 2003, several studies have been performed to assess the millet's increasing demand. This study's objective is to compile marketing mix strategies from those studies and then provide an intelligent perspective on such strategies.

DOI: 10.4018/978-1-6684-9819-4.ch006

Copyright © 2023, IGI Global. Copying or distributing in print or electronic forms without written permission of IGI Global is prohibited.

1. INTRODUCTION

Cereal is an indispensable staple that is consumed by the vast majority of people worldwide. By offering a significant number of calories, protein, and a few specific micronutrients and non-nutrients, cereal grains play a significant part in the diets of people. It also accounts for more than 56% of energy and 50% of protein spent globally. The cereal crops like wheat, maize, rice, barley, sorghum, millet, oat, and rye are all most cost-effective. Cereal grains called millets are divided into two major and minor groups. The primary millet varieties include finger, sorghum, and pearl, while the minor varieties are foxtail, proso, kodo, barnyard, and small millet. Millets are more naturally pest-resistant than common cereals like wheat and rice, which means they require less chemical fertilizers and pesticides to grow. Millets also require a very less water to produce than these common cereals. The ability to combat metabolic disorders, a reduced threat of seditious bowel and heart disease, as well as gastrointestinal issues alike stomach ulcers and colon cancer, have all been related to consuming millet. Additionally, studies have indicated that it promotes bone formation and lowers cholesterol, blood sugar, blood lipids, blood pressure, and hypertension. The first grains, millets, were domesticated at the beginning of human civilization thousands of years ago. Evidence from Northern China suggests that 4000 years ago, noodles were produced from two species of millet, proso and foxtail millet. Small-seeded grains from several tribes and genera that are taxonomically part of the Poaceae family are referred to as millet. Around 90% of the millet produced worldwide is consumed in underdeveloped nations (Shahidi & Chandrasekara, 2013). According to estimations given by the Food and Agriculture Organization (FAO), millet output reached 28.33 million metric tonnes in 2019 and 30.08 million metric tonnes in 2021. Africa is considered as the largest area in the world for millet production and consumption. Further 55% of the world's productivity comes from this area, with Asia coming in another with 40%. Only Africa consumes additional 40% of the world's millet, particularly in Nigeria, Niger, Mali, Burkina Faso and Sudan. The frequent use of millet in underdeveloped countries like India and Africa are important the challenges for food and nutritional safety. High yield and excellent drought resistance of the crop in semi-arid areas of Africa. As a result, initiatives are being undertaken to increase millet production as a response to the region's problems with food security and climate change. African countries began enacting laws in May 2021 to promote millet planting and consumption, employing farmer networks, seed fairs and public seed banks. The development and production of millet seed is the only area of public and private investment where millet entrepreneurs have a substantial impact (FAOSTAT, 2023). But according to Shah et al. (2023), the major barriers to selling millet products include ignorance, unfamiliarity, difficulty in preparation, physical attributes, lack of affordability, loss of trustworthiness, client mistrust, and millets' protracted gestation period. According to Saxena, Saini, and Sasmal (2002), 45.3% of youths did not know how to prepare millet due to the rapid shift towards eating ready-to-eat meals. The majority of customers are aware of the advantages of millet intake, but they have not yet begun doing so. In spite of the fact that 80.6% of women surveyed and believed millet to be nutritious, only 62.7% of them reported eating (Rizwana et al., 2023).

More than two billion people are at risk of iron insufficiency, 1.75 billion are at threat of zinc deficiency, and 127 million teenagers are vitamin A deficient, making micronutrient deficit a severe worldwide health issue. Consuming pearl millet that has had its iron biofortified is one of the best methods to combat the serious health issue of micronutrient deficiencies (Ongudi, Ngigi, & Kimurto, 2017). As part of therapeutic dietary adjustment and variety, offering traditional whole grain and multigrain alternatives to processed carbs might be particularly beneficial. Millets are being revived with increasing attention in India and around the world because of their nutritional value and capacity to thrive in arid climates owing to their climate-smart characteristics. The Indian government developed a national millet mission in addition to designating 2023 as a National Year of Millets and suggested an International Year of Millets for the UN Food & Agriculture Organization (FAO). In India, several state governments created state millet missions in a similar manner. All of these projects understood the need and incorporated elements to interact with customers to generate demand as well as investments in agriculture productivity and production. The demand for millet-based food items with additional value that are being marketed as healthy foods is continuously rising in international markets. It is important to evaluate current knowledge, perceptions, and practices about millets at this time of transition when attitudes towards them are shifting and there is an increase in health awareness. This assessment will set the groundwork for a strategy to effectively promote millets as a common staple (Kane-Potaka et al., 2021). Investigators, nutrition workers, public health personnel, and food producers need to have a thorough understanding of current knowledge and practices in order to propose millet-based food products, interventions, and promotional schemes to improve the nutritional status & common health of the public and for businesses to plan nutritious foods a financially successful venture.

Millets, once considered to be the poor man's food, are nowadays progressively moving how they are perceived to become "trendy" and "cool." Marketers interested in capitalizing on the growing millet movement in the healthy foods category have seen changing consumer food choices. Today, it's usual to discover millets-based products like pizza, noodles, and mixes for cookies, cakes, pancakes, and other baked goods. The shift from traditional to contemporary ready-to-eat meals has changed customer purchasing habits. These Generation-Z customers now place a higher value on nutritional benefits than an unhealthy diet. This paper is limited to a systematic review of 67 research papers from various databases, including science direct, WoS, research gate, google scholar, and willy online library, covering a period of two decades since 2003 on sustainable marketing mix strategies of millets. This paper was motivated by the latest attention in culinary, local feasting, tourism, socializing, wellness, and consumerism towards millets. Further looking into the perspective developments of millets, this paper is divided into 8 sections where section 2 discussed the objective of research; section 3 formulated the methodology of research; section 4 articulated on the sustainable marketing mix strategies of millets; section 5 exhibited the promotional strategies of millets; section 6 discussed on critical analysis; section 7 articulated on the major findings of this research and in section 8 conclusion, discussion & future direction presented.

2. OBJECTIVES OF THE STUDY

This chapter deals with these objectives as follows:

- To study the sustainable marketing mix strategies of millets.
- To understand the level of awareness and consumerism towards millets.
- To analyses the relevant papers and its outcomes on millets sustainable marketing practices.
- To provide necessary findings and suggestions on marketing strategies for future.

3. METHODS OF RESEARCH

A research method is a system to sketch exactly how a researcher aims to convey their investigation. It considers a sequential & orderly method to an effective research problem. A systematic methodology explains how & when a researcher would examine the study in order to acquire consistent, correct results that fulfil their aims and objects. It also includes the data collection and analysis. Experimental research methods have a long history in several domains, including consumer research. Experiments stand out from other research methods despite the fact that they are the most reliable means of establishing causation. As a result, experiments are an effective tool for researchers when they need to demonstrate cause-and-effect connections (Stoner, Felix & Stadler Blank, 2023). Since the beginning of the COVID-19 epidemic, social scientists have had to develop novel methods for conceptualizing and conducting research. This is mainly true for qualitative researchers, as most of them had to rethink how to examine the processes involved in meaning-making. This rethinking has resulted in many qualitative researchers employing state-of-the-art digital tools to collect, analyze, and involve with fresh sorts of data in a variety of digital environments. While qualitative researchers have traditionally utilized digital resources, some academics have started to write about and think on how this past moment may affect how we anticipate conducting qualitative research itself in the future (Lester, 2023). This is why the present chapter chooses, creates, and applies data sets from multiple sources in a logical manner (see Figure 1).

Figure 1. Flow chart of methodology
(Authors' own contribution)

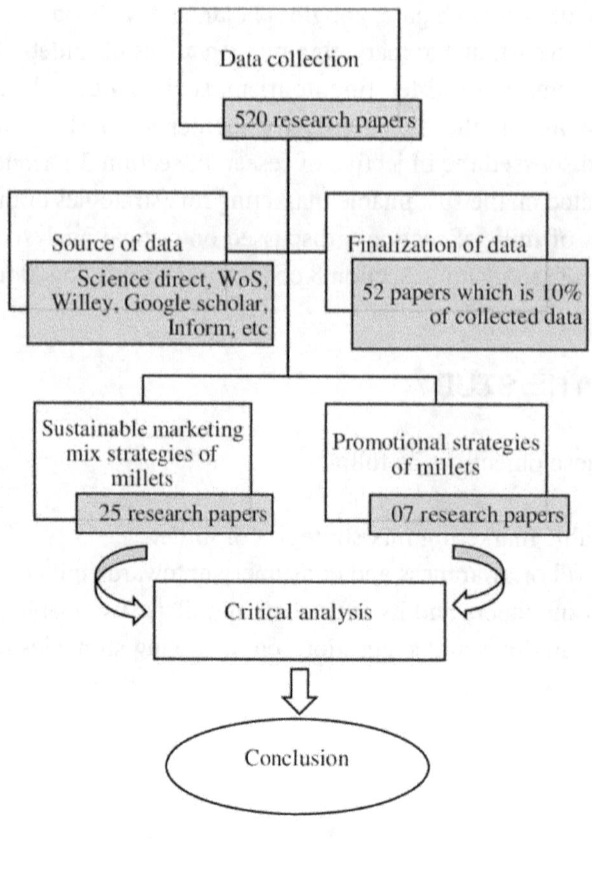

3.1 Research Questions

RQ.1 Are the current marketing techniques suitable for the market for millets' sustainable development?
RQ.2 Are business owners using specific marketing techniques to promote millet products?
RQ.3 Does the government have any regulations in place for the production and export of millet goods?
RQ.4 Does the campaign extend sufficiently to increase awareness among millet users?

4. SUSTAINABLE MARKETING MIX STRATEGIES OF MILLETS

Urban areas have witnessed a rise in the demand for products that contain nutri-cereal bases, which presents several potentials for entrepreneurs that are new to the sector. Entrepreneurship prospects in nutri-cereals are essential for socioeconomic development and promoting rural economic growth, both of which are required to address the mounting properties of accelerated climate transformation on rural livelihoods in less developed nations. Consumer socioeconomic position and health can be improved by promoting these value-added products (Gupta et al.,2023). To resolve the existing matters of undernourishment, food uncertainty, resource reduction and poor routine fitness, global agriculture systems must use sustainable farming techniques. Millets are one such option, outperforming the two most common staple grains in India in terms of economics, sustainability, and nutrition. The effectiveness of the value chain must be improved in this regard, and a comprehensive framework must be provided for millet campaign that takes socio-economic, eco-friendly and well-being sustainability into account (Pandey & Bolia, 2023). The sustainable marketing mix strategies such as product, place, price and promotion of millets and its related cereals could have enhanced the production and consumption limit within the buyers. The pricing strategy before and after millets are harvested has an impact on farmers' earnings. Therefore, along with increasing millets production and consumption, new technologies, government initiatives, and strategies for marketing may also increase farmer income (Abdoulaye & Sanders, 2006). The production process, product improvement, value proposition, selling and use of finger millets might all benefit from greater technological advancements that are currently possible. The finger millet, a novel crop, may improve food safety in Africa and the world by taking benefits of these prospects (Opole, 2019). Millets are remarkable for being an abundant source of protein, carbs, lipids, dietary fibre, and antioxidants. Individuals' daily diet could potentially be improved by including millet, which may aid in the treatment of a variety of health issues (Priya et al., 2023). Food products that are useful and healthful are becoming more and more in demand. A promising option for customers seeking health is buying naturally coloured breakfast cereals with the addition of beneficial components (Sebastião et al.,2023). Researchers are now developing trendy nutri-cereal food products such as extruded, flaked, malted, popped, puffed, and weaning foods. The protein digestibility of extruded weaning food products is improved. For the entrepreneur and market, the emergence of processed nutri-cereals in ready-to-eat or ready-to-cook foods offers excellent opportunities (Gaikwad et al., 2023). Millions of small farmers in East Africa earn their livelihoods by growing finger millets. However, it was severely neglected by the donors of agricultural research on the production and utilization of millets, including unproductive weed management, subpar grain value, ineffective seed schemes, and the production & supply issues, particularly by "spill-in" a revision of pertinent tools established somewhere (Lenne et al., 2007). Understanding these aspects of the study on the marketing strategies of various academic studies, this chapter offered a marketing mix model that could gradually increase the consumption and utilization of millets.

Figure 2. Marketing mix model of millets
(Authors' own contribution)

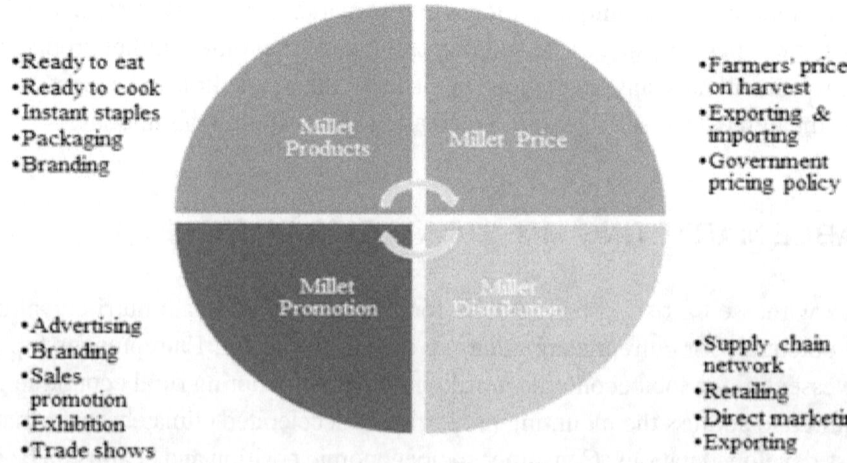

Since 2003, numerous academic studies have utilized the product, pricing, place, and promotion elements of the marketing mix techniques for millets. Understanding the transition from habits to attitude creation towards millets and its products is an amazing stage when sustainable practices like millet promotion or millets missions would be helpful to increase the demand for millets. In Table 1, the marketing strategies employed in the various studies are presented in a logical manner.

Table 1. Contribution of authors on marketing mix strategies

Sl No.	Authors' contribution	Outcome of research	Year of study	Author (s)
1	Millet product mix strategy	Dietary supplement with low glycemic index	2020	Geetha et al.
2	Millet product mix strategy	Dietary value & nutritional content	2013	Krishnamoorthy et al.
3	Finger millet mix strategy	Nutritional value with taste	2012	Pragya & Rita
4	Product attributes of millets	Technology implementation & health benefits	2023	Jindal & Nikhanj
5	Product packaging of millet	Nutritional value in plastic pack	2002	Thathola & Srivastava
6	Product attributes of millets	Nutritional value of finger millet	2013	Shobana et al.
7	Product attributes of millets	Nutritional value of millets among children	2019	Durairaj et al.
8	Product attributes of millets	Reduction of type 2 diabetics, cardiovascular dieses	2022	Amadou
9	Product variety of millet	Innovation in millet products	2019	Alavi, Mazumdar & Taylor
10	Pricing policies of millet	Pearl millet attracted by high income groups	2018	De Groote et al.
11	Pricing of millet	Harvesting price of pearl millet	2008	Gahukar
12	Pricing strategies of millet	Seasonal income of farmers	2016	Roudier et al.
13	Pricing of millet	Reduction in finger millet production due to famers income	2013	Kerr
14	Expensive local seeds	Biological diversity of millets	2007	Nagarajan et al.
15	Price of harvesting	Higher price and lower yields of pearl millet	2001	Rohrbach & Kiriwaggulu
16	Investment in fertilizer	High productivity with less investment	2007	Aune et al.
17	Promotion of millet	Complementary food formation of minor millet	2020	Sharma & Gujral
18	Promotion of pearl millet	Seed treatment for high productivity	2003	Raj et al.
19	Promotion of organic inputs	Soil fertility management	2004	Schlecht & Buerkert
20	Promotion of millet	Health & nutritional benefits	2021	Mbinda & Masaki
21	Distribution strategy of millet	Diversity distribution of pearl millet	2007	Bhattacharjee et al.
22	Diversified distribution	Climate resilient millet	2022	Abdullah et al.
23	Distribution strategy of pearl millet	Sustainable cultivation	2012	Haussmann et al.
24	Diversified cultivation of millet	Seasonal variability of crops	2011	Marteau et al.
25	Distribution of finger millet	Spatial distribution of seeds	2013	Nath et al.

5. PROMOTIONAL STRATEGIES OF MILLETS

To successfully market millets, one must create a USP (unique selling proposition) that promotes them as a practical and healthy decision. A fruitful, sustainable and replicable value-chain approach has been offered, ensuring sustainable nutrition and dietary retreat that is helpful to both the underprivileged dryland agriculturalist and health-conscious urban customers (Dayakar et al., 2021). This model also includes innovations in on farm interventions, treating variation, nutritional assessment, marketing, entre-preneurship growth and awareness raising. In this regard, creative marketing strategies could elevate the dissemination of bi-fortified seeds and products (Foley et al., 2021). With investments in crop breeding, mixing into the public distribution system, nutritional involvement programmes, and commercial seed and food firms with robust mainstreaming nutritional policies, demand of the market is only expected to increase (Govindaraj et al., 2019). In light of the sustainable promotion strategies for millet, governments and policymakers are implementing a number of procedures worldwide to raise awareness of the production and consumption of millet. There are several different promotional tactics listed in Table 2 below.

Table 2. Contribution of authors on promotional strategies

Sl No.	Authors' contribution	Outcome of research	Year of study	Author (s)
1	Advertisement strategy of millet	Publication of isotopic information	2013	Lightfoot et al.
2	Advertisement strategy of millet	nutritional traits of millet	2016	Muthamilarasan et al.
3	Advertisement strategy of millet	nutritional traits of finger millet	2011	Upadhyaya et al.
4	Sales promotion of millet	Adoption of millet through G20	2023	Pandey et al.
5	Public relation strategy of millet	Millet development in climatic conditions	2020	Choudhary et al.
6	Public relation strategy of millet	Stress tolerance techniques of cultivation	2021	Numan et al.
7	Public relation strategy of millet	Plant protection	2021	Singh et al.

6. CRITICAL ANALYSIS

Now people are more conscious towards their health for which there is a growing demand of nutritional food across the globe. Particularly, the behavior of consumers has changed drastically after Covid-19 pandemic. People are now more aware about their health. Consequently, they changed their food habits from fiber based food to nutrition content of food. As a result, the production and consumption proportions of millets are analyzed in the systematic way to understand the impact of marketing strategies of millets for two decades. In this chapter around 520 academic research papers have thoroughly analyzed and relevant papers on marketing and promotional strategies are selected for this analysis. By observing the implementation of various marketing strategies such as product, price, distribution and promotion of millet research (see Table 1) a quantified data exhibited in figure 3.

Figure 3. Analysis of marketing mix strategies of millet research
(Authors' own contribution)

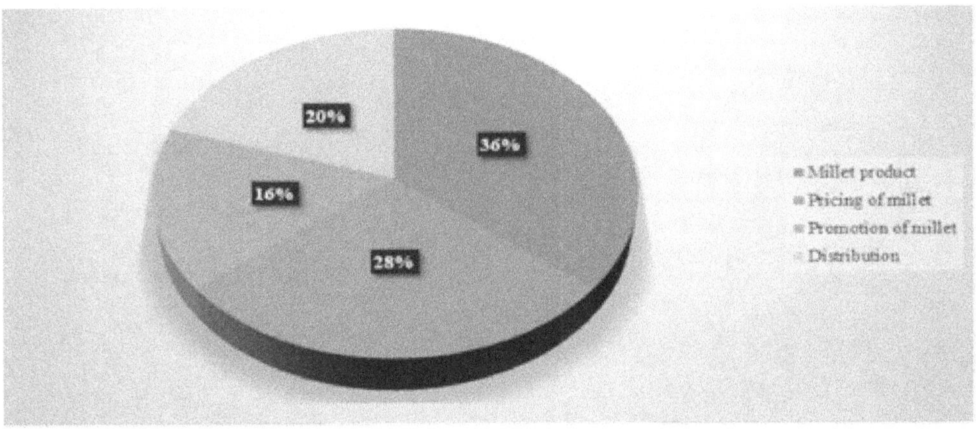

Figure 3 exhibited the marketing mix strategies applied by various authors for two decades where 36% of researches have conducted in product specific outcomes; 28% of authors contributed their study on pricing policies and practices; 20% researches contributed on millet distribution and finally only 16% papers have found specific on promotional parameter. Therefore, it is evidently proved that promotion of millet market severely neglected by the millet producers.

Promotion is an important area of marketing mix where the millet & its byproducts could be promoted through government initiatives and public awareness. Both the major and minor millets could create a sustainable environment for branding, packaging and distribution to change the consumerism. The promotional strategies like minimum selling price, ready to eat foods, staple packaging and dietary supplement for health could accelerate the consumption. The exporting of millets could also be an important area of research. Recent lifestyle food habits like heavy consumption of junk food could be replaced with nutritional food. The promotional strategies such as nutritional benefits, health consciousness, dietary supplement, crop protection, climatic conditions etc. are the major contribution found in this research. Consequently, the millet producers should focus more on the promotional programmes such as advertising, public relations, personal selling or branding of millet based products. The adverting messages, content, media and messages should focus on the nutrition benefits of the millet products so as to attract the attention of the customers. Further, the packaging, branding, personal selling or public relations should also enhance the features of millets. In this connection, the sequence of contribution on promotional strategies are presented (in Table 2) and broader analysis exhibited in figure 4 below.

Figure 4. Analysis of various promotional strategies of millet research
(Authors' own contribution)

Figure 4 exhibited the promotional strategies adopted by various academic research. Among those research contribution 43% of contribution were from adverting research on millet; 43% of authors' contributed on the sales promotion techniques whereas only 14% of academic contribution were on public relation practices. Therefore, it could be observed that public relations are the real challenge for the marketer or entrepreneurs of millet business. As we understand the public relations are very important area where the public awareness could be enhanced properly. Therefore, the millet producers or the government supposed to take a serious measure on sponsorship, exhibitions, trade fairs or export promotions of millets. Various sponsorship programmes like workshops, seminars, conferences, games or symposium should be organized in rural and urban areas for awareness building and penetration on benefits of millets among the consumers. Further, the domestic and international exhibitions have to organize to strengthen the market potential of millet products. The trade fairs should organize at various levels to develop the trading activities among the retailers, wholesalers, or jobbers. The export promotion houses need to be strengthen to internationalize the millet products.

7. FINDINGS

From the above critical analysis, it has been observed several findings of this study as follows:

1. Among the 25 research papers, majority of contributions are based on millet products and pricing of millets whereas few research article found on distribution and promotional strategies of millets.
2. Majority of authors' contributions have observed on advertising and sales promotion of research whereas less contribution found on public relations research.

8. DISCUSSION, CONCLUSION, AND FUTURE DIRECTIONS

Since millets considered as the precise nutritious cereals among all, the promotion and marketing activities of millet produces are still in infancy stage. In this connection, the buying behavior of consumers and the marketing activities of millets could bring the sustainable development of millets. Again it also observed that the consumption and production of millet products are not similar across the world. Particularly the African countries are considered as the highest producers and consumers of millets though the marketing activities are negligible. However, the Asian and European nations have more scope of marketing the millet based products. Similarly, in these areas a rigorous marketing activities such as distribution, promotion etc. could bring a sustainable development of millet products. Therefore, a seamless supply chain network with integrated marketing communication should be implement to develop the marketing and consumption of millets.

The FAO and UN assembly simultaneously proclaimed 2023 as the International Year of Millet by the specific aim of increasing consciousness about the sustainable growth of millets. The establishment of a millet mission, government programmes for neighborhood improvement, the imposition of a minimum selling price, and the socialization process may all help to increase millet's marketing and promotion. Strategies for marketing and advertising such as branding, packaging, distribution, and exporting may be long-term initiatives to support consumption. International trade shows, SHG participation, and cluster development for millets promotion might all help all stakeholders flourish. Farmers and business owners need to be involved in order to fulfil the current marketing practices in order to build a sustainable marketing strategy (see RQ1). Due to poor management and a lack of understanding of the advantages of millets, the business owners' promotional efforts are insufficient (see to RQ2). While significant regions are still in the infancy of millet production and consumption, various academic academics have discussed the regulations implemented in Africa and a small portion of east Asia (see RQ3). Only a small portion of the research contributions are from millet promotion and public relations, according to both analyses (see RQ4).

In order to meet the problems of the future, advertising campaigns, branding, awareness raising, etc. are falling behind. The current study includes a number of limitations in terms of understanding the nutritional benefits of millet products. Numerous topics, including new product creation, the product life cycle, and consumer behavioral research, are not covered in the current chapter. Future studies on millet and its products might concentrate on forecasting methods with accurate estimates of output and demand. Future studies could also focus on the statistical applications on consumer buying behavior towards millets, millets market proposition or supply chain analysis of millet products.

REFERENCES

Abdoulaye, T., & Sanders, J. H. (2006). New technologies, marketing strategies and public policy for traditional food crops: Millet in Niger. *Agricultural Systems*, *90*(1-3), 272–292. doi:10.1016/j.agsy.2005.12.008

Abdullah, K., Abukari, A. B. T., & Abdulai, A. M. (2022). Testing the climate resilience of sorghum and millet with time series data. *Cogent Food & Agriculture*, *8*(1), 2088459. doi:10.1080/23311932.2022.2088459

Alavi, S., Mazumdar, S. D., & Taylor, J. R. (2019). Modern convenient sorghum and millet food, beverage and animal feed products, and their technologies. In *Sorghum and millets* (pp. 293–329). AACC International Press. doi:10.1016/B978-0-12-811527-5.00010-1

Amadou, I. (2022). Millet Based Functional Foods: Bio-Chemical and Bio-Functional Properties. *Functional Foods*, 303-329. doi:10.1002/9781119776345.ch9

Aune, J. B., Doumbia, M., & Berthe, A. (2007). Microfertilizing sorghum and pearl millet in Mali: Agronomic, economic and social feasibility. *Outlook on Agriculture*, *36*(3), 199–203. doi:10.5367/000000007781891504

Bhattacharjee, R., Khairwal, I. S., Bramel, P. J., & Reddy, K. N. (2007). Establishment of a pearl millet [Pennisetum glaucum (L.) R. Br.] core collection based on geographical distribution and quantitative traits. *Euphytica*, *155*(1-2), 35–45. doi:10.100710681-006-9298-x

Choudhary, S., Guha, A., Kholova, J., Pandravada, A., Messina, C. D., Cooper, M., & Vadez, V. (2020). Maize, sorghum, and pearl millet have highly contrasting species strategies to adapt to water stress and climate change-like conditions. *Plant Science*, *295*, 110297. doi:10.1016/j.plantsci.2019.110297 PMID:32534623

Dayakar Rao, B., Bhat, V., Niranjan, T., Sujatha, M., & Tonapi, V. A. (2021). Demand creation measures and value chain model on millets in India. *Millets and millet technology*, 381-411.

De Groote, H., Kariuki, S. W., Traore, D., Taylor, J. R., Ferruzzi, M. G., & Hamaker, B. R. (2018). Measuring consumers' interest in instant fortified pearl millet products: A field experiment in Touba, Senegal. *Journal of the Science of Food and Agriculture*, *98*(6), 2320–2331. doi:10.1002/jsfa.8722 PMID:28990670

Durairaj, M., Gurumurthy, G., Nachimuthu, V., Muniappan, K., & Balasubramanian, S. (2019). Dehulled small millets: The promising nutricereals for improving the nutrition of children. *Maternal and Child Nutrition*, *15*(S3), e12791. doi:10.1111/mcn.12791 PMID:31148399

Foley, J. K., Michaux, K. D., Mudyahoto, B., Kyazike, L., Cherian, B., Kalejaiye, O., Ifeoma, O., Ilona, P., Reinberg, C., Mavindidze, D., & Boy, E. (2021). Scaling up delivery of biofortified staple food crops globally: Paths to nourishing millions. *Food and Nutrition Bulletin*, *42*(1), 116–132. doi:10.1177/0379572120982501 PMID:33593095

Gahukar, R. T. (1988). *Problems and perspectives of pest management in the Sahel: A case study of pearl millet*. Taylor & Francis. doi:10.1080/09670878809371202

Gaikwad, P. S., Reddy, N. B. P., Sandhya, K., & Yadav, B. K. (2023). Traditional vs. Modern Usage of Nutri-Cereals. In *Nutri-Cereals* (pp. 27–56). CRC Press. doi:10.1201/9781003251279-2

Geetha, K., Yankanchi, G. M., Hulamani, S., & Hiremath, N. (2020). Glycemic index of millet based food mix and its effect on pre diabetic subjects. *Journal of Food Science and Technology*, *57*(7), 2732–2738. doi:10.100713197-020-04309-5 PMID:32549623

Govindaraj, M., Rai, K. N., Cherian, B., Pfeiffer, W. H., Kanatti, A., & Shivade, H. (2019). Breeding biofortified pearl millet varieties and hybrids to enhance millet markets for human nutrition. *Agriculture*, *9*(5), 106. doi:10.3390/agriculture9050106

Gupta, R., Singh, H., Kumar, R. P., & Meghwal, M. (2023). Economic, Social, and Market Feasibility of Nutri-Cereals. In *Nutri-Cereals* (pp. 345–360). CRC Press. doi:10.1201/9781003251279-16

Haussmann, B. I., Fred Rattunde, H., Weltzien-Rattunde, E., Traoré, P. S., Vom Brocke, K., & Parzies, H. K. (2012). Breeding strategies for adaptation of pearl millet and sorghum to climate variability and change in West Africa. *Journal Agronomy & Crop Science*, *198*(5), 327–339. doi:10.1111/j.1439-037X.2012.00526.x

Jindal, P., & Nikhanj, P. (2023). A review on processing technologies for value added millet products. *Journal of Food Process Engineering*, *14419*. doi:10.1111/jfpe.14419

Kane-Potaka, J., Anitha, S., Tsusaka, T. W., Botha, R., Budumuru, M., Upadhyay, S., Kumar, P., Mallesh, K., Hunasgi, R., Jalagam, A. K., & Nedumaran, S. (2021). Assessing millets and sorghum consumption behavior in urban India: A large-scale survey. *Frontiers in Sustainable Food Systems*, *5*, 680777. doi:10.3389/fsufs.2021.680777

Kerr, R. B. (2014). Lost and found crops: Agrobiodiversity, indigenous knowledge, and a feminist political ecology of sorghum and finger millet in northern Malawi. *Annals of the Association of American Geographers*, *104*(3), 577–593. doi:10.1080/00045608.2014.892346

Krishnamoorthy, S., Kunjithapatham, S., & Manickam, L. (2013). Traditional I ndian breakfast (I dli and D osa) with enhanced nutritional content using millets. *Nutrition & Dietetics: the Journal of the Dietitians Association of Australia*, *70*(3), 241–246. doi:10.1111/1747-0080.12020

Lenne, J. M., Takan, J. P., Mgonja, M. A., Manyasa, E. O., Kaloki, P., Wanyera, N., Okwadi, J., Muthumeenakshi, S., Brown, A. E., Tamale, M., & Sreenivasaprasad, S. (2007). Finger millet blast disease management: A key entry point for fighting malnutrition and poverty in East Africa. *Outlook on Agriculture*, *36*(2), 101–108. doi:10.5367/000000007781159994

Lester, J. N. (2023). Introduction to special issue: Qualitative research methodologies and methods for theory building in human resource development. *Human Resource Development Review*, *22*(1), 7–14. doi:10.1177/15344843221146871

Lightfoot, E., Liu, X., & Jones, M. K. (2013). Why move starchy cereals? A review of the isotopic evidence for prehistoric millet consumption across Eurasia. *World Archaeology*, *45*(4), 574–623. doi:10.1080/00438243.2013.852070

Marteau, R., Sultan, B., Moron, V., Alhassane, A., Baron, C., & Traoré, S. B. (2011). The onset of the rainy season and farmers' sowing strategy for pearl millet cultivation in Southwest Niger. *Agricultural and Forest Meteorology*, *151*(10), 1356–1369. doi:10.1016/j.agrformet.2011.05.018

Mbinda, W., & Masaki, H. (2021). Breeding strategies and challenges in the improvement of blast disease resistance in finger millet. A current review. *Frontiers in Plant Science*, *11*, 602882. doi:10.3389/fpls.2020.602882 PMID:33488650

Muthamilarasan, M., Dhaka, A., Yadav, R., & Prasad, M. (2016). Exploration of millet models for developing nutrient rich graminaceous crops. *Plant Science*, *242*, 89–97. doi:10.1016/j.plantsci.2015.08.023 PMID:26566827

Nagarajan, L., Smale, M., & Glewwe, P. (2007). Determinants of millet diversity at the household-farm and village-community levels in the drylands of India: The role of local seed systems. *Agricultural Economics*, *36*(2), 157–167. doi:10.1111/j.1574-0862.2007.00195.x

Nath, M., Roy, P., Shukla, A., & Kumar, A. (2013). Spatial distribution and accumulation of calcium in different tissues, developing spikes and seeds of finger millet genotypes. *Journal of Plant Nutrition*, *36*(4), 539–550. doi:10.1080/01904167.2012.748072

Numan, M., Serba, D. D., & Ligaba-Osena, A. (2021). Alternative strategies for multi-stress tolerance and yield improvement in millets. *Genes*, *12*(5), 739. doi:10.3390/genes12050739 PMID:34068886

Ongudi, S. O., Ngigi, M. W., & Kimurto, P. K. (2017). Determinants of Consumers' Choice and Willingness to Pay for Biofortified Pearl Millet in Kenya. *East African Agricultural and Forestry Journal*, *82*(2-4), 175–187. doi:10.1080/00128325.2017.1391489

Opole, R. (2019). Opportunities for enhancing production, utilization and marketing of finger millet in Africa. *African Journal of Food, Agriculture, Nutrition and Development*, *19*(1), 13863–13882. doi:10.18697/ajfand.84.BLFB1004

Pandey, A., & Bolia, N. B. (2023). Millet value chain revolution for sustainability: A proposal for India. *Socio-Economic Planning Sciences*, *101592*, 101592. doi:10.1016/j.seps.2023.101592

Pandey, N. L., Rath, R., Datta, M., Choudhury, B. R., & Kandpal, R. (2023). *Mainstreaming Millets in G20 Countries Through Integrated Behavioural and Policy Shifts*. ORF.

Pragya, S., & Rita, S. R. (2012). Finger millet for food and nutritional security. *African Journal of Food Science*, *6*(4), 77–84. doi:10.5897/AJFSX10.010

Priya, V., Verma, R. K., Lakhawat, S., Yadav, V. K., Gacem, A., Abbas, M., Yadav, K. K., Park, H.-K., Jeon, B.-H., & Mishra, S. (2023). Millets: Sustainable treasure house of bioactive components. *International Journal of Food Properties*, *26*(1), 1822–1840. doi:10.1080/10942912.2023.2236317

Raj, S. N., Deepak, S. A., Basavaraju, P., Shetty, H. S., Reddy, M. S., & Kloepper, J. W. (2003). Comparative performance of formulations of plant growth promoting rhizobacteria in growth promotion and suppression of downy mildew in pearl millet. *Crop Protection (Guildford, Surrey)*, *22*(4), 579–588. doi:10.1016/S0261-2194(02)00222-3

Rizwana, M., Singh, P., Ahalya, N., & Mohanasundaram, T. (2023). Assessing the awareness of nutritional benefits of millets amongst women in Bangalore. *British Food Journal*, *125*(6), 2002–2018. doi:10.1108/BFJ-05-2021-0593

Rohrbach, D. D., & Kiriwaggulu, J. A. B. (2001). Commercialization Prospects for Sorghum and Pearl Millet in Tanzania. (Working Paper Series no. 7). *Journal of SAT Agricultural Research*, *3*(1), 1–28.

Roudier, P., Alhassane, A., Baron, C., Louvet, S., & Sultan, B. (2016). Assessing the benefits of weather and seasonal forecasts to millet growers in Niger. *Agricultural and Forest Meteorology*, *223*, 168–180. doi:10.1016/j.agrformet.2016.04.010

Saxena, S., Saini, S., & Sasmal, S. (2022). General Public Awareness Survey Drive on Impact of Indian Culinary Practices on Nutritional Profile of Food: Special Emphasis on Millet Awareness. *Journal of Culinary Science & Technology*, 1–11. doi:10.1080/15428052.2022.2073936

Schlecht, E., & Buerkert, A. (2004). Organic inputs and farmers' management strategies in millet fields of western Niger. *Geoderma*, *121*(3-4), 271–289. doi:10.1016/j.geoderma.2003.11.015

Sebastião, V. G., Batista, D., Rebellato, A. P., Macedo, J. A., & Steel, C. J. (2023). Sustainable Production Of Naturally Colored Extruded Breakfast Cereals From Blends Of Broken Rice And Vegetable Flours. *Food Research International*, *113078*, 113078. doi:10.1016/j.foodres.2023.113078

Shah, P., Dhir, A., Joshi, R., & Tripathy, N. (2023). Opportunities and challenges in food entrepreneurship: In-depth qualitative investigation of millet entrepreneurs. *Journal of Business Research*, *155*, 113372. doi:10.1016/j.jbusres.2022.113372

Shahidi, F., & Chandrasekara, A. (2013). Millet grain phenolics and their role in disease risk reduction and health promotion: A review. *Journal of Functional Foods*, *5*(2), 570–581. doi:10.1016/j.jff.2013.02.004

Sharma, B., & Gujral, H. S. (2020). Modifying the dough mixing behavior, protein & starch digestibility and antinutritional profile of minor millets by sprouting. *International Journal of Biological Macromolecules*, *153*, 962–970. doi:10.1016/j.ijbiomac.2019.10.225 PMID:31759014

Shobana, S., Krishnaswamy, K., Sudha, V., Malleshi, N. G., Anjana, R. M., Palaniappan, L., & Mohan, V. (2013). Finger millet (Ragi, Eleusine coracana L.): A review of its nutritional properties, processing, and plausible health benefits. *Advances in Food and Nutrition Research*, *69*, 1–39. doi:10.1016/B978-0-12-410540-9.00001-6 PMID:23522794

Singh, S., Sharma, R., Chandra Nayaka, S., Tara Satyavathi, C., & Raj, C. (2021). Understanding pearl millet blast caused by Magnaporthe grisea and strategies for its management. *Blast Disease of Cereal Crops: Evolution and Adaptation in Context of Climate Change*, 151-172.

Stoner, J. L., Felix, R., & Stadler Blank, A. (2023). Best practices for implementing experimental research methods. *International Journal of Consumer Studies*, *47*(4), 1579–1595. doi:10.1111/ijcs.12878

Thathola, A., & Srivastava, S. (2002). Physicochemical properties and nutritional traits of millet-based weaning food suitable for infants of the Kumaon hills, Northern India. *Asia Pacific Journal of Clinical Nutrition*, *11*(1), 28–32. doi:10.1046/j.1440-6047.2002.00251.x PMID:11890635

Upadhyaya, H. D., Ramesh, S., Sharma, S., Singh, S. K., Varshney, S. K., Sarma, N. D. R. K., Ravishankar, C. R., Narasimhudu, Y., Reddy, V. G., Sahrawat, K. L., Dhanalakshmi, T. N., Mgonja, M. A., Parzies, H. K., Gowda, C. L. L., & Singh, S. (2011). Genetic diversity for grain nutrients contents in a core collection of finger millet (Eleusine coracana (L.) Gaertn.) germplasm. *Field Crops Research*, *121*(1), 42–52. doi:10.1016/j.fcr.2010.11.017

Chapter 7
Millet Farming and Tribal Livelihood:
An Analysis of Odisha Millet Mission in Koraput District, Odisha, India

Minati Sahoo
https://orcid.org/0000-0002-6061-8035
Central University of Odisha, India

Karishma Rana
https://orcid.org/0000-0001-8699-6301
Central University of Odisha, India

Pradeep Kumar Samantray
Central University of Odisha, India

ABSTRACT

Millet is one of the oldest cultivated crops which provide food for human and animal consumption, along with its nutritional and health benefits. It is also a pro-environment crop which can grow and sustain with extreme environmental conditions; it has low water requirements and less requirements of soil fertility, nutrients, and moisture. Hence, to increase production, boost consumption, improve processing mechanisms, and improve marketing chain of millet Government of Odisha has introduced a flagship programme called Odisha Millet Mission, to revive and promote millet farming. With these backdrops, this chapter intended to examine the role of Odisha Millet Mission to improve tribal livelihood. This study found that the OMM has remarkable contribution for reviving and encouraging the production, consumption and processing of millet. This is also helpful for improving the employment and livelihood among the tribal in the Koraput district. But the government also needs to endure the tribal people reading the benefits of such potential programmes to further enhance the livelihood opportunities.

DOI: 10.4018/978-1-6684-9819-4.ch007

Copyright © 2023, IGI Global. Copying or distributing in print or electronic forms without written permission of IGI Global is prohibited.

1. INTRODUCTION

The focus of the policy makers, government and economist are now shifting to modifying the agriculture system towards, highly nutritious, climate smart crops that can be grow with most extreme climatic condition, low water requirement, less requirement of chemical fertilizer etc. Such kind of crops and cereals help the countries especially less developed countries to deal with hunger, malnutrition and scarcity of food in affordable price. One of such climate smart crop is millet. Millet is one of the oldest cultivated crops which provide food for human consumption, animal food along with its nutritional and health benefits. It is also a pro-environment crop which can grow and sustain with extreme environmental condition, low water requirements and less requirements of soil fertility, nutrients and moisture. Millet is a comprehensive term used for indicating a wide variety of small-seeded grasses that are mainly cultivated on hilly lands on dry and temperate areas under tropical and sub-tropical regions. India is the largest producer of different varieties of millets. There are different varieties of millet crops grown in India namely Sorghum (Great millet), Ragi (Finger millet), Bajra (Pearl millet), and small millets viz., Korra (Foxtail millet), Little millet, Proso millet, Barnyard milletKodo millet.But recently the contribution of millets in total foodgrain production of India reduced from 22.17% to 6.94% over the last six decades from 1950-51 to 2011-12 due to globalization, urbanization and rising income and standard of living etc. Millet is traditionally serving as one of the sole cultivated crops among the tribal dominated and non-affluent societies for various food andbeverages. Most of the tribal people consumed millet in their daily diet twice or thrice per day for their food, nutrition and water needs. Millets are rich in nutritional values with non-glutinous and not acid forming properties. Hence it is very soothing and light on the palate which makes iteasy to digest grains available (Michaelraj & Shanmugam, 2013). Millets are rich in various nutrients like iron, potassium magnesium, phosphorous and what not. Finger millet is the richest source in case of calcium content compared to 10 times of rice or wheat which has great potential to improve tribal livelihood, nutritional need. Long ago, India is the second largest producer of various kinds of millet but recently there is remarkable reduction in the area, production and productivity of millet due to urban life style, changing food habits, distribution of cheap rice and wheat through public distribution system etc. Millet is one of the pro-poor and pro-nutrition crops that is not only climate smart in nature but also deeply related to rural and tribal life. But in the recent past the area, production and productivity decreased due to green revolution which focused on the production of rice and wheat, distribution of cheap rice and wheat through the fair price shop etc. Hence, to increase the production, consumption and improve supply chain of millet Government of Odisha has introduced Odisha Millet Mission to revive and promote millet.

Odisha Millet Mission is one of the flagships programme in the year 2017 to revive millet from farm to plate to increase the production, boost consumption, improving processing mechanism and improve marketing chain of millet. This initiative prioritizing local crops like millet which helps for shifting our agriculture system towards sustainable cropping practices with diversity in our diets along with respecting the biodiversity in nature rather than forcefully changing cropping patterns to grow wheat and paddy everywhere. This is a comprehensive programme which focuses on the improvement in agronomic practices for production and post production processing activities, improvement in the marketing channel by providing direct marketing linkage to the farmers through millet procurement system, popularizing consumption. To encourage the consumption this project has two fold initiatives. Firstly; provision of providing millet to the households under PDS to increase food and nutritional security within the farming communities. Secondly; inclusion of millet based ready to eat and cooked food in various state level programmes through women SHGs which popularize millet outside the farming communities. To improve marketing strategies there is provision of direct marketing linkage through Millet Procurment Automation System (MPAS) where farmers can easily sell their millet grain at nearest PACS / Mandis. This programme is also helpful for reduction of drudgery of women by developing women friendly tools for post-production processing activities (Jeeva, 2019). Hence, it is essential to look into the role of this programme for the promotion of tribal livelihood. Thus, an attempt has been made in the present study to examine the role of Odisha Millet Mission on millet producing tribal livelihood in Koraput district of Odisha.

2. VARIOUS TYPES OF MILLET

India is the largest producer of different varieties of millets. There are different varieties of millet crops grown in India namely Sorghum (Great millet), Ragi (Finger millet), Bajra (Pearl millet), and small millets viz., Korra (Foxtail millet), Little millet, Proso millet, Barnyard millet and Kodo millet. Each variety of millet contains some unique characteristics for health, environment and nutrition. Like finger millet is the richest sourcing of calcium content, about 10 times that of rice or wheat.

2.1 Finger Millet

Finger millet is the most popularly cultivated crop in the hilly and rural areas. It is an excellent source of various vitamins and nutrients like calcium, iron, protein and amino acids. Finger millet is the richest source of calcium i.e 10 times higher than rice and wheat. It is popularly consumed among various tribes for different foods and beverages like cake, pudding, porridge and other cooked items. This can also used for some alcoholic beverages related to their cultural and traditional food and drinks. It can have potential to improve food and nutritional need among the tribal and rural non-affluent communities. As finger millet farming is a major cultivated crop among the tribes this can also leads to improve rural income and livelihood opportunities. The residue of the finger millet can also used as fodder. Hence, finger millet can helpful for the all-round development of the small holder tribal and farming communities.

2.2 Little Millet

Little millet is one of the small millets commonly known as kutki in hindi, suan in odia. This variety of millet is the powerhouse of various vitamins, minerals and nutrients like protein, carbohydrates, fiber, iron, magnesium, phosphorus and niacin. The nutritional and health benefits of little millet ensure the food and nutritional need of the rural and tribal belts in many parts of India. It contains low glycemic index and it is also very easy to digest compare to other crops like rice and wheat. This property makes it an ideal option for the people with diabetics. Little millet is widely used for the

2.3 Pearl Millet

Pearl millet (Pennisetum typhoides) is another popularly cultivated variety of millet. This variety is popularly known as bajra and can grow in sandy soil with less requirements of water. Pearl millet is normally used as bajra flour for making roti and chapatti etc in many parts of India. Pearl millet is rich with carbohydrates that make it a healthy option for the diabetic. The richness in dietary fibers and cholesterol lowering properties is helpful for a healthy heart. Pearl millet is one of such rare food which reduces frequent acidity and stomach ulcers. The presence of insoluble fibre in bajra helps to prevent the problem of constipation.

2.4 Kodo Millet

Kodo millet (Paspalum scrobiculatum) is historically originated in west Africa and later on carried out to many countries like India, Philippines, Indonesia, Vietnam, Thailand, etc. In India kodo millet is grown in Madhya Pradesh, Tamil Nadu, Karnataka, Gujurat and Chhatishgarh. Kodo millet is very hard in nature. Along with that it also required least water and moisture and highest drought resistance compared with other crops. It contains a good amount of carbohydrates, protein and dietary fibers. The richness of calcium, phosphorus, iron and anti-oxidant properties helps for lowering blood sugar level, reducing blood pressure and lipid level. It is also rich with anti-allergic and anti bacterial properties.

2.5 Barnyard Millet

Barnyard millet is characterized with a number of nutrients like protein, carbohydrates, fiber along with iron and zinc. There are different varieties of barnyard millet but the most popular varieties are Indian barnyared millet (Echinochloa frumentacea) and Japanese barnyard millet (Echinochloa esculenta). Barnyard millet is the fastest growing cereals which can produce harvest within 45-50 days of cultivation with normal weather conditions. The Indian barnyard millet or Swarna dhan is eaten during the fasting time or during any festivals which proves the religious importance of this variety. Barnyard millet is also a good source of protein with low calories which makes it very light on the palate at the same time energetic for the body. It is also rich with fiber which helps in preventing constipation, excess gas, bloating and cramping. The low glycemic index with gluten-free property makes it suitable for people with cardiovascular diseases and diabetic symptoms.

2.6 Foxile Millet

Foxile millet (Setaria italic) is one of the primitive cultivated crops originated from china. It can be cultivated in dry and rain-fed uplands. In India foxile millet is widely cultivated in Karnataka, Andhra Pradesh, Maharashtra and Tamil Nadu. Foxile millet is a great source of nutrients like healthy carbohydrates, iron, magnesium and calcium. This is a great food for balancing the blood sugar and reducing diabetes. The presence of vitamin B12 makes it a healthy option for maintaining healthy heart. It is also rich with natural iron which helps to reduce muscular pain and restlessness. Not only health benefits foxile millet is also related to cultural beliefs in the southern parts of India. Foxile millet has a sweet and nutty flavor and used for various cooked food and ready to eat instant food.

2.7 Sorghum

Sorghum or great millet is one of the widely cultivated cereals in the world used for food, fodder and production of ethanol. It originated in Africa and now widely cultivated in various tropical and subtropical regions. Sorghum is rich in a number of nutrients including Vitamin B which posses an important factor for good metabolism, healthy nerve, hair and skin. It also contains minerals like magnesium and anti-oxidants like phenolic acid which is helpful for strengthening bone health, reducing stress and inflammation etc. Sorghum is also used for making sorghum syrup which is used as a sweetener in food industry. It is lower in total sugar contains but higher in fructose which makes it sweeter than molasses.

2.8 Proso Millet

The culture of cultivation of proso millet (Panicum miliaceum L) goes around 10000 years ago originated from china. Proso millet is the most climate smart crop with least water requirement, low requirement of soil fertility and moisture. Along with climate smart properties it is also rich with various nutritional properties and health benefits. It is also less vulnerable to the attack of disease and pest. This is one of the best options for the people with gluten intolerance because it is completely gluten free. It is also helpful for the nervous system health for their smooth functioning. The loaded magnesium with it also helps for a healthy heart. The niacin and anti-oxidants present in proso millet helpful for slower down the process of ageing by dealing with skin diseases like pellagra, dryness and dullness in the skin etc.

3. REVIEW OF LITERATURE

Millet is traditionally served as one of the sole cultivated crops among the tribal dominated and non-affluent societies for various food and beverages. Millet grains are Millets are rich in nutritional values with non-glutinous and not acid forming properties. Hence it is very soothing and light on the palate which makes iteasy to digest grains available (Michaelraj and Shanmugam, 2013). Millet is not only providing income and livelihood opportunities to the farmers but also fulfil the nutritional and health requirements among the deprived and poor tribal communities (Kumar et.al, 2018). Behera (2017) also found that millet cultivation can improve food and nutritional security has a great role not only for meeting the food and nutritional requirement of many rural poor but also persevered tribal culture and major source of income for them. Millet cultivation has significant role not only for tribal livelihood, consumption but also the socio-cultural aspects of tribal life. With an aim to tackle malnutrition and to increase production of millet based cereals, Government also introduced various policies and programmes in tribal regions. Considering the nutrient contents, finger millet is the richest source of calcium content, about 10 times that of rice or wheat and millet farming has the great potential to improve tribal livelihood, nutritional need and resource management in the tribal and hilly areas and serve as a staple food, weaning food and also a cash crop which provides better opportunities for tribal livelihood (Riley et al., 1992). Along with that millet is also helpful for ensuring food and nutritional security among the tribal community because of the various nutritional and health benefits of millet in affordable price. Mulualem & Melak (2013) found that the nutritional benefits of finger millet have the potential contribution on reducing food scarcity among the households in Metekel Zone, North Western Ethiopia. But the production of these crops are quite less because the farmers are using low yielding local varieties and traditional farming techniques for the production of millets. Michaelraj & Shanmugam (2013); Kumar et.al (2018) found that allocation of budget for food security and tribal development through promotion of local crops like through integration of millet in existing food and nutritional security programmes like existing PDS, ICDS, school mid-day meals, welfare hostels and such other international, national and state-level feeding programs will encourage farmers to grow millets and also help to overcome the existing nutrient deficiencies. With a view to increase production government of Odisha has introduced a special programme for promotion of millet in tribal areas of Odisha in the year 2017 popularly known as Odisha Millet Mission to revive millet on farm and on plate. Government also started millet procurement under OMM based on KMS 2019-20 (Annual report, 2020-21).

4. STUDY AREA & RESEARCH DESIGN

Koraput is purposively used for this study because it is the highest producer of millet-based cereals in Odisha and rich with various tribal cultures.It has highest area coverage i.e. 2709.8 ha for the cultivation of millet (Annual report, OMM, 2020-21). Koraput is one of the poverty-stricken pockets of southern Odisha state in India. The district has the 3rd largest district in terms of geographical area with 8807 square kilometers with 14 blocks and 2 subdivisions. The total population of the district is 1,379,647 with 6,78,809 male population and 7,00,838 females (Census, 2011). One of the surprising facts of the district is that, in spite of ignorance, illiteracy and poverty the sex ratio is 1032 than the national average of .This district is also the heart land of various tribal where around 50% of the population belongs to various tribes like kandha, paraja, gadaba etc. Poverty and low standard of living is another major road blockage for the development of this district with 83% population living below the poverty line as compared to the state wise figure of 47%. As it is one of the rural based economy so the employment opportunities are very less and agriculture is the main occupation in this area. This study is based on secondary data collected from various published sources to examine the role of OMM on variables like area and production of millet, number of farmers enrolled, consumption status, various processing mechanism have been examined. This study is both analytical and descriptive in nature. Various graphs, diagrams and percentage method have been used to analyse and interpret the data.

5. RESULTS AND DISCUSSION

5.1. Area Wise Programme Outreach in Koraput District

Historically, Millet is one of the sole cultivated crops among the tribes of the Koraput district.They are using millet as one of the staple foods for their daily consumption. But due to climatic changes, irregular monsoon cycles and new variety of pests along with resource degradation of natural resources like quality of soil also negatively affected the crop productivity, production cycle and gross output. Over the years, fragmentation of land holdings and diversion of agricultural land into commercial plantations have further affected their production and productivity. But due to the implementation of Odisha Millet Mission there is increase in the production and area under millet farming.

Figure 1. Area wise program outreach in Koraput district (NCDS, 2022)

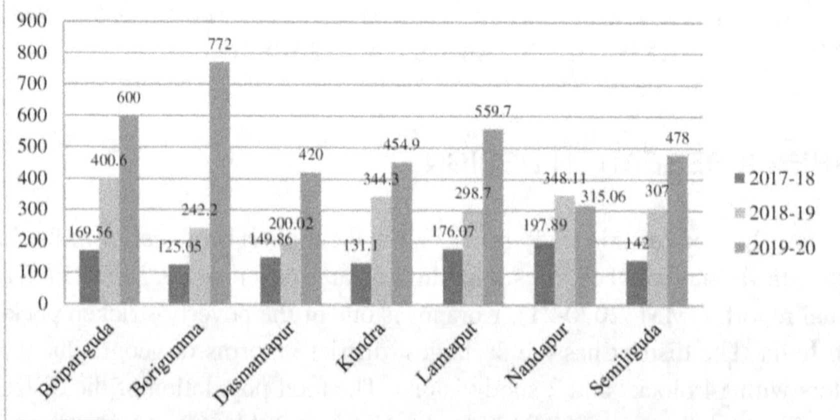

Figure 1 reveals the extension of the program in the 7 blocks that are predominantly grown millet in the study area which are taken for the first phase of the OMM from2017-18 to 2019-20. It can also be seen that there is increase in the number of farmer coverage in all the blocks2017-18 to 2019-20. Maximum coverage of area can be found in the Nandapur block and minimum in the Kundra block.

5.2. Production of Different Kinds of Millet

In Koraput district mainly two types of millet are cultivated finger millet (Ragi), Little millet (Suan) and very small quantity of Kangu, Janha and Kodo. In the figure 2, the average production per farmer has been taken. It can also be seen that there is remarkable increase in the per farmer production of finger millet. Before the OMM the average production was 1.4 qntls but with the implementation of the project there is sharp increase to 5.5 qntls. There is also improvement in the production of other millets like kodo, janha,kangu 0.75 to 2.25 followed by suan 2.2 to 2.6 qntls.

Figure 2. Production per farmer of different kinds of millet (in quintal) (NCDS, 2022)

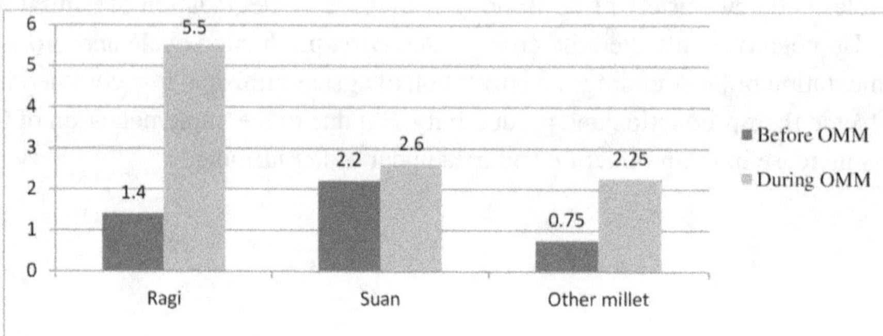

5.3. Cost- Benefit Per Acre of Different Kinds of Millet

Cost-benefit analysis is an important method to find out the profitability and sustainability of the production. The cost components include various costs like seed cost, fertilizer cost, tractor cost, bullock cost, human labour cost both male labour and female labour cost and other costslike interest paid on the leased land, interest paid for the money and loan etc. The return from millet includes the money from the total sale produce like millet grain, by products from millet like straw etc. It ensures that the farmers are earning profit and motivate them to continue their production in future. The table 1 and 2 compare the before and after scenario of the cost structure and profit earned to examine the effectiveness of the project for the farmers profit.

Table 1. Cost and return structure of millet before OMM

Particulars	Ragi	Suan
Total Cost/ Acre (Rs.)	2093	1968
Total Cost/ Quintal of marketable surplus (Rs.)	3128	1295
Total Sales Proceeds/ Acre	1148	2097
Total Sales Proceeds/ Quintal of marketable surplus (Rs.)	1560	2076
Net Income/ Acre (Rs.)	-945	129
Net Income/ Quintal of marketable surplus (Rs.)	-1568	781

Source: NCDS, 2022

The table1 reveals the cost and return of the millet cultivation in the study area. Here the two main cultivating millet variety ragi and suan has been taken. Before the implementation of millet mission the farmers are using traditional agronomic practices and seed which leads to low production and marketable surplus. Before OMM, the total cost per acre is much higher than the revenue earned from sale of the millet grain which leads to loss in the net income of Rs 945 and loss of Rs 1568 in case of net income per quintal of marketable surplus.

Table 2. Cost and return structure of millet during OMM

Particulars	Ragi	Suan
Total Cost/ Acre (Rs.)	4650	4371
Total Cost/ Quintal of marketable surplus (Rs.)	987	2537
Total Sales Proceeds/ Acre	17692	3886
Total Sales Proceeds/ Quintal of marketable surplus (Rs.)	2960	4646
Net Income/ Acre (Rs.)	13042	-485
Net Income/ Quintal of marketable surplus (Rs.)	1973	2109

Source: NCDS, 2022

Table 2 explains the cost and return picture of the two main varieties of millet ragi and suan. It also can be seen that there is increase in the total cost due to use of improved seed variety and mechanization of the agronomic practices but alongside the revenue from the sale proceed also increased which leads to increase in the net income per acre with Rs 13,042 and net income per quintal of marketable surplus. There are mainly two reasons for the increase in the net income. Firstly, due to increase in the production quantity use of improved seed variety and mechanization of the agronomic practices; secondly, due to sale through the PACS/ Mandis with the MSP, which is double than the market pricing. But if we consider suan the net return per acre is -485 which shows loss to the farmer. The reason is that the processing activities of suan is much difficult because of the outer layer of little millet (suan) is very hard. So, the farmers are more interested to invest on the farming of finger millet (ragi) than little millet (suan).

5.4. Selling and Procurement of Millet

Another important provision under this OMM project is providing better marketing and supply chain linkage directly to the farmers. This helps the farmers to get better price and protect them against distress sale at very low price. This helps the ragi and suan producing farmers to sell their crops at better price in the Mandis/PACS under this program through Tribal Development Cooperative Corporation of Odisha Limited (TDCCOL).The government of India has announced the minimum support price for Ragi to be Rs. 3295 only per quintal which is double then the market pricing confirming to fair average quality norms for KMS 2020-2021.Under this scheme 6271 farmers are registered and 71401.84 qunitals are procured in Koraput district in 2020-21. And the procured millet is used for various welfare programs like PDS, ICDS, Mid-Day Meal programs.

Figure 3. Selling of millets before and after OMM
(NCDS, 2022)

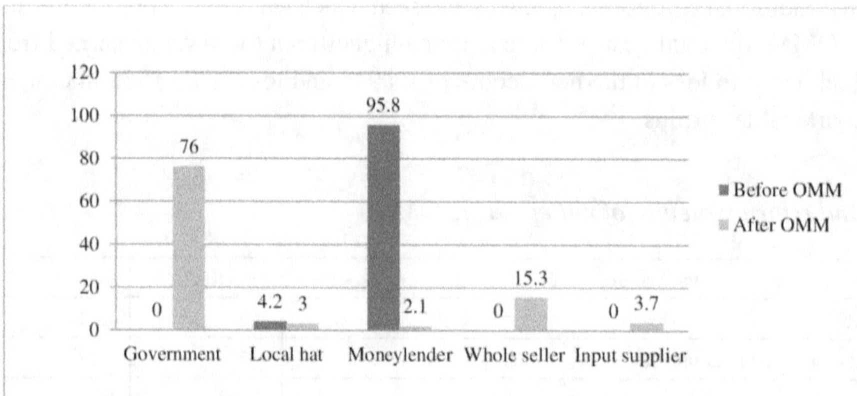

The above table shows that, before the implementation of this project most of the millet producing farmers i.e. sold their produce to the money lender for the loan they have taken before the farming and only 4.2% of the farmers sold in the local hats. But after the implementation of Odisha Miilet Mission most of the farmers are selling under millet procurement system through mandis/PACs. Around 15.3%

of the farmers sells to the whole seller and 3.7% sell to the input suppliers for the purchase of various inputs, 3% to local hats and 2.1% to the money lender.

5.5. Household Purchase of Millet Than Own Production

Millet is one of the staple foods among the tribal people in the Koraput district. They use millet for their daily dietary pattern even trice in a day. Their diet pattern also varies seasonally based on summer, winter and rainy season. In the summer season they prefer millet twice to trice to maintain their whole day water requirement.

Figure 4. Household purchase of average quantity of millet purchased for household use (Quintal) *(NCDS, 2022)*

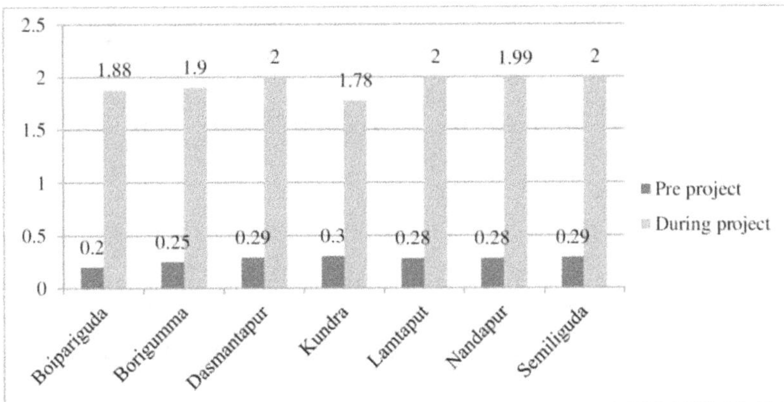

Figure 4 explains the average quantity of millet purchase than their own production of millet in quintals. It shows that before the project the purchase of millet for household use is less varies from 0.2 qntls to 0.3 qntls. But during the project there is increase in the quantity of millet purchase. The purchase includes all the purchase outside their own production like from local hat, barter, wage good and PDS. As one of the objectives of this mission is to include the procured millet into various welfare programs like PDS which increase the quantity of millet purchase at the household level.

5.6 Processing and Marketing of Millet

After harvesting the post production activities or processing like threshing, cleaning and drying of grain, selection and management of seed, winnowing and storage etc are important task for the milet farmers before sending the final produce to the market or storing the produce for self consumption. There are two ways of doing these processing activities, by machine or by manually. In most of the blocks in the koraput district these processing activities are done manually. But doing these activities involve lots of pain and drudgery as most of these processing activities are performed by women manually.

Figure 5. Millet processing methods
(NCDS, 2022)

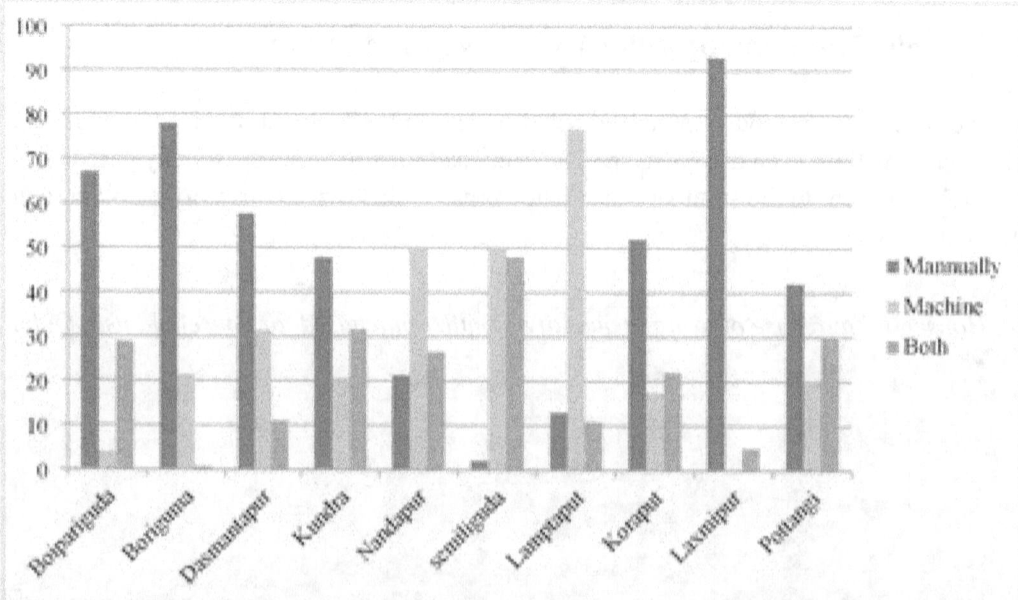

Figure 5 suggests that most of the blocks performed their post production and processing activities manually. In Laxmipur block 93% of the processing activities are done manually followed by Boriguma 77.9% and Boipariguda 67%. In two blocks namely Lamptaput (76.3%) and Semiliguda (52%) followed by Nandapur (50%) major post production processing activities are done through processing machines. In Laxmipur block not a single household performed major processing activities through machine.

5.7 Method Wise Yield of Millet Under OMM

Odisha Millets Mission improved the productivity of millets by adopting standard and improved agronomic practices such as System of Millets Intensification (SMI), Line transplanting (LT) and Line Sowing (LS). These improved agronomic practice benefited the farmers by increasing the crop yield, and by increasing profitability of millet cultivation. This not only improved agronomic practices but also improved the soil quality with efficient usage of water thereby promoting better crops and overall cropping system that is environment friendly.

Figure 6. Method wise yield of millet
(OMM, 2017-18)

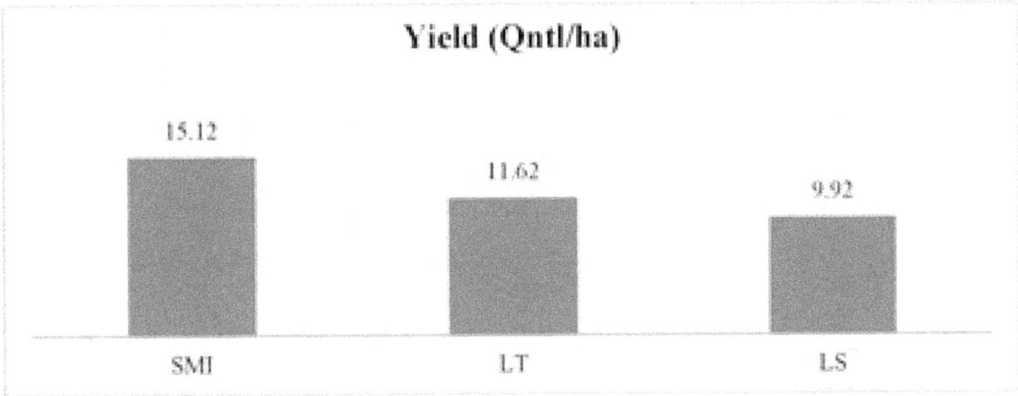

The above figure shows that method wise yield rate of millet and it also reveals that the SMI method provide highest yield i.e. 15.12 quintal per hector to the farmers and in Koraput district 4598.5 hectors are cultivated in 10 blocks in this method under OMM.

5.8 Enhance the Value of Millet Under OMM

Odisha millet mission also helps the farmers by providing supply chain linkage to the market. This helps the ragi and suan producing farmers to sell their crops at better price in the Mandis/PACS under this programme. Government of Odisha has appointed Tribal Development Cooperative Corporation of Odisha Limited (TDCCOL) for the procurement of millet in the tribal dominated areas. The TDCCOL procured ragi with the support of PACS/LAMPCS/FPO from the farmers provided that the farmers has to follow the FAQ specification suggested by the government for procurement issued by the Agriculture and Farmers Empowerment department from time to time under the scheme. The TDCCOL procured ragi from all the blocks under Odisha Millets Mission for development of the marketing channels of millet. The government also purchase the millet from te farmers with the minimum support price for Ragi to be Rs. 3295.00 only per quintal assuring the quality guidelines prescribed by the for KMS 2020-2021. Under this scheme 6271 farmers are registered and 71401.84 qunitals are procured in koraput district in 2020-21.

Figure 7. Value of millet (per ha & per hh)
(OMM, 2017-18)

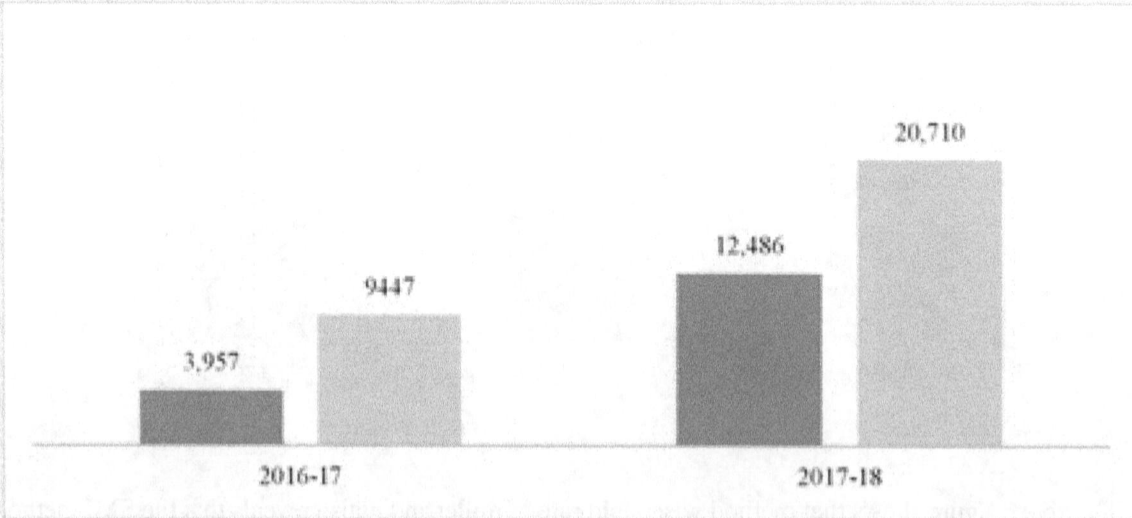

Figure 7 shows comparison of the value of the millet produce before the mission and during the Odisha Millet Mission. This also reveals that the farmers income value per household increased from 3957 to 9447 i.e. by 3 times. Similarly, the value per hector increased from 12,486 to 20,710 i.e. by 2 times. Because before the OMM they are normally preferred to sale to the middleman or at the local hat which forced the farmers for distressed sale at lower price. But after the program the farmers get higher price per quintal which increased their gross value.

5.9 Role of OMM for the Women Farmer

Millet is considered as pro-poor and pro-women in nature because most of the labor intensive farming activities like on farm agricultural activities such as field preparation, seed selection, varietal selection, manuring, irrigation, intercultural operations and hand weeding, harvesting, drying, threshing, winnowing, storage and grain management and marketing are performed by women. Majority of the women which is around 70 percent are also involved in the marketing of their product in the local hat or to the middlemen (Jeeva, 2019). Manual post-harvest operations of millets are difficult mainly due to its small grain size as well as complicated process involved in threshing on the ground. This creates lots of drudgery for the women farmers. But this program also mechanism the agronomic practices such as thresher, pulverizer, cleaner-grader-stoner etc. These improve efficiency and reduce drudgery of the farmers.

Table 3. Enterprises with WSHGs under OMM

Enterprises with WSHG	2020-21	2021-22	2022-23	Total
Millet Shakti Tiffin centre	15	81	27	123
Millet Shakti on Wheel	0	1	0	1
Mission Shakti Cafe	0	2	0	2
Millet Shakti Outlet	0	1	1	2
THR Units (under ICDS)	59	0	0	59
Total	74	85	28	187

Source: Odisha Millet Mission Dashboard, 2022

This initiative encourages not only production of millets but also provide opportunity for varieties of millet-based food. Local SHGs and specifically women are empowered economically and socially after getting some income generating sources due to millet based value chain. The WSHGs with the help of mission shakti, under OMM established millet-based tiffin centre to serve highly nutritious millet-based ready to eat food.

6. CONCLUSION AND SUGGESTION

As most of the rural and tribal livelihood are depends on the cultivation of the locally cultivated cops and cereals which are normally un-recognized and un-popular. Millet is one of the pro-poor and pro-nutrition crops that is not only climate smart in nature but also deeply related to rural and tribal life. It is a nutritional power package for the tribal and poor section of the communities for ensuring food and nutrition among them. But in the recent past the area, production and productivity has decreased due to green revolution, urbanization, rising income, changing food habits and most importantly focus on the production of rice and wheat and distribution of cheap rice and wheat through the fair price shop etc. So, to ensure food and nutritional security and to ensure rural employment and livelihood Government of Odisha introduced a comprehensive programme to revive the production and increase the household level consumption of millet mainly in the tribal dominated area for their food and nutritional requirements in affordable price. Odisha Millet Mission. This programme helps the farmers to increased their production through improved agronomic practices increased the consumption and nutrition through distribution in the PDS. This mission also helps to improve the post production processing by mechanization and also provide market linkage to the farmers thereby improve their level of income. The facility of selling their grain in the mandi/PACS in the MSP helps to improve the value chain system. The MSP provided by the government is the highest price in which the farmers can sell their product compare to the market price. This all together lead to more availability of millet which increased the consumption. Along with that this program also helpful to improve the ease of doing farming for the women farmers especially by reducing the drudgery of women by improving the agronomic practices through mechanization of production and processing activities. Hence, this program is no doubt an innovative initiative to improve economies of millet in one hand and to promote the livelihood status and the tribal and rural on the other hand. But the government need to do lot more in the way of creating awareness and provision for the remote and tribal farmers to make this program reach out to the mass. The value chain system in Odisha is very much limited to millet-based flour only. With this program various millet based tiffin centers are established through SHG but this needs to be emphasized more to encourage millet based ready to eat food outside the farming communities. Under OMM, with the help of District Mineral Fund (DMF) the SHG women in the Keonjhar district are engaged for ragi laddoo flour mix for the preschool and anganwadi children. This needs to be encouraged district wise so that not only SHG women get more financial benefit but also to improve nutrition level of the children.

REFERENCES

Behera, M. K. (2017). Assessment of the State of Millets Farming in India. *MOJ Ecology and Environmental Sciences*, *2*(1), 16–20. doi:10.15406/mojes.2017.02.00013

Jeeva, J. C. (2019). Engendering finger millet-based value chains for livelihood and nutritional security of women in agriculture. *Current Science*, *116*(11), 1893–1896. https://www.currentscience.ac.in/Volumes/116/11/1893.pdf. doi:10.18520/cs/v116/i11/1893-1896

Kumar, A., Tomer, V., Kaur, A., Kumar, V., and Gupta, K. (2018). Millets: A solution to agrarian and nutritional challenges. *Agriculture and Food Security*, *7*(1), 1–15. doi:10.1186/s40066-018-0183-3

Michaelraj, P. S. J., & Shanmugam. (2013). A Study on Millets Based Cultivation and Consumption in India. *International Journal of Marketing, Financial Services and Management Research*, *2*(4), 49–58. www.semanticscholar.org/paper/A-Study-on-Millets-Based-Cultivation-and-in-India-Michaelraj-Shanmugam/0973456 81c669b828c2fb8d8bbee94d92414388f

Mulualem, T., & Melak, A. (2013). A survey on the status and constraints of Finger Millet (Eleusine coracana L.) production in Metekel Zone, North Western Ethiopia. *Direct Research Journal of Agriculture and Food Science*, *1* (5), 67-72. https://directresearchpublisher.org/drjafs/abstract/a-survey-on-the-status-and-constraints-of-finger-millet-eleusine-coracana-l-production-in-metekel-zone-north-western-ethiopia/

NCDS. (2022). *Mid Term Evaluation of Special Programme for Promotion of Millets in Tribal Areas of Odisha*. Phase-I Blocks, Koraput District. https://ncds.nic.in/?q=node/744

Chapter 8
An Assessment on Marketing Promotions and Strategies Adopted by Retailers Towards Millet-Based Products in Hyderabad

Ashita Waghray
GIET University, India

Saumendra Das
https://orcid.org/0000-0003-4956-4352
GIET University, India

Shehbaz Ahmed
Independent Researcher, India

ABSTRACT

Today, millet has been rebranded as "Nutri-cereals." It became a choice for the taste of many consumers. Its nutri-rich sources and benefits are creating a fascination for many consumers and apparently there will be further increase in demand for the future. On the other hand, millet also attracted attention due to lower investments by farmers requirements such as water fertilizers, pesticides, etc., with good income and market potentiality. Moreover, it is increasing minimum support price (MSP), and millets can also be considered as positive cultivation for the farmers to make decisions on agriculture. The eventual declining trend of the natural resource availability also identifies the need for crop shift as of paramount importance. Millets are generally categorized into major and minor millets, ICAR-IIMR has been leading their efforts for a decade now.

DOI: 10.4018/978-1-6684-9819-4.ch008

Copyright © 2023, IGI Global. Copying or distributing in print or electronic forms without written permission of IGI Global is prohibited.

INTRODUCTION AND A REVIEW ON THE PERFORMANCE OF RETAILING SECTOR

Retail is probably the most important industry because it has a direct relationship with consumers. This includes all stores that sell personal and household products and services to end consumers, from kiosks and small grocery stores to supermarket chains and shopping malls. The objective of this article is to conduct a survey on the retail industry and its type of business. To collect data, a second resource was used. An in-depth analysis of the literature was carried out. According to the survey, the retail industry is generally organized into two main groups: in-store retail and off-store retail.

India's retail segment is generally controlled by conventional and advanced shapes of organized retail since a decade. The conventional retail format was sensed by Kirana stores and there are still Kirana stores. Getting a charge out of administration and dominance within the retail segment in India. Indeed, in little towns and urban zones, Find the control of little, family-owned autonomous stores advertising a wide assortment of items in one put customer's ask. In expansion, organized retailers, both conventional and modern, contribute financial development of the nation. Concurring to the Worldwide Retail Improvement Index-2017, India secured the primary position within the worldwide retail list. Hence, the final decade seen dynamism within the Indian retail sector. Some urban zones have ended up centers of fascination with the rise of numerous diverse sorts of organizations.

In-store retail is classified agreeing to various characteristics, but the foremost critical classifications are based on proprietorship, stock, and price. According to the classification based on proprietorship, the foremost critical sorts are free stores, chain stores, establishments and rental office stores. Concurring to the commodity-based classification, the foremost vital retail sorts are office stores, supermarkets, forte stores, comfort stores, mass retail stores, and retail administrations. Concurring to price-based classification, the foremost critical retail sorts are discount stores, outlet stores, off-price stores, off-price stores, distribution center clubs, and hypermarkets.

Marketing makes up the majority of trade operations. There's gigantic competition to draw in shoppers. Pay consideration to the item or benefit to persuade them. As a result, each producer should construct one Promoting technique and activity plan more alluring than current competitors. A celebrated instrument for fascination Shopper consideration to an item or service is the special blend. The limited time blend is one of the most key component within the promoting blend and it plays a crucial part in the victory of the showcase and accomplishing its objectives. As a rule, advancements are applied and used to guarantee that customers are aware of the items the company offers. That's action plan to set up a communication relationship between the advertiser and his target gather Company. Promoting advancement is to a few degrees unmistakable from mass communication, in which a company organization targets an undifferentiated mass audience for non-commercial purposes through implies such as: publications on radio, daily papers and tv. Beneath the components of promoting advancement, an organization will endeavor to target an gathering of people that's purposely separated for commercial reasons and will utilize implies such as publicizing methodology, deals advancement, individual offering, open relations and publicizing. These promotions These components are exceptionally overwhelming procedures to pull in and draw in open consideration to the product traders offer to sell

THE FACTORS THAT ARE INFLUENCING ADOPTING THE MILLETS BASED PRODUCTS BY THE RETAILERS IN HYDERABAD

Several factors currently determine millet consumption in India. Urbanization has caused a shift in the consumer mindset and food consumption patterns, which is the primary cause. Modern consumers are increasingly fitness conscious and want to focus on health and wellness, especially in today's environment. But it's important to provide them with healthy foods, like millet, in a product form that they can easily integrate into their lifestyle.

Additionally, increased disposable income and the push for e-commerce allow consumers to experiment with their food choices by choosing healthy premium foods. Millet has several important health benefits. As well as being sustainable and having high nutritional value; They have 3-5 times more nutritional content than the rice products and wheat; rich in fiber content and lower glucose index provides health benefits.

The explosion of brands working with millets aims to provide nutrition, not just a product. Consumers are becoming more aware of the benefits of millet products, leading to increased demand for this product.

MILLETS AND ITS NUTRITIONAL VALUES

Millet is a rich source of nutrients, providing many powerful health benefits (often called "nutritious grains"). They contain three times more fiber than wheat and ten times more than rice. Because of their reduced glycaemic index, they helps in regulating the glucose levels, which stands them ideal for diabetics. Those who want to consume them without gluten are also affording the perfect meal replacement.

TARGET GROUP

Food consumers who prioritize health and convenience are the intended beneficiaries of Millet products. From 6 months to 60 years old, consuming millets can benefit your health in many ways.

THE OUTBREAK CAUSED A SURGE IN THE DEMAND FOR MILLET

Traditional millet is gaining more popularity as a super grain due to growing concerns about health and immunity. We are definitely seeing an increase in demand for millet products. People are becoming more health conscious during this pandemic and consumption of millets has increased which is one of the reasons why there is a demand and this has become a boon for the farmers. People have increased the area under millet cultivation. Millet consumption has become popular among individuals who want to improve their immunity due to this situation.

INNOVATIONS ARE MADE TO MILLET PRODUCTS

Brands continuously focus on innovation. The appeal of millets to consumers can be enhanced by creating products that intelligently blend them, while maintaining their flavor and nutritional value.

According to Meghana Narayan and Shauravi Malik, the company has introduced over twelve new products in the last year or so, including baking mixes, jaggery powder, chia seeds, and coconut oil.

"We deliver our products packed with nutrients and flavor, and we also make portioning easy by providing each serving in individual packets," says Prabhu Gandhikumar.

PUSH YOUR BRAND DIGITALLY

Since the pandemic hit, brands have been marketing to consumers primarily through digital audiences. Their approach involves both sharing campaigns on social media and their website to share the story of their brand, as well as educating people about their values and beliefs.

"We have made our brand digitally accessible to the masses using various platforms like Instagram, Facebook etc," said Prabhu. People can easily contact us through these websites and place orders.

Our strategy is similar to that of any new-age brand: We will be concentrating on the internet and making the most of online channels. We will continue to prioritize this capability for the sake of our business, given the rapid pace at which omnichannel is becoming mainstream. We are currently selling on online marketplaces like Flipkart, Amazon and BigBasket. There has been a lot of interest in our D2C service, which allows us to engage directly with consumers and create highly informative conversations about brands. Prashant remarked that our campaigns in recent months have been entirely digital.

Explaining their digital initiative, Meghana and Shauravi said we have also run several thematic campaigns linking our products with current health issues. When air pollution in various parts of India reached alarming levels during winter, we initiated a campaign to clarify the benefits of jaggery for the health of the lungs. This was done through education. We also published a detailed article on the impact of the pandemic on children's mental health, a topic that urgently needs attention.

FUTURE PROSPECTS OF MILLET USE IN INDIA

With the increasing number of consumers in India who are changing their lifestyle, there is no reason to doubt the future prospects for millets. We firmly believe that it will grow into a popular genre. These cereals are expected to gain more acceptance as people embrace and utilize products that suit their needs and lifestyle.

- Urban population's growing preference for nutritious eating habits
- Iron, calcium, and fibre found in millets aid to reinforce vital nutrients for children's healthy growth. Millets are being used in baby food and nutrition products, and several producers are growing their businesses.
- The production of rice and wheat, both of which require a lot of water, is becoming increasingly unsustainable as freshwater supplies are being depleted worldwide. Millet requires little water, grows well in dry climates, and has a short harvesting season.

- Rise in the prevalence of diseases including diabetes, obesity, and heart disease, as well as other conditions like arrhythmias and coronary artery disease. Millets are rich in proteins and minerals like calcium, iron, and others that can help prevent such illnesses.

Figure 1. Marketing strategies

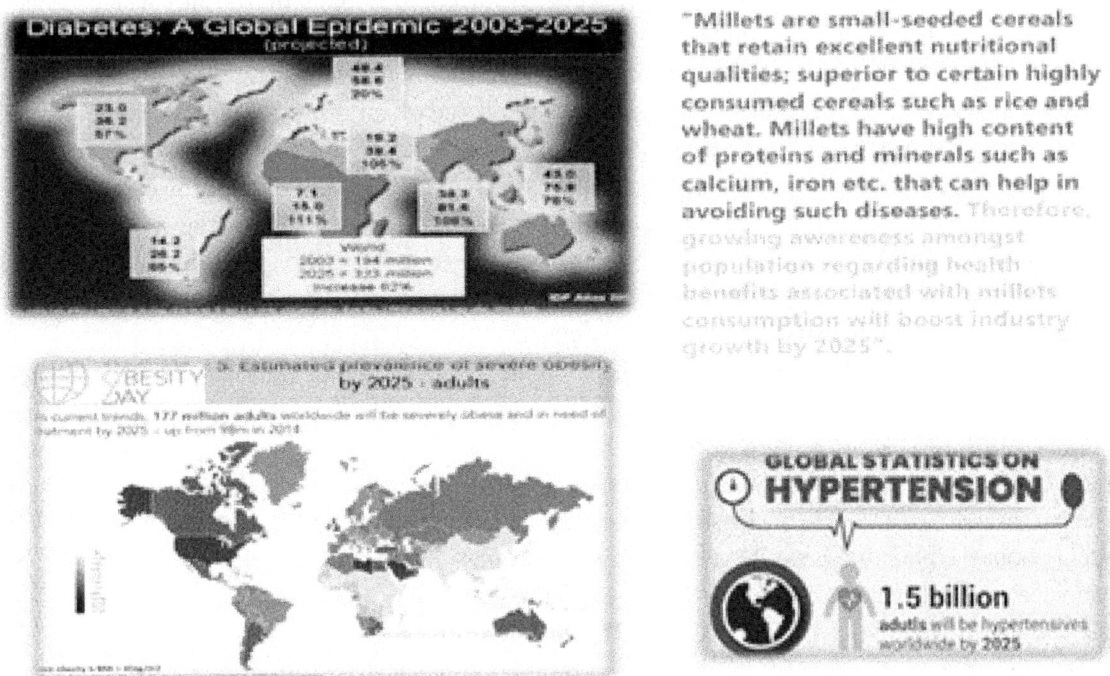

"By 2030, there will be more than 23.6 million cardiovascular disease patients, which will help increase in millet consumption."

TO INVESTIGATE ON THE MARKETING STRATEGIES AND PROMOTION TOOLS FOR BUSINESS SUSTAINABILITY

Sustainable marketing is an approach that aims to benefit both the environment and business by producing goods and services that are socially responsible, environmentally friendly, and have eco-friendly properties. These positive attributes help to build brand loyalty and promote best practices. Sustainable marketing is not tied to specific products or short-term campaigns. It represents the company's overall goals. However, it can be emphasized in the short term for specific reasons, such as promoting a company's specific selling points (White et al. 2019)

This idea was led to much debate during 1970s and 1980s, but was implemented by some companies. When the sustainability concept was introduced in the late 1990s, it got good familiarity but many of the companies reluctant to adopt it because their benefits were not known to the companies (Obermiller et al. 2008). However, in recent times, the companies had realized that traditional or conventional models do not develop sustainable competitive advantages and the new ways are required (García-Rosell 2013). So, so the marketing in social platforms has become more sustainable and played an prominent role in the spread of marketing uses,formation and development of the marketing. Marketing and its Sustainability is a mixture of green, social and environmental marketing. Therefore, the researchers of academics and the industry consider marketing as an improved method. It is a process of strategic planning, implementing, developing and controlling pricing strategies, even the promotional and distribution strategies helps to reach the consumers with products that could meet their expectations and requirements which helps in achieving the organizational goals. In this way, it allows the company to adopt to the changing environment (Belz and Peattie 2009). Feasible promoting guarantees a long-term introduction toward relationship marketing, which recognizes it from the short-term objectives of present day promoting. To turn advanced marketing into feasible promoting, social and natural concerns must be integrated with routine promoting issues. Thus, this alters the presents a modern system for breaking up the showcase and catering to buyers. Feasible promoting requires an unused and inventive thoughts in four key ranges: Firstly, social and natural concerns, customer practices being examined in detail, the promotion in blend being re-evaluated, and after that utilizing proactive showcasing exercises to construct commerce connections (Ingenbleek and Frambach,2010)The correlation between sustainable marketing strategy and business value By adopting sustainability initiatives at the strategic, business and operational levels, companies can achieve great business value by the integrating sustainable marketing research design into business operations will lead to comprehensive business innovation, as these strategies require working with multiple stakeholders (Lacy et al. 2019). Investors and companies still have a gap between themselves and the potential for sustainable business practices to generate net income, as per the Story Investors are skeptical about the potential benefits of sustainability because they are not straight involvement in the intricate dealings between stakeholders that have become part of business as usual.

BRANDING THE MILLETS AS HEALTHY & FUNCTIONAL FOOD: A recent scientific case of millet consumption was done through research by ICRISAT and NIN, others who have found that regular consumption can help reduce risks treat diabetes and obesity while improving cholesterol and promoting reduced risk of cardiovascular disease.

Handling shelf life and rancidity: Current shelf life of millets is lower compared to other major cereals. Brings good taste and comfort a longer shelf life of about six months has shown promise technologies. A shelf life of 8-10 months is crucial for the export market. In the pearl millet, the aim is to reduce rancidity by standardizing processing and storage, moreover, it is solved by biochemical and molecular interventions.

Developing versatile products: providing versatile foods according to traditional and modern taste, requirements and comfort the consumer is a key factor to increase the consumption of millet in the food industry, more than 50 versatile value-added millet processing techniques (blowing, baking, blasting, shaking, cold and hot extrusion, grinding) is used to prepare millet food in a convenient way.

Campaigns: Planning and implementation of campaigns with public status ambassadors (celebrities, social media influencers) of the public it would be useful to include them in popularizing social media, fairs, Exhibitions, festivals and campaigns are organized to promote millets, National level "The Eat Millet Campaigns" helping to raise awareness of the issue about the benefits of millet and reaching larger consumer groups to create a demand for millet and thus increase production and sales.

Mission in millet: is very important to generate additional demand. The initiatives taken by the Government of India, has to include millets in national food security Mission. In addition, the Government of India has declared 2018 National Year Millet to increase the production and consumption of millet.

Thanks to its nutritional value due to economic and climatic conditions, the Government of India started functioning mainstreaming of millet in the country and around the world, these formulations helping retailers of millet as a strategy to increase the sales.

Product sampling and market testing undertaken by the retailers: before taking the initiative in the millet business the retailers are adopting test marketing strategies by providing free sample and demonstrating the ready to eat methods to the customers to improve sales and awareness

TO ANALYZE THE MARKET POTENTIALITY AND PROS AND CONS IN DEALING WITH MILLET-BASED PRODUCTS IN INDIA WITH SPECIFIC REFERENCE TO HYDERABAD

Millet Mission In India

The ministry is on a mission to increase the production and consumption of millets in partnership with other central ministries, state governments, and other concerned organizations. A change in the public distribution system's basic calorie distribution programs needs to be redirected to an expanded food basket, including millets, which will improve the nutritional status of preschool children and middle-aged women. childbearing age. The Indian government announced millets as a nutritious cereal in April 2018. Millets were also included in the Poshan Mission campaign.

India has set a target of achieving a 20% blend of ethanol and petrol within five years, viz.

from 2030 to 2025, as the country seeks to reduce its dependence on costly oil imports. The government wants to use it to achieve this goal. Integrate this policy change, we expect millet usage in distilleries to grow at a CAGR of 7.05% during the year period 2019-20 to 2029-30

Figure 2. Combined millets demand break up in MMT

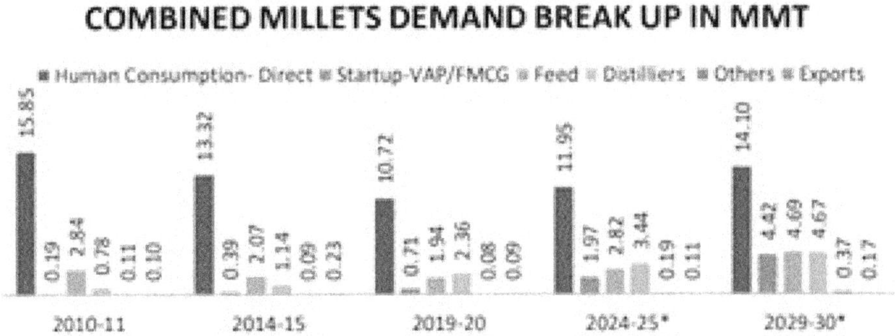

Table 1. Nutritional composition of millets in comparison with rice and wheat

Millets	Carbohydrates(g)	Protein (g)	Fat (g)	Energy (Kcal)	Dietary Fibre (g)	Ca (mg)	P(mg)	Mg(mg)	Zn(mg)	Fe(mg)	Thiamine(mg)	Riboflavin(mg)	Niacin (mg)	Folic Acid (mg)
Sorghum	67.68	9.97	1.73	334	10.2	27.6	274	133	1.9	3.9	0.35	0.14	2.1	39.4
Pearl millet	61.8	10.96	5.43	347	11.49	27.4	289	124	2.76	6.42	0.25	0.2	0.86	36.11
Finger millet	66.82	7.2	1.92	320.75	11.18	364	210	146	2.5	4.6	0.37	0.17	1.3	34.7
Kodo millet	66.19	8.92	2.55	331	6.39	15.27	101	122	1.65	2.34	0.29	0.2	1.49	39.49
Proso millet	70.4	12.5	1.1	341	-	14	206	153	1.4	0.8	0.41	0.28	4.5	-
Foxtail millet	60.1	12.3	4.3	331	-	31	188	81	2.4	2.8	0.59	0.11	3.2	15
Little millet	65.55	10.13	3.89	346	7.72	16.1	130	91.41	1.82	1.26	0.26	0.05	1.29	36.2
Barnyard millet	65.5	6.2	2.2	307	-	20	280	82	3	5	0.33	0.1	4.2	
Wheat	64.7	10.6	1.47	321	11.23	39.36	315	125	2.85	3.97	0.46	0.15	2.68	30.1
Rice	78.24	7.94	0.52	356	2.81	7.49	96	19.3	1.21	0.65	0.05	0.05	1.69	9.32

Source: 1. Indian Food Composition Table 2017 – National Institute of Nutrition,
2. Nutritive value of Indian foods, 2004 - National Institute of Nutrition

With the emergence of the several young players, the retail industry of India has become one of the fast-growing segments. It represents the country's gross domestic product (GDP) more than 10% and the employment reserves for 8%.In the world the India has ranked as the 5th largest retailing destination. India was ranked 73rd in the United Nations Conference on Trade and Development's 2019 Business-to-Consumer (B2C) E-Commerce Index. India is the fifth largest retailing destinations of the world and ranks as 63rd in the World Bank's report 2020 for its business.

Figure 3. Types of millets available in grocery shops

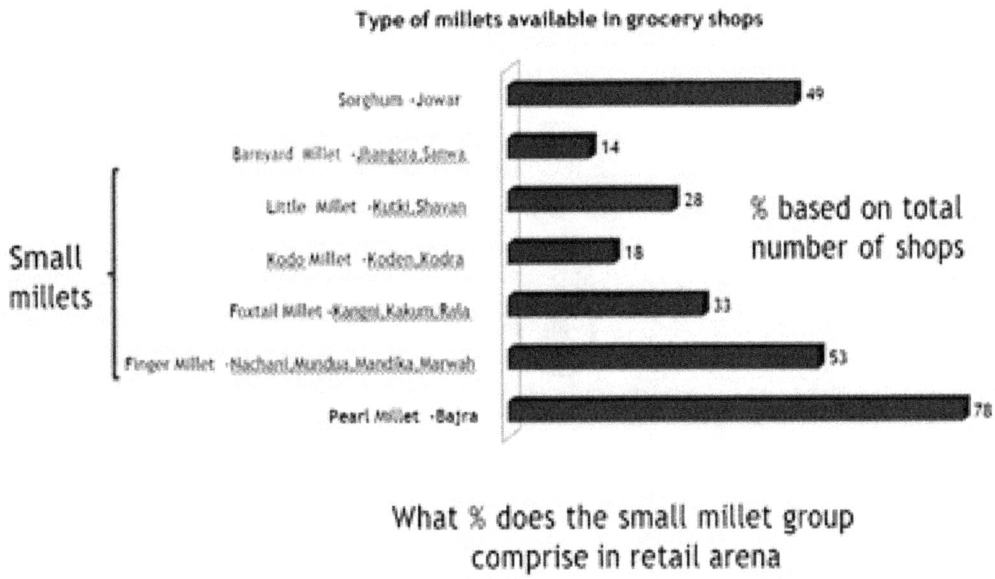

The above diagram shows the availability of Millets in the retail stores and helps to access the market potentiality of specific millet and its maintenance, higher the availability of stock of specific millet higher will be the potentiality of the market, Pearl millets are fast moving and scored high percentage in stock maintenance and results good sales followed by finger millets, jowar and foxtail millets.

Figure 4. Secondary processing units

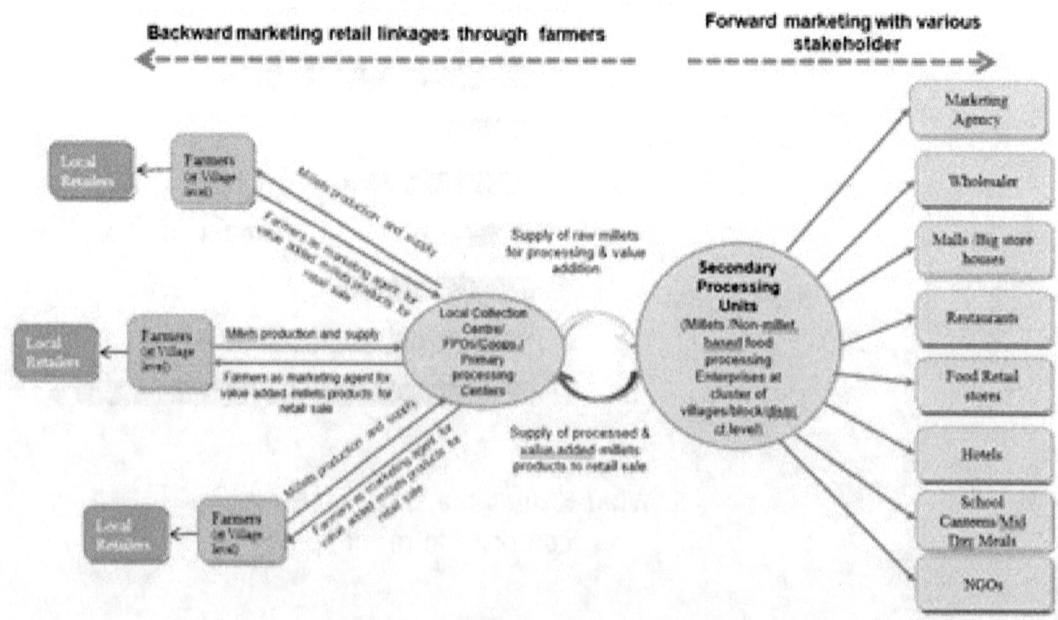

From the above source how the retailers or startups work on assessing the potentiality of millet market by backward and forward marketing integration in dealing with millet-based products in Hyderabad and it also shows how millets sales are reaching local to international markets

RETAILERS' RESPONSE ON MILLETS SALES OR MARKET POTENTIALITY OF CUSTOMERS DUE TO THE FOLLOWING FACTORS

Table 2. Millets sales

parameter	Most influencing	Influencing	Neutral	Less Influencing	Not influencing
Price	66%	10%	15%	4%	5%
Packaging	25%	16%	8%	27%	24%
Brand	58%	21%	11%	6%	4%
Taste	23%	16%	18%	32%	11%
Superiority	28%	25%	12%	20%	15%
Availability	72%	23%	2%	1%	2%
Health benefits	86%	11%	1%	1%	1%

Figure 5. Retailers' response on millets sales or market potentiality of customers

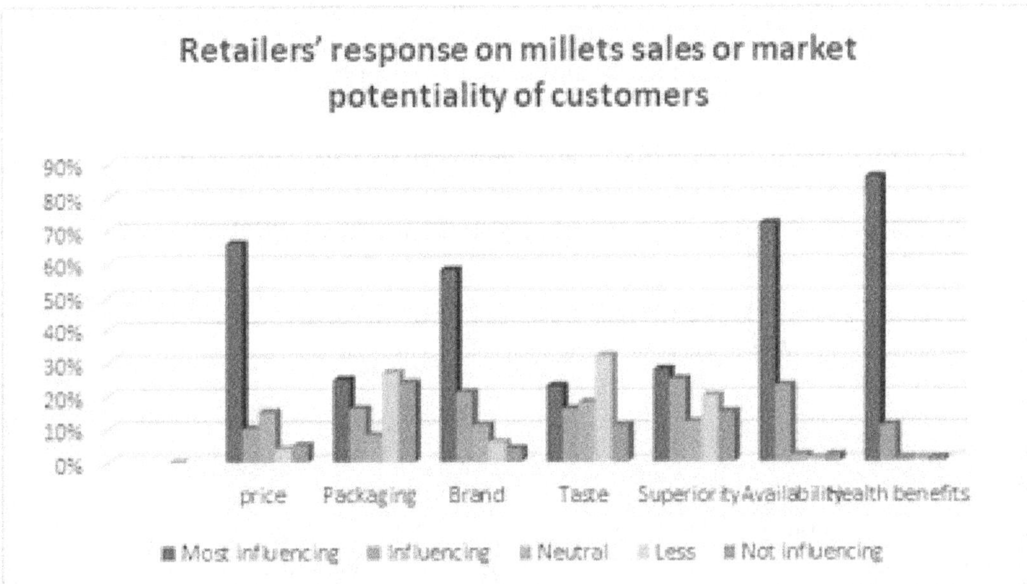

From the above analysis of Retailers' response on millets sales or market potentiality of customers and the factors effecting are identified where the factor of health benefits, availability and Price are playing dominant role and most influential in finding the potentiality of sales, and next followed by the brand, superiority, packaging and taste.

CHALLENGES TO THE RETAILERS IN THE MILLET SECTORS

- **Poor infrastructure:** Poor infrastructure such as roads and transport services can make it difficult for farmers to reach the markets and buyers. This can impact farmers' incomes and retailer accessibility, Inadequate market access: Inadequate market access is another limitation.
- **Low prices:** The low prices offered by millet buyers can make it difficult for farmers to make a profit from their crops. This can lead to decreased motivation for millet production.
- **Unfavorable government policies**: Unfavorable government policies such as high taxes and subsidies can have a negative impact on millet production. This can make it difficult for farmers to benefit from their crops.
- **Lack of agricultural extension services:** Agricultural extension services are essential to provide farmers with access to up-to-date information and advice on millet production, but these services are often inadequate in India. This can lead to poor crop management and reduced yields

ISSUES RELATED TO REGULATIONS

1. Millet products do not fall within the scope of standard foods and will therefore be subject to regulation approval process for FSSAI to review and launch these innovative products with standards.
2. Quality standards and certification remain major constraints on millet exports. Product claims must be in the list of claims approved by FSSAI for companies Interested people can use it in public relations.
3. Insufficient awareness of export policies and understanding markets across countries.

Nation. 'Integrated solutions are necessary to address these challenges. millet by diversifying production technology, establishing upstream and downstream linkages, nurturing the startup ecosystem and bringing millets to everyone's plate. In this Successive governments have so far taken a series of measures to promote the millet industry in Indiaand abroad.

CONCLUSION

From the paper we can conclude that the retailers must choose the strategies and promotions for the millet-based products in such lines, whereas the developed, ready for eating and ready for cooking products must be available at various stores to make consumers aware of this existence and also innovative products must reach the requirement for 5-year-old consumers to 20 years to create a special market segment. Some of the traders have experience in military operations for years, they expressed that there is no awareness of processed products in millets category available as healthy food for soldiers, the government can make a step forward in organizing meetings and information programs for millet traders and develop at the state level entrepreneurship and can promote for the increase in consumption of millets.

Businesses can engage in location-based marketing strategy to stand out in the market choose a store greater number of walks per day, must always focus on consumer changing tastes and preferences towards millet products .where most of retailers are expressed that they need to act double to the consciousness for the market demand, in the while as an immediate approach, the public funded business incubators should come out with the awareness programs for the sensitization of urban consumers in the both domestic & International markets.

REFERENCES

Accenture. (2020). How COVID-19 will permanently change consumer behavior. *Accenture.* https://www.accenture.com/us-en/insights/consumer-goods-services/coronavirus-consumer-behavior-research

Chaudhury, R. (2010). Determinants of Consumer Behavior in Buying RTE Foods. *The Journal of Business and Retail Management Research*, *5*(1), 76–86.

Dayakar Rao, B., Patil, J. V., & Hemavati, T. V. (2014). *Terminal report of NAIP sub-project on Creation of Demand for Millet Foods through PCS Value Chain.* Directorate of Sorghum Research, Rajendra nagar, Hyderabad. https://naip.icar.gov.in/download/79347/207601.pdf/207601.pdf

John, M. (2020). The coronavirus pandemic and US consumption. *Vox EU& CEPR Coverage of the Covid-19 Global Pandemic.* https://voxeu.org/article/coronavirus-pandemic-and-us- consumption

Sharma, R., & Sharma, A. (2008). *Assessment of market potential and marketing strategies of Kangra tea. P.G. Dr.* Yashwant Singh Parmar University of Horticulture and Forestry.

Sirisha, S., & Lakshmi, J. (2018*). A study on consumption pattern of millets and millet-based products in Guntur city, Andhra Pradesh PG.* [Thesis, Ranga University, Rajendra nagar, Hyderabad].

Swamy, B. M., Kumar, T. A., & Rao, K. S. (2012). Buying behavior of consumers towards instant food products. *International Journal of Research and Computational Technology.*, *2*(2).

Chapter 9
Prospective of Millets in the Rejuvenation of the Bakery Industry

Sanjay Indani
MITADT University, India

Anjali Bhoite
MITADT University, India

Vasant N. Pawar
MITADT University, India

Bhagwan Sakhale
Dr. Babasaheb Ambedkar Marathwada University, India

ABSTRACT

Considering the growing market for bakery products, and healthy food products rising in consumer perception, new functional food product development in bakery industry is a worthwhile focus. In search of appropriate nutritional and functional ingredients, the application of millet flours as alternative ingredients is critically assessed by the researchers. The present review highlighted the India's potential (19% productivity share at global level) in millet production, its nutritional goodness, as well as recent traits in bakery product development. Not only the nutrient goodness, but also the phenolic and bioactive components associated with these nutrigrains, could lead the researchers to categorizing it as superfood. Non-gluten nature associated with these arid gain is the worthy source for nutrient enrichment and fortification in various processed bakery products; its low glycemic index (GI) based nature also confirms millets significant role in curing various disorders, making it to categorised as healing grain.

DOI: 10.4018/978-1-6684-9819-4.ch009

Copyright © 2023, IGI Global. Copying or distributing in print or electronic forms without written permission of IGI Global is prohibited.

1. BAKERY PRODUCTS GENESIS AND SCOPE FOR REJUVENATION AS DOMINANT COMMERCIAL SECTOR

Baking has a long and illustrious history that predates written records. Although the Egyptians had ovens that were well-designed and operated by skilled professional bakers, there is no evidence to indicate exactly when or where baking with an oven was first discovered. Baking and milling may have started at the same time, as they moved across the savannas, early nomadic tribes were gathering grass seeds, which are the precursors to oats, barley, rye, and wheat. They eventually came to the conclusion that milling was the result of rubbing off some of the husks, which made it easier to chew and enjoy the seeds. Maybe some meagre slop was splattered on a hot stone by accident to make a fresh, delicious tidbit. Alternately, wild yeast and lactic acid bacteria might have ended up in a pot of porridge that had been set by someone else. It's possible that the owner drank it that way or allowed it to dry out in the pot over the fire, producing a flavorful, hard cake. The dried cake was not damaged. It might be carried on the hunt, cooled in the water, and eaten without having to set aside some time to make grains. When the prey was scarce, it could even be kept in the cave to consume. In any case, the processing, baking, and fermentation of cereal grains appear to have a common ancestor and are still closely related. (Walker and Eustace, 2016).

At the point when we take a look at the historical backdrop of baking and heated products, we track down that baking as a "modern" movement has been around for millennia. The idea that a baker has specialized skills and can deliver goods that could theoretically have been made at home started it all. Baking has always been and will always be subject to evolution, and always will be. Since bakers have always been looking for new ways to improve product quality and processing, the development of baking technology has been driven by innovation. The production and distribution of baked goods were fundamentally altered by the industrialization of equipment and manufacturing procedures in the nineteenth century. The seeds of present-day baking were planted during this time span when the size of metropolitan population started to emphatically grow and there was a rising interest in additional viable techniques for fulfilling purchaser interest for baked products. Over the course of about 200 years, we have transitioned the production of baked goods from largely small, local baking businesses into the era of mass production, and this trend continues in many parts of the world. (Cauvain, 2013).

Even though the Indian bakery industry has been around for a long time, it didn't really get a boost until the latter half of the 20th century (Manley, 2011). One of the factors was urbanization, which increased the demand for ready-to-eat products at a reasonable price. Bakery products made with whole, natural grains and other natural ingredients have become increasingly important (Saranraj and Geetha, 2012).

Bakery products are used to incorporate a variety of nutrient-dense ingredients. The amino acid profile of non-wheat proteins added to wheat flour enhances protein quality. Due to their superior flavor and desirable sensory and nutritional characteristics, multigrain cakes have gained market share. Multigrain cakes are the result of a combination of factors and offer a variety of flavors and textures to consumers (Heenan et al., 2009; Rashmi et al., 2022; Singh et al., 2017). It has been reported that baking raises the amount of resistant starch in the product. However, the baking conditions, such as time and temperature determine the structural changes that reduce starch digestibility (Mitelut et al.,2021).In view of these properties the nutraceutical efficacy through bakery products could be properly embedded to satisfy the consumer aspiration. Figure 1 shows the variety of bakery products with respect to fat to sugar ratio to upgrade critical thinking of researchers for drawing strong road map to rejuvenate the bakery industry as dominant functional food processing sector.

Figure 1. Plot of baked product groups according to ratios of fat to flour and sugar to flour (Cauvain, 2013)

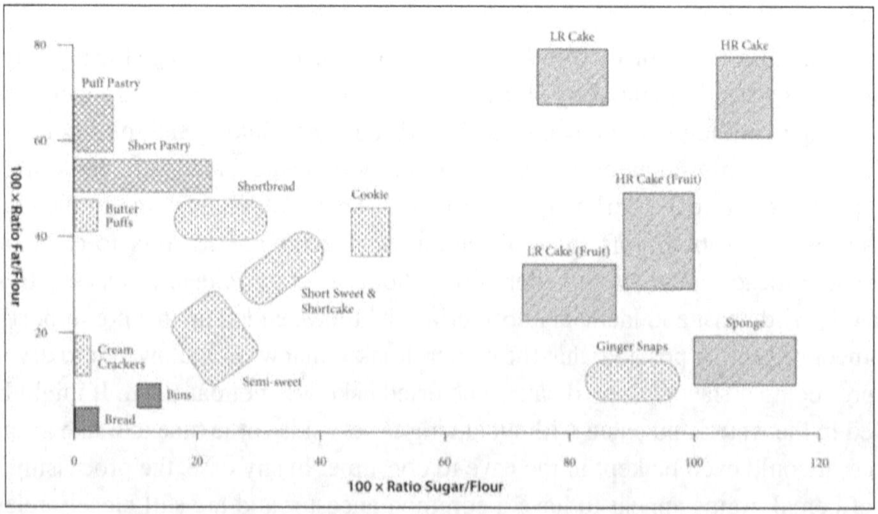

2. MILLETS AS A DISTINCTIVE RAW MATERIAL FOR FUNCTIONAL BAKERY PRODUCT FORMULATIONS

Ongoing trend of research has underlined the need of variety of flours in formulation of functional bakery products. In response to India's request to the Food and Agriculture Organization (FAO), the United Nations has declared 2023 the "International Year of Millets." The use of millet flours could be encouraged for novel food products especially bakery products. The purpose of revelling International Year of Millets is to bring the nutritional advantages of various millets to the attention of all stakeholders (FAO, 2023). Since ancient times, many millets have been used in various cuisines in various parts of the world. Till date their use in contemporary processed food products was not investigated. However, recently, numerous millet-based products have been introduced to the market. Finger millets, sorghum, and others are among them have a lot of potential for baking goods (Mishra et al., 2022; Rashmi and others, 2022).Figure 2 is the reflection of lots of a core considerations behind the acknowledgement of these grains because of these distinctive magnetism the millet could be the appreciable raw material for functional food development.

2.1 Reported Potentials of Millets as a Metaphor in Functional Food Product Development

Millet grains are a superior decision for superfoods or nutria-cereals than cereal grains since they contain a more noteworthy amount of minerals like calcium, potassium, magnesium, iron, manganese, zinc, and the B complex nutrients. They also have a lot of good quality fat, protein, carbohydrates, and dietary fiber. Table 1 depicted the nutritional values obtained in various studies also figure 2 summarizes the health benefits associate with various components, accorded in most of the previous studies. Based on which potential in curing diseases and disorders are also aggravated. (Dayakar Rao et al., 2017; Ramashia et al., 2018; Taylor and Duodu.,2018).

Figure 2. Health benefits of bio actives in millets
(Anderson and Hanna, 1999; Yousaf et al., 2021,Dayakar Rao et al., 2017).

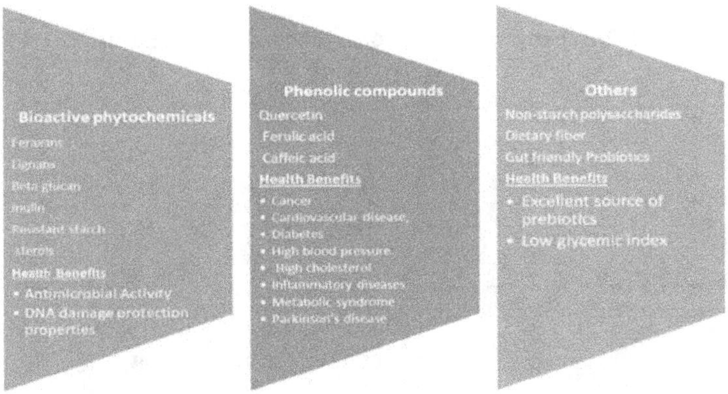

Millets, which live in close proximity to trillions of *Lactobacillus acidophilus, Lacticaseibacillus rhamnosus,* Actinobacteria, and Bifido species, are a great source of fiber and slow-digesting starches that are good for the gut (Amritha et al., 2018). Millets are non-starch polysaccharides that are a major source of dietary fibre and are excellent prebiotics, and the fermentation of resistant starch results in the production of short-chain fatty acids. Millets are a good source of probiotics that are good for the gut because the fermentation process with different cultures makes Gram-negative bacteria grow. Millets have lower glycemic properties because they contain more fiber. Although the idea of the handling, food networks, fixing arrangement, food structure (grain or flour), and so forth., The majority of millet recipes can be made low on the glycemic index due to the fact that all of these factors affect the glycemic properties of foods.

2.2 Millets Production Scenario in India and at Global Level

As depicted in figure 3 India is the largest producer of millets in the world. In India, millets are grown on 12.45 million hectares, resulting in a yield of 1247 kg/ha and 15.53 million tons. India produces 40.51 percent of the world's pearl millet, followed by 8.09 percent of the world's sorghum. 19% of the world's Millets will be manufactured in 2020. India's most important millets-producing states are Rajasthan, Karnataka, Maharashtra, Uttar Pradesh, Haryana, Gujarat, Madhya Pradesh, Tamil Nadu, Andhra Pradesh, and Uttarakhand. In 2020 and then some, these ten states will represent around 98% of India's millets creation. Six states account for more than 83% of millet production: Rajasthan, Karnataka, Maharashtra, Uttar Pradesh, Haryana, and Gujarat. Sorghum is India's fourth most important food grain in terms of area, following rice, wheat, and maize. hector and output (4,31 percent Mn. MT). Bajra, which also produces nearly the same amount, produces more than half of the country's millets (7.05 million ha). It is intriguing to take note of that India delivers the most barnyard (99.9%), finger (53.3%), kodo (100 percent), little millet (100 percent), and pearl millet (44.5%) from 8.87 million ha, creating around 12.46 million metric tons. (*Dhruvendra Sachan* et al., 2023)

Figure 3. Country wise production of Millet (2020)
(Agricultural and Processed Food Products Export Development Authority (APEDA), 2023)

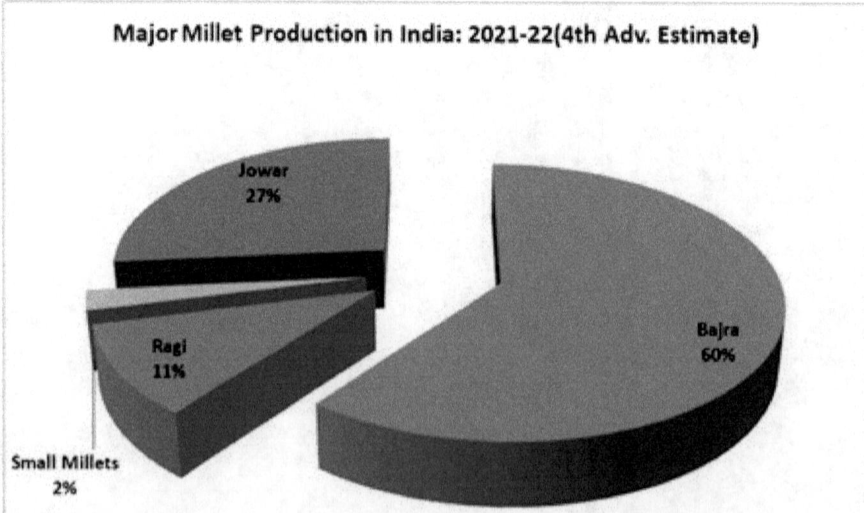

Maharashtra (20.67 percent), Karnataka (13.46%), Uttar Pradesh (8.06%), Madhya Pradesh (6.11%), Gujarat (3.94%), and Tamil Nadu (3.74%) are the other states with the highest percentage of millets under cultivation (29.05 percent). In recent years, the areas covered by millets in Gujarat and Madhya Pradesh have grown. However, Andhra Pradesh had the highest yield at 2626.58 kg/ha, followed by Tamil Nadu at 2153.22 kg/ha, Haryana at 1906.78 kg/ha, Gujarat at 1762.05 kg/ha, and Madhya Pradesh at 1729.70 kg/ha. When compared to their counterparts, states like Gujarat and Andhra Pradesh have demonstrated higher levels of productivity.

Figure 4. State wise millet production year 2021-22
(Agricultural and Processed Food Products Export Development Authority (APEDA), 2023)

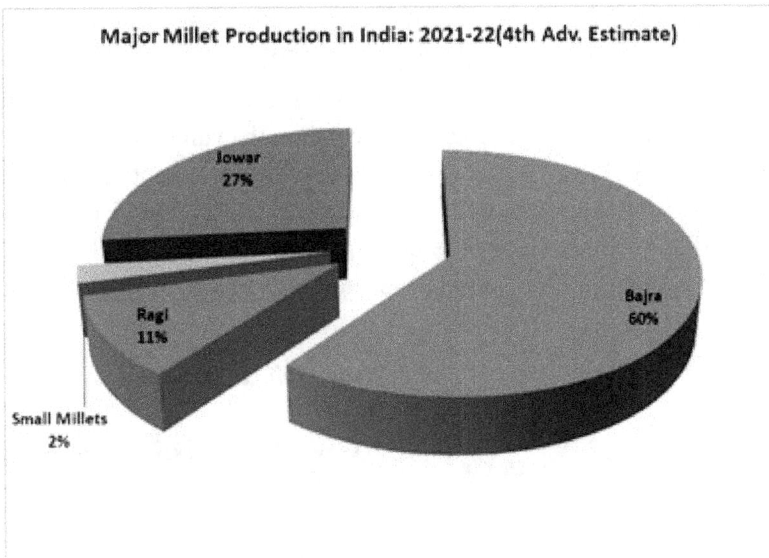

Figure 5. Major millet production in India (2021-22)
(Agricultural and Processed Food Products Export Development Authority (APEDA), 2023)

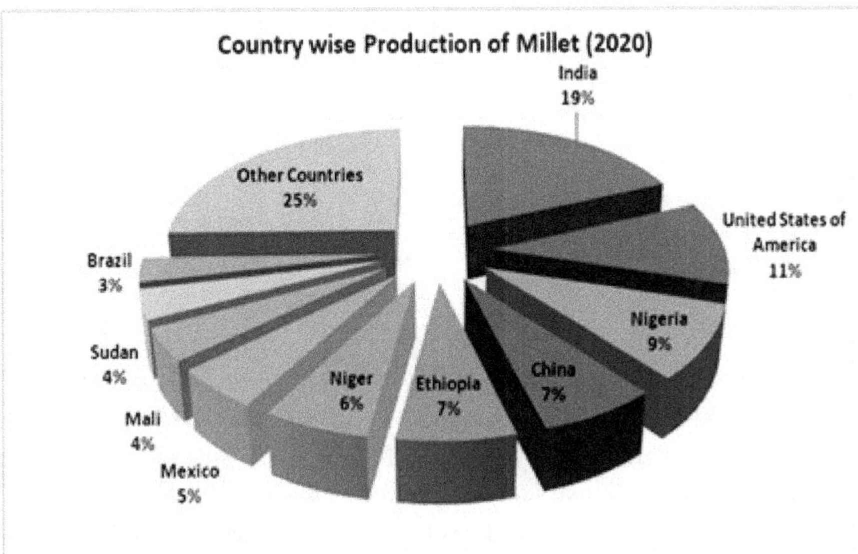

Table 1 summarizes the nutritional value of diverse millets and in table 2 vernacular names of cultivated millets are depicted .Finger millet also known as *Eleusine coracana*, is a underutilized gluten –free grain. It has a lot of bioactive and phenolic compounds, a lot of nutritional value, and the potential to be a healthy food ingredient (Ramashia et al., 2018). It is an exceptionally esteemed crop because of its high dietary substance, dry season resilience, and reasonableness for development in peripheral soils. The traditional Indian crop of finger millet is grown in Jharkhand; Karnataka is India's largest producer. As indicated by Peng and Yao (2017), finger millet is otherwise called ragi, nachni, or nageli. Finger millet, one of the important millets, is round and brick red (Bhatt et al., 2022). Finger millet contains 8.5% carbohydrates, 9.8% protein, and 4.3% crude fiber, making it a good source of minerals like calcium and iron. The fact that finger millet has ten times more calcium than any other cereal crop is particularly noteworthy (Ceasar et al., 2023). Finger millet has a number of health advantages. It contributes to weight loss. It helps lower cholesterol levels in the body because it has a lot of fiber. It can be used as a medicine to treat conditions like diabetes, anemia, brittle bones, and osteoporosis (Selladurai et al., 2023) and frequently found in baked goods, non-alcoholic beverages, and porridges (Taylor and Kruger, 2019).

Table 1. Nutritional composition of millets in comparison with rice and wheat

Millets	Carbohydrates(g)	Protein (g)	Fat (g)	Energy (Kcal)	Dietary Fibre (g)	Ca (mg)	P(mg)	Mg(mg)	Zn(mg)	Fe(mg)	Thiamine(mg)	Riboflavin(mg)	Niacin (mg)	Folic Acid (mg)
Sorghum	67.68	9.97	1.73	334	10.2	27.6	274	133	1.9	3.9	0.35	0.14	2.1	39.4
Pearl millet	61.8	10.96	5.43	347	11.49	27.4	289	124	2.76	6.42	0.25	0.2	0.86	36.11
Finger millet	66.82	7.2	1.92	320.73	11.18	364	210	146	2.5	4.6	0.37	0.17	1.3	34.7
Kodo millet	66.19	8.92	2.55	331	6.39	15.27	101	122	1.65	2.34	0.29	0.2	1.49	39.99
Proso millet	70.4	12.5	1.1	341	-	14	206	153	1.4	0.8	0.41	0.28	4.5	-
Foxtail millet	60.1	12.3	4.3	331	-	31	188	81	2.4	2.8	0.59	0.11	3.2	15
Little millet	65.55	10.13	3.89	346	7.72	16.1	130	91.41	1.82	1.26	0.26	0.05	1.29	36.2
Barnyard millet	65.5	6.2	2.2	307	-	20	280	82	3	5	0.33	0.1	4.2	
Wheat	64.7	10.6	1.47	321	11.23	39.36	315	125	2.85	3.97	0.46	0.15	2.68	30.1
Rice	78.24	7.94	0.52	356	2.81	7.49	96	19.3	1.21	0.65	0.05	0.05	1.69	9.32

Source: 1. Indian Food Composition Table 2017 - National Institute of Nutrition,
2. Nutritive value of Indian foods, 2004 - National Institute of Nutrition

Sorghum (*Sorghum bicolor*) as a grain (*Linn.) Moench*) is a worldwide crop of cereal. Despite the fact that it has been assessed that somewhere around 75% of the complete world creation is devoured as human food (Taylor and Duodu, 2018), the nations of the Western Side of the equator utilize a large portion of their grain sorghum for animal feed. According to a review of the research on baked goods made from composite flours containing wheat and sorghum, composite flour products do not meet North American standards for quality. Researchers have attempted to characterize the harmful components or come up with plausible explanations for the differences in the baking properties of composite flours, but they have had difficulty (Xu et al., 2020).

In light of the development of low-gluten food products, food researchers face the challenge of finding an alternative to wheat flour. Sorghum is one of the cereal grains that is used to make gluten-free foods, along with other grains like rice and maize. Sorghum contains the unique protein kafirin and it is categorized as non-gluten protein. Although researchers shows good interest in sorgum as a non-gluten flour source but it is documented that, sorghum shows some drawbacks and limitations as an ingredient in different food systems. Some cultivars of sorghum have a light flavor and color, while others have distinctive colors and strong flavors that remain in the finished product (Ndimba et al., 2017). The majority of composite flour research that has been published has been centered on gluten-based foods like cakes and other similar foods. Okpala et al., 2013) utilized aged sorghum, cocoyam, and ground pigeon pea to make treats. The highest ash content, 2.73 percent, was found in cookies made with 100 percent fermented sorghum flour (100FSF), indicating that sorghum cookies are a good source of minerals. The protein content of the treats went from 4.85% to 19.89% with 100 percent cocoyam flour.

Pearl millet (*Pennisetum glaucum* (L.)) is a type of millet that is popular in India and the drier tropics. It originated in Central tropical Africa. It became laid out as a minor search in the Southeast and Bay Coast states in the wake of being acquainted with the Western state during the 1850s. After sorghum, rice, and wheat, pearl millet, or bajra, is India's fourth most important cereal. It is a crucial Kharif food crop that grows in areas of western India, Gujarat, Rajasthan, and Haryana that are either semi-arid or arid. It is regarded as the healthiest and least expensive food for animals and the poor. Pearl millet production in Africa and Asia is around 56% and 41% of complete world production (Shahidi and Chandrasekara, 2013).

Proso millet, also known as *Panicum miliaceum* (L.), is a grass that grows back from seeds every year. It has a long and illustrious history, as it is thought to have grown in the Central regions of Europe as early as 2000 B.C. This plant does well in dry places like Central Russia, the Middle East, northern India, Africa, Manchuria, and the Great Plains of North America. Proso millet was first brought to Canada in the 17th century, but it was only used sparingly as a forage crop up until the early 1900s. (Sheahan et al., 2014)

Kodo millet (*Paspalum scrobiculatum* (L.)) is common in damp habitats all over the world's tropics and subtropics. It is a native Indian cereal that is currently grown in Kerala, Tamilnadu, and Uttar Pradesh in the north, respectively. This cereal is also known as varagu, haraka, and arakalu. It provides the majority of food's nutritional requirements. It has 11 percent protein, low fat (4.2%), and a lot of fiber. Kodo millet is very easy to digest, has a high lecithin content, and is great for strengthening the nervous system. Minerals like calcium, iron, potassium, magnesium, zinc, and B vitamins like niacin, B6, and folic acid are abundant in kodo millets. Kodo millets are great for people who don't like gluten because they don't contain any gluten. Kodo millet consumption greatly benefits postmenopausal women with cardiovascular disease symptoms like high blood pressure and cholesterol. (Organic Farmers Market, 2022)

One of the oldest crops ever grown, foxtail millet (*Setaria italica* (L.)) is thought to have originated in China. Foxtail millet keeps on assuming a huge part in worldwide horticulture, delivering roughly 6,000,000 tons of nourishment for a great many individuals, basically on poor or negligible soils in southern Europe and calm, subtropical, and tropical Asia. In terms of global millet production, it is second. (Aidoo et al., 2016)

Echinochloa crusgalli (L.), also known as barnyard millet, is a crop grown for both food and feed .Additionally, it goes by the following moniker: These Japanese barnyard millets are called ooda, oodalu, sawan, sanwa, and sanwank. It is a good source of both soluble and insoluble fractions, as well as a good source of highly digestible protein and dietary fiber. Since barnyard millet has low starch content and is challenging to process rapidly, it is a gift from nature for present-day individuals who accomplish stationary work. (Roopashree et al., 2011). Linoleic acid, palmitic acid, and oleic acid are the next most abundant fatty acids in barnyard millet. Furthermore, it shows a lot of amylase retrogradation, which makes it more straightforward to create more prominent amounts of safe starches. (Dayakar Rao et al.,

2017). Diabetes mellitus and cardiovascular disease patients may benefit from it as a result. The best way to lower blood glucose and lipid levels is with barnyard millet. (Surekha et al., 2013)

In India, the little millet (*Panicum sumatrense*) was tamed. A small-grained, annual cereal crop of the Poaceae family, it is also known as samai or kutki. In India, Nepal, and other parts of Asia, it is grown extensively. It only grows to a small extent up to 2100 meters in height in India, but it is of little significance elsewhere. Compared to regular millet, the seeds of little millet are smaller. Although it is smaller, the proso millet has a habit that is similar to that of this cereal species. (Shweta et al., 2018)

Table 2. Vernacular names of millets

English	Sorghum	Pearl Millet	Finger millet	Little millet	Kodo millet	Foxtail/ Italian millet	Barnyard millet	Proso millet
Marathi	Jowari	Bajri	Nachni	Sava	Kodra	Kang, Rala	Shamul	Vari
Hindi	Jowar	Bajra	Mandua	Kutki	Kodon	Kangni, Kakum	Sanwa, Jhangon	Barre
Sanskrit	-		Nandimukhi, Madhuli	-	Kodara	Kanguni	Shyama	China
Kannada	Jola	Sajjai	Ragi	Same	Harka	Navane	Oodalu	Baragu
Tamil	Cholam	Kamboo	Kelvaragu	Samai	Varagu	Tenai	Kuthiravaali	Panivaragu
Telugu	Jonna	Sajjalu	Ragulu	Samalu	Arikelu, Arika	Korra, Korralu	Udalu, Kodisama	Varigulu, Varagalu
Malayalam	Cholam	Kamboo	Moothari	Chama	Varagu	Thina	---	Panivaragu
Gujrati	Juar	Bajri	Nagli, Bavto	Gajro, Kari	Kodra	Kang	Sama	Cheno
Bengali	Juar	Bajra	Mandua	Kangani	Kodo	Kaon	Shamula	Cheena
Punjabi		Bajra	Mandhuka, Mandhal	Swank	Kodra	Kangni	Swank	Cheena

Source: Indian Institute of Millet Research, 2020

2.3 Indulge in Baking Manufacturing Traits for Aligning Millet Flour Application as a Functional Component

Bakery industry is always focused whenever there is functional product development as it is an ideal vehicle for nutrient bioavailability, ease of product development and radial available to meet the product availability with high level of frequency. Studies have underlined the following are the main trends in baking product development:

1. Allergen-free products
2. Enriched products
3. Products with reduced content.

2.3.1 Allergen-Free Products

An allergen is a substance or product that causes an allergic reaction in the body. The minimum amount of gluten that can be considered an allergen in bakery products has been reduced. Gluten is made up of gliadin and glutenin, two proteins in flour that are not soluble in water. Products that are gluten-free contain less gluten than 20 parts per million (ppm), whereas products that are gluten-reduced contain less gluten than 100 ppm. The primary consumers of allergenic baking products are celiac or gluten-intolerant individuals. Although the symptoms of celiac disease are more severe, they are similar in both cases (Di Cairano et al., 2018). The disease's damage to the small intestine is reflected in symptoms such as diarrhea, abdominal distension, weight loss, and abnormalities in digestion and nutrition. The disorder of fat-soluble vitamin absorption, the disorder of bone metabolism, and the anemia all occur in the latter scenario.

According to Xu et al., 2020, gluten intolerance, also known as gluten sensitivity, frequently develops following a digestive disorder in which the intestinal wall becomes more porous, making it difficult to tolerate certain amounts of gluten. Gluten sensitivity can be treated with basic healing and an individual gluten-free diet. Celiac disease, also known as gluten-sensitive enteropathy, is a multifactorial autoimmune condition that is characterized, according to David A. Nelsen (2002), primarily by issues with the absorption of nutrients and gastrointestinal symptoms brought on by antibodies produced against cereal gluten proteins. Although gastrointestinal symptoms can be alleviated by following a gluten-free diet, celiac disease cannot be completely cured. In allergen-free baking products, wheat and rye flour are substituted for flour, soy, maize, rice, and other grains. (Stantiall and Serventi, 2018).

2.3.2 Enriched Products

Overall, enhanced bakery products are functional foods with basic nutritional benefits that have been demonstrated. During nourishment, it is important that the interaction doesn't adjust the item's basic organoleptic properties (Heenan et al., 2009). Baked goods are made more nutritious by adding basic nutrients like proteins and carbohydrates, supplementary nutrients like vitamins and minerals, and accompanying substances like fibers.

2.3.2.1 Vitamin and Minerals Enriched Products

The body requires vitamins, which are essential biological compounds. Vitamin enrichment aims to provide the body with more of the vitamins it needs from the food. Baking products contain vitamins B, such as B1, B2, B3, B6, and B9. For this purpose, B-complexes are typically utilized (Kaur et al., 2022). The daily dose, the vitamin's reactivity and stability, and the amount of vitamin lost during storage all need to be taken into account when selecting the appropriate dosage ratio for a vitamin. According to personal experience, these vitamins typically require an additional 10 to 20 percent dose to ensure that, throughout its shelf life, the product contains the intended quantity. The amount of nutrients to be controlled is very low, so the batter should be mixed and sprinkled with a transporter like starch or sucrose (Kamali Rousta et al., 2021). It is important to keep in mind when developing the technology that the vitamins are extremely reactive and, as a result, unstable.

Enzyme and stimulus transmission processes are aided by minerals in our bodies. Minerals like Fe, Ca, and P are often added to baking products. The human body only needs a small amount of iron, but it is necessary for the enzymes that make hemoglobin, cytochrome, peroxidase, and catalase. Ca and P are the most abundant minerals, with a daily intake of 800 mg. The Ca: P proportion is 1: 2. The same principles that govern the advancement of nutrients apply to the measurement levels of minerals.

2.3.2.2 Protein-Enriched Products

Proteins are our fundamental building blocks, aid in the transfer of nutrients and water, take part in metabolic processes, and serve as an important source of energy. Protein-enriched bakery products are becoming increasingly popular and necessary in today's market. Because fewer or fewer amino acids are present in the majority of plant-derived proteins than are required by the human body, their consumption alone causes life-threatening conditions. To prevent this, more and more technological processes are being used to complete the protein content of the products. Amino acid preparations or naturally occurring proteins with a favorable amino acid arrangement can be used to complete the process, with the latter typically being preferred. As a substitute for lysine and threonine in wheat flour, various soy preparations are typically utilized as a plant-derived supplementary additive for this purpose. Using whole eggs in baking products can increase their biological value even more because, in comparison to breast milk and other foods, they have the highest nutritional value determined by their protein and amino acid composition.

2.3.2.3 Carbohydrate-Enriched Products

Carbohydrates, including mono- and disaccharides, are an important source of energy for our body's system due to their ease of digestion and quick absorption. Adding carbohydrates to bakery products is essential for patient nutrition and feeding phenylketonuric patients. Carbs go through energy rapidly during exercise. The absence of the essential amino acid dissociating phenylalanine hydroxylase enzyme, phenylalanine, is a genetic condition known as phenylketonuria (PKU). As a result, severe and frequently permanent brain damage occurs as the amino acid builds up in the blood and brain. There is currently no treatment for the disease; the only thing that can be done is to prevent further damage. The brain needs more glucose from carbohydrate-enhanced foods because of disease-related damage to the brain. (MacLeod et al., 2010)

2.3.2.4 Fiber-Enriched Products

Carbohydrates that are nearly impossible to digest make up food fibers. Such materials integrate cellulose, hemicellulose, gelatin, and other set aside polysaccharides. The digestion of cellulose in foods with a lot of fiber helps the intestines move faster, which reduces the amount of time it takes for food to go through the digestive system. Fibers can prevent a number of diseases and abnormalities. Most importantly, they help prevent constipation, obesity, and colon cancer, stabilize blood glucose levels, and lower cholesterol levels in the blood. Hence, it is essential for the body to get fiber to work with its not startling movement. A product must have at least 3 grams of fiber for every 100 grams of solids; or on the other hand 1.5 g of fiber per 100 ml for fluids; or on the other hand 1.5 g of fiber per 100 kcal. (Food Safety & Standard (Advertising and Claims) Regulations, 2018). Fiber-enriched baking products have emerged to encourage healthy eating. Due to their high fiber content and economics, apples and oats are used most frequently to increase fiber content. The technology must take into account the fact that the dough's capacity to absorb water changes when dietary fibers are used. (Liu et al ., 2019)

2.3.3 Products With Reduced Content

Foods with a reduced content are functional foods that contain substantially fewer substances that can be harmful to one's health when consumed in large quantities. In the baking industry, foods with fewer carbohydrates, salt, or fat have emerged for this purpose.

2.3.3.1 Carbohydrate-reduced products

Products with less carbs are preferred by people who have problems with their metabolism of carbs, but dieters also like them. There are two types of diabetes, the most severe form of carbohydrate metabolism disorder: types 1 and 2. The pancreas makes insulin, which helps the glucose units in the blood get into the cells. As blood glucose levels decrease, insulin release also decreases. The liver gives the typical scope of blood glucose levels. This process fails in diabetes, resulting in blood sugar accumulation. In Type 1 diabetes, the pancreas does not produce enough insulin to keep blood glucose levels normal, whereas, in Type 2 diabetes, cells develop insulin resistance. Type 1 can only be treated because it is likely caused by a genetic predisposition. In contrast, type 2 diabetes can be cured by curing another disease, like obesity, which causes it. A customized diet that restricts and minimizes carbohydrate intake can be used to treat diabetes in the latter scenario. When compared to the original, a carbohydrate-reduced product contains at least 30% less carbohydrates than the original. In practice, protein filtration is typically used to accomplish this. It is important to keep in mind that the dough's water absorption capacity will decrease as a result of the quantitative reduction in carbohydrates during technology. Hydrocolloids like guar gum and locust bean flour are utilized in the industry to address this issue. (Maningat et al., 2005).

Table 3. GI of various food products

Glycemic Index	Value	Sources	Reference
High	≥70	White rice, Whole wheat bread, Cornflakes, Instant oat porridge, Watermelon,	Swathi Handoo, (2023)
Medium	56-69	Grapes, Pineapple, Raisins, French fries, Couscous, Brown rice, Muesli, Millet porridge, Popcorn	Swathi Handoo, (2023)
Low	0-55	Grapefruit, Apple, Mango, Carrots, Green peas, Oatmeal, Barley, Sweet corn, Quinoa, Soy milk, Lentils, Cashew, Chickpeas,	Howard E. LeWine, (2021)

Source- Compiled from the information available with references (Swathi Handoo, 2023; Howard E. LeWine, 2021)

2.3.3.2Salt-reduced products

In India, FSSAI has initiated HFSS program (High- fat, sugar and salt) and has recommended ≤ 5 g/day of salt. Therefore salt reduction is now a national program with the primary goal of curbing the population's intake of too much salt, thereby lowering the prevalence of high blood pressure and the risk of heart attack and stroke that comes with it. It is essential to keep in mind that salt has techno-functional properties, which means that it will alter the method by which dough is made, when lowering the salinity of bakery products. (WHO, 2023)

2.3.3.3 Fat-Reduced Pastries

Fats give our body calories, energy, and the plan of the movies, building materials for synthetic compounds and supplements. Our bodies will accumulate an excessive amount of food, which may result in obesity and its complications. Today, innovative foodstuffs with lower fat content are on the market to prevent this. Various low-fat bakery products are being developed by various researchers, but the baking industry is still in the early stages of product development, giving it scope for the industry. (Wickramarachchi et al., 2015)

2.4 An Impression on Cookies and Cakes for millet Based Functional Bakery Product Formulation

Different items that are made with a combination of flour (normally wheat), sugar, and fat as the fundamental fixings and have a low moisture content (ordinarily under 5%) are alluded to as "cookies" contingent upon the language utilized. In this gathering, we can find sweet and sour treats that are explained using various strategies and plans. As a result, general advice is extremely difficult to provide. However, it should be noted that the majority of cookies do not contain the gluten network, and even when they contain wheat flour, they should not be eaten. This is due to the fact that some of these cookies lack flour and contain a lot of sugar and fat. However, there is less mechanical work required during the process because doughs are extremely liquid or because mixing times are shorter. All of these scenarios will make it simple to produce gluten-free products, and it may be sufficient to substitute another suitable gluten-free flour or their mixture for wheat flour (Gómez, 2022).

The majority of studies on gluten-free cookies are based on the creation of sugar snap cookies or formulations similar to them. According to Adedara and Taylor (2021), gluten-free cookies made with sorghum flour and wheat flour are very similar.

They also noted that the low moisture content of the starch prevents it from gelatinizing during the cookie-making process. However, the sorghum cookies' breaking strength is decreased when pregelatinized starch is added to the recipe (Adedara and Taylor, 2021).

Worldwide, people of all ages consume a lot of cake, a sweet baked dessert. Cakes have a sweet flavor and a moist, tender crumb. The global cake market is anticipated to grow at an annual rate of 3.3% and reach USD 75 billion by 2023, according to Research and Markets. Global Cake Market Growth, Trends, and the Future, 2018–2023). Godefroidt et al. claim that, Cake typically has a final moisture content of 18 to 28% in 2019, which is lower than that of bread but higher than that of cookies. Cakes can be divided into a few different categories based on their structure or potential method of making them, such as a high-proportion cake (where the ratio of sugar to flour is higher than one) or a low-proportion cake (where the ratio of sugar to flour is lower than one) or an abbreviated cake (a clammy, delicate, and smooth bread shop product made with fat and raising specialists like baking powder or pop) or an unshortened cake (known as a heavenly messenger or wipe cake without fat and raised via air caught in beaten eggs and steam) (Wilderjans et al., 2013).

Two methods for determining the quality of a cake are sensory analysis and physical analysis (weight, symmetry, volume, cake height, color of crust and crumbs, crumb structure and texture, and crumb structure). High volume, uniform morsel structure, staling resilience, and a long period of realistic usability are all examples of great cake quality. According to Cauvain (2016), cake typically loses most of its quality due to staling and microbial spoilage after a one to four week shelf life. Due to their delicate surface and sweet flavor, cakes are frequently consumed as breakfast or evening snacks (Tsatsaragkou et al., 2015).

According to Ronda et al., (2011) cake batter is like an oil-in-water emulsion with a continuous aqueous phase containing dissolved or suspended dry ingredients. The effectiveness of air retention in cake batter is inversely proportional to its viscosity, according to Sahi and Alava (2003). Prior to the formation of the cake structure, the primary factors in cake production are the aeration and stability of the gas bubbles during baking (Sahi and Alava, 2003). The quality of the cake is also determined by the volume and texture of the crumb, which are related to the gas bubbles that are incorporated into the batter while it is being mixed. The cake's volume is significantly affected by the temperature at which the starch becomes gelatinous. Higher temperatures of gelatinization resulted in better batter expansion (Avila et al., 2017). The grain milling technique is one of many factors that affect the quantity of gluten-free cake (De la Hera et al., 2012), flour type, and the size of the flour particles (De la Hera et al., 2013; De la Hera et al., 2012; Ronda et al., 2011), and additional cake batter components and additives, such as fiber, gum, and an emulsifier (Avila et al., 2017).

A new strategic area for product development has recently been established within the food industry. These days, shoppers are searching for food sources with two primary properties: The first is about the food's traditional nutritional qualities, and the second is about the expected added health benefits of eating it often. In many cases, these kinds of foods are referred to as useful food sources. Over the past few years, functional foods have received a lot of attention all over the world as a result of changes in healthy lifestyles. One interesting reason to live a healthier lifestyle is the rising number of people with cardiovascular diseases (CVDs), high blood pressure, obesity, diabetes, and other related diseases. A poor diet with foods high in saturated fatty acids (SFAs) and a lack of physical activity are typically the root causes of these conditions. Because of its constructive outcomes on wellbeing, absolute dietary fiber (TDF) has arisen as a fundamental piece of the everyday eating regimen. One of the biggest challenges facing the baking industry is making gluten-free products. Gluten, which is necessary for dough development, is frequently associated with both gluten intolerance and gluten sensitivity, which have recently been classified as two distinct medical conditions linked to the consumption of gluten protein. With regards to integrating gluten- free sources into starch-rich bread kitchen items, (Rizkalla et al., 2002; Tsatsaragkou et al., 2015).

Gluten- free bakery products have opened up on the lookout, ready out of non-wheat flour like rice, ragi, sorghum, green gram, and wheat. Beta-glucans and inulin, two types of fiber, were found to be beneficial to gluten-free bakery products in a number of studies. The majority of whole grains, such as corn, rice, sorghum, buckwheat, green gram, and ragi, are excellent sources of fiber, iron, and vitamin B, so manufacturers of gluten-free foods are investing in their use.

Over the course of the past two decades, a number of studies have demonstrated that foods with a low glycemic index (GI) reduce the risk of developing chronic diseases like cardiovascular disease (CVD) and type 2 diabetes (T2D). For better management and prevention of chronic diseases like T2D and CVD, the FAO/WHO report from 1998 suggests that the GI of foods and information about the composition of the food be used together to guide food choices. India already has the world's highest prevalence of T2D. In order to prevent and manage T2D and its complications, it is important to control blood sugar after eating.

There is a lot of evidence to suggest that the GI (or quality) of the carbohydrate is just as important as how much of it you eat if you want to reduce postprandial glycaemia as part of a plan to prevent and treat diabetes and cardiovascular disease. Low-GI foods may help prevent cardiovascular events linked to diabetes by lowering insulin demand, improving blood glucose control, lowering blood lipid concentrations, and lowering body weight. In the Asian Indian diet, carbohydrates (cereal-based foods), particularly rice and wheat (60–65 percent), provide the majority of the energy. The consumption of wheat is higher in northern India, while rice consumption is higher in the country's southern and eastern regions. Refined grains account for nearly half (46–9 percent) of the South Indian population's daily energy intake (mean intake: 333 g/d), with white rice accounting for 253–4 g/d. For the proper selection of foods, it is necessary to know the GI of such staples. Indians, who are more likely to be insulin resistant, may particularly benefit from this information. Although the GI of most products has not been tested, whole grain-based foods from traditional Indian diets have only recently been reintroduced to the Indian market. There is subsequently a need to decide the GI of nearby food items utilizing normalized philosophy. In the current review, we look at the GI of rotis (unleavened flatbread) made of entire wheat flour and with a recently evolved atta blend containing Bengal gram, psyllium husk, and debittered fenugreek flour in solid non-diabetic subjects (Ganesan et al., 2010).

Celiac disease is an enteropathy of the small intestine that is caused by an immune process and is caused by an intolerance to gluten that lasts a lifetime. It can start in childhood or later in life (Murch; Walker-Smith,1995). Gluten is a group of proteins that are found in wheat, barley, rye, oats, and a few other bowls of cereal. Because of its unique cohesive and elastic properties, gluten makes dough, which are important technological feature for making cookies, bread, cakes, and other bakery products (Wall; Huebner, 1981). However, because the small intestine does not properly absorb nutrients, celiac patients cannot consume foods that contain this protein. As a result, many symptoms, including weakness, weight loss, and diarrhea, can occur. Food fortification, a method for balancing the nutritional profile or restoring nutrients lost in the process, represents one of the most significant steps in the improvement of food quality over the past forty years (Credidio, 2005).

Additionally, food fortification can be a cost-effective, adaptable, and socially acceptable method of increasing the nutrient intake of at-risk groups in order to guarantee dietary sufficiency (Hoffpauer; Wright, 1994) as a means by which milled or processed food can be obtained (Mason et al., 2001). Numerous foods have been enriched with substances that have some positive effects on health: cookies made from Saccharomyces cerevisiae extract (Santucci); high in folate and other B vitamins Alvim; Cookies enriched with retinyl acetate (Butt et al., Faria, 2003), Casein from milk (Kruger et al.,, 2003), dietary fibers that assist in regulating the flow of food through the digestive tract (Matias et al., 2005; Zürk et al. 2002; Silva, 1997), wheat germ and vitamin C, both of which possess antioxidant properties (Bilgiçli; İbanog; 2007 Herken; Torres; Lobo; Vegetative proteins (Martin et al., 2007 Sato, 1995) and hydrolyzed proteins (Silva et al., 2007) that boost the nutritional value of numerous food items.

However, the addition of minerals to flours and starches for the production of cookies and bread has also been investigated (Torres; Lobo; Sato, 1996; Scabio; Granato; Ferrari, 2005). Lack of iron is the most widely recognized nourishing problem on the planet, being predominant in emerging nations like Brazil, and furthermore in affluent nations. Children, adolescents, and women are the most affected groups due to their nutritional requirements (WHO, 2008). Iron is found in relatively few foods; Moreover, the body only takes in a trace amount. As a result, food fortification has been found to be one of the most efficient and long-lasting methods for controlling populations' iron deficiency (Inacg, 1977).

2.5 Low Gluten (LG) or Gluten-Free Bakery Products Based Consumer Demand as a Driving Force for Bakery Industry Advancement

People who have celiac disease, are allergic to gluten, or are concerned about their health are increasingly looking for gluten-free bakery products. Synthetically raised gluten- free items like treats, cakes, biscuits, and wafers, notwithstanding bread items, are acquiring notoriety because of their accommodation and particular flavor and surface. In recent years, numerous research papers have been conducted on gluten-free bakery products. Table 4 and 5 summarizes few of the significant research abstracts. The Food Safety and Standards (Advertising and Claims) Regulation, 2018, states that a food's glycemic index (GI) indicates the rate at which the food's carbohydrate is broken down into glucose and absorbed into the bloodstream from the intestine. A GI of less than 55 is regarded as low, while a GI of greater than 70 is regarded as high. Between 56 and 69, moderates fall somewhere in between.

In accordance with the low gluten (LG) standards for international trade, the Codex Alimentarius Commission concluded that LG foods cannot contain wheat, rye, barley, oats, or their crossbred varieties unless they have been specially processed to reduce the gluten level to less than 20 ppm. The 2009 report of the Codex Alimentarius Commission states that gluten-free foods sold in the European Union may be referred to as having "very low gluten." According to a final regulation issued by the Food and Drug Administration (Food and Drug Administration, 2013), the definition of the term "gluten-free" for the labeling of foods in the United States is comparable. This implies that any undeniable presence of gluten in the food should be under 20 sections for each million.

The Commission Regulation (EC) No. 41/2009 sets the rules for gluten-free and low-gluten food products in the EU. As indicated by this guideline, bakery items with under 100 milligrams of gluten for each kilogram can be named as "exceptionally low gluten" or "gluten- free". However, it is essential to keep in mind that gluten-free and low-gluten products are not the same thing. Low-gluten products still contain some gluten, so they may not be suitable for people with celiac disease or gluten intolerance whereas, gluten-free products contain no gluten at all.

As previously stated, the primary obstacle in LG cakes and cookies is the replacement of gluten network functionality. LG flours produce batters rather than doughs when combined with water, making the production of cakes and cookies a technological challenge. The repellent forces between starch granules are largely to blame for the system's lack of stability and coherent structures to entrap the air added by kneading the batter and the carbon dioxide produced by yeast fermentation. LG cakes and cookies are rigid and crumbly due to the irregular and unstable cells produced by a weak capacity for gas retention (Onyango et al.,, 2010). In the traditional technological approach to the production of LG foods, starches, dairy products like milk powder, milk protein isolate, albumin, caseinate, and whey protein concentrate, gums and hydrocolloids like methylcellulose derivatives, galactomannans, and xanthan gum, and other nongluten proteins like soy protein isolate, pea protein isolate, and egg proteins are used to improve LG bakery products' structure, mouthfeel, acceptability, and shelf life.

LG cakes and cookies are made in an essentially unexpected manner in comparison to standard wheat cakes and cookies. Mixed, bulk fermented, divided/molded, proofed, and baked are the traditional steps for wheat dough. The majority of LG doughs contain more water and have a structure that is more fluid. According to Duodu and Taylor (2012), mixing, proofing, and baking them also take less time than wheat counterparts. Moore and others, In subsequent studies on LG cakes and cookies, this strategy was successfully utilized (Hussain et al., 2004) to develop a method for producing high-quality LG cakes and cookies through baking, proofing, and mixing. 2019; Naseer and other, 2021). The authors made

LG cakes and cookies and compared their quality to that of wheat cakes and cookies or LG cakes and cookies made from a commercial mix made from wheat starch, with or without dairy. Nondairy LG cakes and cookies contained corn starch, brown rice flour, soy, buckwheat flour, and xanthan gum. The primary ingredients in the dairy LG cakes and cookies were brown rice flour, skim milk powder, whole eggs, potato and corn starches, soya flour, xanthan, and konjac gums.

2.6 Scenario of Low GI Bakery Food Product Investigations and Acceptability

Research advancement of Low GI food products rather than low Gluten or gluten free product has encouraged to focus on GI value of prepared food product. Cakes and cookies are two major bakery products focused with this aspect in various studies. Table 6 summarizes the scientific contribution made with low GI food preparation. There is very little written about millet-based low-GI cookies and cakes, but there is written about low-GI bakery products made with resistant starch or sugars that are low in calories. Millets are essentially wealthy in safe starch and fiber content, these extraordinary properties of millets make them an optimal element for the advancement of low GI food sources.

Dangi et al., (2019) mixed PMS with the hydrolysate produced by the breakdown of guar gum by acid. Acid hydrolysis improved PMS's solubility, WHC, and SP. When PMS-acid hydrolyzed gums were used, the paste's viscosities increased and the paste's temperatures decreased. Acid-hydrolyzed starches are used as gelling agents, fat substitutes, and RS in the cookies and gum candies.

Bakery products are used to incorporate a variety of nutrient-dense ingredients. The amino profiles of non-wheat proteins added to wheat flour improve the protein's quality. Due to their appealing sensory and nutritional qualities as well as superior taste, multigrain cakes have gained market share. According to Staticall and Serventi (2017), multigrain cakes can be found in a variety of flavors and textures for customers to choose from.

CONCLUSION

India's largest share in millet production is encouraging to boost the research on the new product development. The dominant state like Rajasthan and Maharashtra could be the leader in initiating the new start-ups of new category functional food. There need to be proper channel form farm to fork coling around nutritional goodness preservation and, processing, preservation. As Millets are identified as an ideal source of fortification and enrichment of nutrients. Its distinct functional ingredients such as bioactive, phenolic and non-gluten nature would be exploited to devlopmpment functional foods. Popularity of bakery sector for section of food has upsurge and bakery products are recognized as ideal vehicle in delivering the food nutrient inside the body. Considering the availability, cost effectiveness research on millet based bakery products could be widely acknowledged. Subsequent investigations about apposite functional constituent to mitigate the challenge of texture and sensorial properties of millet based food products has to be addressed by next generation researchers.

Table 4. Effect of millet flour on properties of dough and product with special emphasis on cookie dough and cookies characteristics alterations

Flour	Other additives	Dough properties	Cookie properties	Reference
Pearl millet and wheat flour	Fermented flour using probiotic culture	Cookies were prepared by substituting 50 percent cooked fermented pearl millet flour for 100 percent wheat flour.	The sensory acceptability of cookies containing 50% fermented pearl millet was satisfactory. Both the cookies made with 100% wheat flour and 50% cooked fermented pearl millet flour were well received.	Ranasalva and Visvanathan, (2014)
Pearl millet flour (PMF) and wheat flour	N/A	N/A	Addition of PMF resulted in a significant decrease in weight, diameter and spread factors of cookies. Replacement of whole wheat flour with PMF (60%) saw as fundamentally better with respect to general quality attributes over the other blends.	Kulkarni et al., (2021)
Rice flour	Konjac flour (0, 0.25, and 0.4%)	Expanding konjac flour expanded the thickness, sticking temperature, top consistency, breakdown, and hot paste consistency of rice flour.	The diameter, lightness, and greenness of the konjac were reduced, and the spread ratio, specific volume, and hardness were increased.	Akesowan, (2016)
Buckwheat (B), buckwheat/ maize (BC) (50/50), buckwheat/rice (BR) (50/50), buckwheat/maize/rice (BCR) (50/25/25).	N/A	N/A	B, BC, and BR all had wider widths than BR; B was thicker than the other formulation; Ratio of spread: The relative hardness of BC, BCR, and BR was lower than that of B and BR. B was more fracturable than BC, BR, and BCR.	Altindag et al., (2015)
Rice flour/buckwheat flour (100/0, 90/10, 80/20, 70/30)	Carboxymethyl cellulose (CMC) (0.69%)	There were elastic-like characteristics in all treatments; The elastic modulus decreased with an increase in buckwheat, but the tan σ increased. The dough's tenacity and resistance to deformation increased when CMC was added.	The addition of buckwheat flour decreased the fracturability and hardness of the cookies, but it increased their eccentricity and overall acceptability.	Dapčević Hadnadev et al., (2013)
Oat flour, substituted by oat bran (0, 30, 50, 70, and 100%)	N/A	The oat bran's water absorption index increased.	The lightness, greenness, and hardness were diminished by an increase in oat bran.	Duta and Culetu, (2015)
Rice/buckwheat (80/20)	Canola protein disconnects or focuses (3, 6, and 9% replacement of rice flour)	Canola protein concentrates/segregates decreased hardness and expanded springiness.	Lightness, greenness, and yellowness were not significantly affected by canola protein concentrates or isolates.	Gerzhova et al. (2016)
Flour (40 percent rice flour, 20 percent maize flour, and 30 percent maize starch)	Inulin (3, 3.5, 4, and 4.5%)	N/A	The addition of inulin had no effect on the thickness, width, spread factor, color, or sensory properties of the cookies.	Maghaydah *etal.*, (2013)
Gluten-free commercial mix: maize flour, pre-gelatinized rice flour, tapioca starch, integrated with chestnut flour (0, 50, 80, and 100 percent)	N/A	Chestnut flour expanded springiness, cohesiveness, and adhesiveness while diminishing hardness, stickiness, water restricting limit, water ingestion file, consistency, and flexible modulus.	Chestnut flour increased the greenness and yellowness while decreasing the lightness.	Paciulli et al., (2018)

REFERENCES

Adedara, O. A., & Taylor, J. R. N. (2021). Roles of protein, starch and sugar in the texture of sorghum biscuits. *Lebensmittel-Wissenschaft + Technologie*, *136*, 110323. doi:10.1016/j.lwt.2020.110323

Agricultural and Processed Food Products Export Development Authority (APEDA). (2023). *Indian Millets Production*. APEDA. https://apeda.gov.in/milletportal/Production.html

Aidoo, M. K., Bdolach, E., Fait, A., Lazarovitch, N., & Rachmilevitch, S. (2016). Tolerance to high soil temperature in foxtail millet (Setaria italica L.) is related to shoot and root growth and metabolism. *Plant Physiology and Biochemistry*, *106*, 73–81. doi:10.1016/j.plaphy.2016.04.038 PMID:27149034

Akesowan, A. (2016). Influence of konjac flour on foaming properties of milk protein concentrate and quality characteristics of gluten-free cookie. *International Journal of Food Science & Technology*, *51*(7), 1560–1569. doi:10.1111/ijfs.13125

Altindag, G., Certel, M., Erem, F., & Ilknur Konak, Ü. (2015). Quality characteristics of gluten-free cookies made of buckwheat, corn, and rice flour with/without transglutaminase. *Food Science & Technology International*, *21*(3), 213–220. doi:10.1177/1082013214525428 PMID:24613830

Amritha, G. K., Dharmaraj, U., Halami, P. M., & Venkateswaran, G. (2018). Dephytinization of seed coat matter of finger millet (Eleusine coracana) by Lactobacillus pentosus CFR3 to improve zinc bioavailability. *Lebensmittel-Wissenschaft + Technologie*, *87*, 562–566. doi:10.1016/j.lwt.2017.09.024

Anderson, J. W., & Hanna, T. J. (1999). Whole grains and protection against coronary heart disease: What are the active components and mechanisms? *The American Journal of Clinical Nutrition*, *70*(3), 307–308. doi:10.1093/ajcn/70.3.307 PMID:10479191

Ávila, B. P., Braganca, G. C. M., Rockenbach, R., Alves, G. D., Monks, J., Gularte, M. A., & Elias, M. C. (2017). Physical and sensory characteristics of cake prepared with six whole-grain flours. *Journal of Food Measurement and Characterization*, *11*(3), 1486–1492. doi:10.100711694-017-9527-0

R M Beattie, J A Walker-Smith, S H Murch (1995). Indications for investigation of chronic gastrointestinal symptoms. *Archives of Disease in Childhood, 73*, 354-355.

Bhatt, D., Fairos, M., & Mazumdar, A. (2022). Millets: Nutritional composition, production and significance: A review. ~ 1577 ~ Pharma. *The Innovation Journal*, 1577–1582.

Bilgiçli, N., İbanogˇlu, Ş., & Herken, E. N. (2007). Effect of dietary fiber addition on the selected nutritional properties of cookies. *Journal of Food Engineering*, *78*(1), 86–89. doi:10.1016/j.jfoodeng.2005.09.009

Butt, M. S., Arshad, M. U., Alam, M. S., & Nadeem, M. T.BUTT. (2007). Bioavailability and storage stability of vitamin A fortificant (retinyl acetate) in fortified cookies. *Food Research International*, *40*(10), 1212–1219. doi:10.1016/j.foodres.2007.07.002

Cauvain, S. P. (2013). Evolution in the manufacture of baked products. *Cereal Foods World*, *58*(5), 277–278. doi:10.1094/CFW-58-5-277

Cauvain, S. P. (2016). *Bread and other bakery products, The Stability and Shelf Life of Food*. Elsevier Ltd., doi:10.1016/B978-0-08-100435-7.00015-0

Ceasar, S.A., Maharajan, T., Krishna, T.P.A., & Ignacimuthu, S. (2023). Finger millet (Eleusine coracana (L.) Gaertn). *Neglected Underutilized Crop,* 137–149. doi:10.1016/B978-0-323-90537-4.00031-4

Credidio, E. (2005). Alimentos Fortificados. *Boletim Brasileiro de Nutrilogia*, *3*(1), 1–3.

Dangi, N., Yadav, B. S., & Yadav, R. B. (2019). Pasting, rheological, thermal and gel textural properties of pearl millet starch as modified by guar gum and its acid hydrolysate. *International Journal of Biological Macromolecules*, *139*, 387–396. doi:10.1016/j.ijbiomac.2019.08.012 PMID:31381921

Dapčević Hadnadev, T. R., Torbica, A. M., & Hadnadev, M. S. (2013). Influence of Buckwheat Flour and Carboxymethyl Cellulose on Rheological Behaviour and Baking Performance of Gluten-Free Cookie Dough. *Food and Bioprocess Technology*, *6*(7), 1770–1781. doi:10.100711947-012-0841-6

Nelsen, J. (2002). Gluten-Sensitive Enteropathy (Celiac Disease). *More Common Than You Think. Am. Fam. Physician*, *66*, 2259–2266. PMID:12507163

Dayakar Rao, B., Kandlakunta, B., Christina, A., Golla, S. D., & Tonapi, V. (2017). *Nutritional and Health Benefits of Millets [WWW Document].* ICAR_Indian Inst. Millets Res. Rajendranagar. https://www.researchgate.net/publication/318921493_Nutritional_and_Health_Benefits_of_Millets (accessed 3.29.23).

De la Hera, E., Martinez, M., Oliete, B., & Gómez, M. (2013). Influence of Flour Particle Size on Quality of Gluten-Free Rice Cakes. *Food and Bioprocess Technology*, *6*(9), 2280–2288. doi:10.100711947-012-0922-6

De la Hera, E., Ruiz-París, E., Oliete, B., & Gómez, M. (2012). Studies of the quality of cakes made with wheat-lentil composite flours. *Lebensmittel-Wissenschaft + Technologie*, *49*(1), 48–54. doi:10.1016/j.lwt.2012.05.009

Di Cairano, M., Galgano, F., Tolve, R., Caruso, M. C., & Condelli, N. (2018). Focus on gluten free biscuits: Ingredients and issues. *Trends in Food Science & Technology*, *81*, 203–212. doi:10.1016/j.tifs.2018.09.006

Duta, D. E., & Culetu, A. (2015). Evaluation of rheological, physicochemical, thermal, mechanical and sensory properties of oat-based gluten free cookies. *Journal of Food Engineering*, *162*, 1–8. doi:10.1016/j.jfoodeng.2015.04.002

FAO, UN. (2023). *International Year of Millets 2023*. Food and Agriculture Organization of the United Nations. https://www.fao.org/millets-2023/en

FAO/WHO. (1998). Carbohydrates in human nutrition: Report of joint FAO/WHO expert consultation. *FAO Food and Nutrition*, *66*, 1–140. PMID:9743703

Food and Drug Administration- FDA. (2013). *'Gluten-Free' Means What It Says*. FDA.

Gerzhova, A., Mondor, M., Benali, M., & Aider, M. (2016). Incorporation of canola proteins extracted by electroactivated solutions in gluten-free biscuit formulation of rice–buckwheat flour blend: Assessment of quality characteristics and textural properties of the product. *International Journal of Food Science & Technology*, *51*(3), 814–827. doi:10.1111/ijfs.13034

Godefroidt, T., Ooms, N., Pareyt, B., Brijs, K., & Delcour, J. A. (2019). Ingredient Functionality During Foam-Type Cake Making: A Review. *Comprehensive Reviews in Food Science and Food Safety*, *18*(5), 1550–1562. doi:10.1111/1541-4337.12488 PMID:33336911

Handoo, S. (2023). *What Is Glycemic Index & List Of Foods With Their GI.*

Heenan, S. P., Hamid, N., Dufour, J. P., Harvey, W., & Delahunty, C. M. (2009). Consumer freshness perceptions of breads, biscuits and cakes. *Food Quality and Preference*, *20*(5), 380–390. doi:10.1016/j.foodqual.2009.02.008

Hoffpauer. (1994). Enrichment of rice. In W. E. Marshall & A. Wadsworth (Eds.), *Rice science and Technology*. Marshall Rekker.

Howard, E., & LeWine, M. D. (2021). *Glycemic index for 60+ foods*. Harvard Medical School.

Huebner, F. R., & Wall, J. S. (1981). Fractionation and quantitative differences of glutenin from wheat varieties varying in baking quality. *Cereal Chemistry*, *53*, 258–268.

Hussain, S., Mohamed, A. A., Alamri, M. S., Ibraheem, M. A., Abdo Qasem, A. A., Serag El-Din, M. F., & Almaiman, S. A. M. (2004). Wheat–millet flour cookies: Physical, textural, sensory attributes and antioxidant potential. *Food Science & Technology International*, *26*(4), 311–320. doi:10.1177/1082013219894127 PMID:31818128

International Nutritional Anemia Consultative Group - INACG. (1977). *Guidelines for the eradication of iron deficiency anemia Washington: International Nutritional Anemia Consultative Group*. Nutrition Foundation.

Kamali Rousta, L., Bodbodak, S., Nejatian, M., Ghandehari Yazdi, A. P., Rafiee, Z., Xiao, J., & Jafari, S. M. (2021). Use of encapsulation technology to enrich and fortify bakery, pasta, and cereal-based products. *Trends in Food Science & Technology*, *118*, 688–710. doi:10.1016/j.tifs.2021.10.029

Kaur, N., Agarwal, A., & Sabharwal, M. (2022). Food fortification strategies to deliver nutrients for the management of iron deficiency anaemia. *Current Research in Food Science*, *5*, 2094–2107. doi:10.1016/j.crfs.2022.10.020 PMID:36387591

Krüger, C. C. H., Comassetto, M. C. G., Cândido, L. M. B., Baldini, V. L. S., Santtucci, M. C., & Sgarbieri, V. C. (2003). Biscoitos tipo "cookie" e "snack" enriquecidos, respectivamente com caseína obtida por coagulação enzimática e caseinato de sódio. *Food Science and Technology (Campinas)*, *23*(1), 81–86. doi:10.1590/S0101-20612003000100017

Kulkarni, D. B., Sakhale, B. K., & Chavan, R. F. (2021). Studies on development of low gluten cookies from pearl millet and wheat flour. *Food Research*, *5*(4), 114–119. doi:10.26656/fr.2017.5(4).028

Liu, S., Willett, W., Stampfer, M., Hu, F. B., Franz, M., Sampson, L., Hennekens, C. H., & Manson, J. A. E. (2000). A prospective study of dietary glycemic load, carbohydrate intake, and risk of coronary heart dis- ease in US women. *The American Journal of Clinical Nutrition*, *71*(6), 1455–1461. doi:10.1093/ajcn/71.6.1455 PMID:10837285

Maghaydah, S., Abdul-Hussain, S., Ajo, R., Obeidat, B., & Tawalbeh, Y. (2013). Enhancing the Nutritional Value of Gluten-Free Cookies with Inulin. *Advance Journal of Food Science and Technology : AJFST*, *5*(7), 866–870. doi:10.19026/ajfst.5.3174

Maningat, C., Bassi, S., Woo, K., Dohl, C., Gaul, J., & Stempien, G. (2005). Formulation of High-Protein, High-Fiber (Low-Carbohydrate), Reduced Calorie Bread. *Technical Bulletin.*, *27*(4), 1–16.

Manley, D. (2011). Product development in the biscuit industry. Manley's Technol. Biscuits, Crackers Cookies. Woodhead Publishing. doi:10.1533/9780857093646.1.69

Martin-Diana, A. B., Izquierdo, N., Albertos, I., Sanchez, M. S., Herrero, A., Sanz, M. A., & Rico, D. (2017). Valorization of carob's germ and seed peel as natural antioxidant ingredients in gluten-free crackers: Carob antioxidant by-products in gluten-free snack. *Journal of Food Processing and Preservation*, *41*(2), e12770. doi:10.1111/jfpp.12770

Mason. J. B. (2001). The Micronutrient Report: Current Progress and Trends in the Control of Vitamin A, Iodine, and Iron Deficiencies. Ottawa Micronutrient Initiative

Matias, M. F. O., Oliveira, E. L., Gertrudes, E., & Magalhães, M. M. A. (2005). Use of fibers obtained from the cashew (Anacardium ocidentale) and guava (Psidium guayava) fruits for enrichment of food products. *Brazilian Archives of Biology and Technology*, *48*(spe), 143–150. doi:10.1590/S1516-89132005000400018

Mishra, A., Pattnaik, B., Dutta, T., & Baitharu, I. (2022). Nutritional Values and Potential Health Benefits of Millets- A Review. *The Journal of Nutrition*, *8*(1), 9–26. doi:10.18488/87.v8i1.3176

Mitelut, A.C., Popa, E.E., Popescu, P.A., & Popa, M.E. (2021). Trends of innovation in bread and bakery production. *Trends Wheat Bread Mak*, 199–226. doi:10.1016/B978-0-12-821048-2.00007-6

Naseer, B., Naik, H. R., Hussain, S. Z., Zargar, I., Beenish, Bhat, T. A., & Nazir, N. (2021). Effect of carboxymethyl cellulose and baking conditions on in-vitro starch digestibility and physico-textural characteristics of low glycemic index gluten-free rice cookies. *Lebensmittel-Wissenschaft + Technologie*, *141*, 110885. doi:10.1016/j.lwt.2021.110885

Okpala, L., Okoli, E., & Udensi, E. (2013). Physico-chemical and sensory properties of cookies made from blends of germinated pigeon pea, fermented sorghum, and cocoyam flours. *Food Science & Nutrition*, *1*(1), 8–14. doi:10.1002/fsn3.2 PMID:24804009

Onyango, C., Mutungi, C., Unbehend, G., & Lindhauer, M. G. (2010). Rheological and baking characteristics of batter and bread prepared from pregelatinised cassava starch and sorghum and modified using microbial transglutaminase. *Journal of Food Engineering*, *97*(4), 465–470. doi:10.1016/j.jfoodeng.2009.11.002

Paciulli, M., Rinaldi, M., Cavazza, A., Ganino, T., Rodolfi, M., Chiancone, B., & Chiavaro, E. (2018). Effect of chestnut flour supplementation on physico-chemical properties and oxidative stability of gluten-free biscuits during storage. *Lebensmittel-Wissenschaft + Technologie*, *98*, 451–457. doi:10.1016/j.lwt.2018.09.002

Peng, X., & Yao, Y. (2017). Carbohydrates as Fat Replacers. *Annual Review of Food Science and Technology*, *8*(1), 331–351. doi:10.1146/annurev-food-030216-030034 PMID:28245156

Radhika, G., Sumathi, C., Ganesan, A., Sudha, V., Jeya Kumar Henry, C., & Mohan, V. (2010). Glycaemic index of Indian flatbreads (rotis) prepared using whole wheat flour and 'atta mix'-added whole wheat flour. *British Journal of Nutrition*, *103*(11), 1642–1647. doi:10.1017/S0007114509993680 PMID:20100375

Ramashia, S. E., Gwata, E. T., Meddows-Taylor, S., Anyasi, T. A., & Jideani, A. I. O. (2018). Some physical and functional properties of finger millet (Eleusine coracana) obtained in sub-Saharan Africa. *Food Research International*, *104*, 110–118. doi:10.1016/j.foodres.2017.09.065 PMID:29433775

Ranasalva, N., & Visvanathan, R. (2014). Development of cookies and bread from cooked and fermented pearl millet flour. *African Journal of Food Science*, *8*(6), 330–336. doi:10.5897/AJFS2013.1113

Rashmi, S., Dhumketi, K., Chaturvedi, P., Shukla, N., Shukla, R., & Dhumketi, K. (2022). Development and assessment of healthy multigrain biscuits of wheat flour with pearl millet, maize and buckwheat composite flours. *Pharma Innov. J.*, *11*, 1702–1707.

Rizkalla, S. W., Bellisle, F., & Slama, G. (2002). Health benefits of low glycaemic index foods. Such as pulses, in diabetic petients and health individuals. *British Journal of Nutrition*, *88*(S3), 255–262. doi:10.1079/BJN2002715 PMID:12498625

Ronda, F., Oliete, B., Gómez, M., Caballero, P. A., & Pando, V. (2011). Rheological study of layer cake batters made with soybean protein isolate and different starch sources. *Journal of Food Engineering*, *102*(3), 272–277. doi:10.1016/j.jfoodeng.2010.09.001

Sachan, D. S., Kumar, R., Kumar, P., Pal, V., & Yadav, K. K. (2023). Millets Production and Consumption in India. *Just Agriculture*, *3*(5), 275–283.

Sahi, S. S., & Alava, J. M. (2003). Functionality of emulsifiers in sponge cake production. *Journal of the Science of Food and Agriculture*, *83*(14), 1419–1429. doi:10.1002/jsfa.1557

Saloni, S., Sindhu, S., Kumari, S., & Suman, S. (2018). Little millets: Properties, functions and future prospects. *International Journal of Agricultural Engineering*, *11*(Special), 179–181. doi:10.15740/HAS/IJAE/11.Sp.Issue/179-181

Saranraj, P., & Geetha, M. (2012). Microbial Spoilage of Bakery Products and Its Control by Preservatives. *International Journal of Pharmaceutical and Biological Archives.*

Sato, H., Matsumura, T. and Shibukawa, S. 1995. Apparent heat transfer in a forcedconvection oven and properties of baked food. *J. Food Sci. 52*, 185-188.

Scabio, A., Granato, D., & Ferrari, R. A. (2005). Aplicação do bagaço de laranja na produção de biscoitos. In: SEMANA DE ENGENHARIA DE ALIMENTOS. Ponta Grossa: Universidade Estadual de Ponta Grossa.

Selladurai, M., Pulivarthi, M. K., Raj, A. S., Iftikhar, M., Prasad, P. V. V., & Siliveru, K. (2022). *Considerations for gluten free foods - pearl and finger millet processing and market demand.* Grain Oil Sci. Technol. doi:10.1016/j.gaost.2022.11.003

Shahidi, F., & Chandrasekara, A. (2013). Millet grain phenolics and their role in disease risk reduction and health promotion: A review. *Journal of Functional Foods*, *5*(2), 570–581. doi:10.1016/j.jff.2013.02.004

Sheahan, C. M. (2014). Plant guide for proso millet (Panicum miliaceum). USDA-Natural Resources Conservation Service, Cape May Plant Materials Center. Cape May, NJ.

Silva, M. R. (1997). Caracterização química e nutricional da farinha de jatobá (Hymenaea stigonocarpa Mart.): desenvolvimento e otimização de produtos através de testes sensoriais afetivos. Universidade Estadual de Campinas, São Paulo,.

Silva, M. A. (2007). Uso de hidrolisado de pescado como suplemento protéico em biscoito de polvilho. In: Encontro Regional Sul De Ciência E Tecnologia De Alimentos, 9, 475-479.

Singh, M., & Mohamed, A. (2007). Influence of gluten-soy protein blends on the quality of reduced carbohydrates cookies. *Lebensmittel-Wissenschaft + Technologie*, *40*(2), 353–360. doi:10.1016/j.lwt.2005.09.013

Stanley, P., Cauvain, & Young, L. (1999). Bakery Food Manufacture and Quality: Water Control and Effects: Second Edition. Wiley.

Stantiall, S.E. & Serventi, L. (2017). *Nutritional and sensory challenges of gluten-free bakery products: a review*. NIH. https://doi.org/ doi:10.1080/09637486.2017.1378626

Stantiall, S. E., & Serventi, L. (2018). Nutritional and sensory challenges of gluten-free bakery products: A review. *International Journal of Food Sciences and Nutrition*, *69*(4), 427–436. doi:10.1080/09637486.2017.1378626 PMID:28954552

Surekha, N., Chimmad, B. V., & Naik, R. S. (2013). Hypoglycaemic and hypolipidemic efficacy of barnyard millet (Echinochloa frumentacea Link) based health food. *Asian Journal of Home Science*, *8*(2), 383–387.

Taylor, J. R. N., & Duodu, K. G. (2019). *Sorghum and millets: chemistry, technology, and nutritional attributes* (2nd ed.). Woodland Publishing.

Taylor, J.R.N. & Kruger, J. (2019). *Sorghum and Millets: Food and Beverage Nutritional 36 Attributes*. Sorghum Millets Chem. Technol. Nutr. Attrib. doi:10.1016/B978-0-12-811527-5.00007-1

Torres, M. A. A., Sato, K., Lobo, N. F., & Queiroz, S. S. (1995). Efeito do uso de leite fortificado com ferro e vitamina C sobre os níveis de hemoglobina e condição nutricional de crianças menores de 2 anos. *Revista de Saude Publica*, *29*(4), 301–307. doi:10.1590/S0034-89101995000400008 PMID:8729281

Tsatsaragkou, K., Papantoniou, M., & Mandala, I. (2015). Rheological, Physical, and Sensory Attributes of Gluten-Free Rice Cakes Containing Resistant Starch. *Journal of Food Science*, *80*(2), E341–E348. doi:10.1111/1750-3841.12766 PMID:25604540

Ugare, R., Chimmad, B., Naik, R., Bharati, P., & Itagi, S. (2011). Glycemic index and significance of barnyard millet (Echinochloa frumentacae) in type II diabetics. *Journal of Food Science and Technology*, *51*(2), pages392–395. doi:10.100713197-011-0516-8 PMID:24493902

Uyir-Organic Farmers Market. (2022). *The Several Benefits of Kodo Millet*. Uyir-Organic Farmers Market. https://uyirorganic.farm/benefits-of-the-kodo-millet

Walker, C. E., & Eustace, W. D. (2016). *Milling and Baking, History*. Ref. Modul. Food Sci. doi:10.1016/B978-0-08-100596-5.00158-X

Wickramarachchi, K. S., Sissons, M. J., & Cauvain, S. P. (2015). Puff pastry and trends in fat reduction: An update. *International Journal of Food Science & Technology*, *50*(5), 1065–1075. doi:10.1111/ijfs.12754

Wilderjans, E., Luyts, A., Brijs, K., & Delcour, J. A. (2013). Ingredient functionality in batter type cake making. *Trends in Food Science & Technology*, *30*(1), 6–15. doi:10.1016/j.tifs.2013.01.001

World Health Organization. (2022). Ageing and health [WWW Document]. World Health Organization. https://www.who.int/news-room/fact-sheets/detail/ageing-and-health

WORLD HEALTH ORGANIZATION – WHO. (2008). Iron deficiency anaemia: assessment, prevention and control. Geneva: World Health Organization.

Xu, J., Zhang, Y., Wang, W., & Li, Y. (2020). Advanced properties of gluten-free cookies, cakes, and crackers: A review. *Trends in Food Science & Technology*, *103*, 200–213. doi:10.1016/j.tifs.2020.07.017

Yousaf, L., Hou, D., Liaqat, H., & Shen, Q. (2021). Millet: A review of its nutritional and functional changes during processing. *Food Research International*, *142*, 110197. doi:10.1016/j.foodres.2021.110197 PMID:33773674

Chapter 10
Smart Security System for Women's Safety

Sangeetha Ganesan
R.M.K. College of Engineering and Technology, India

ABSTRACT

The issue of security is paramount in society, especially to the women who are being illtreated in this prevailing era. Thus, the authors intend to aid security by bringing up a smart security system for women safety. This system involves an EEG sensor, heartbeat sensor, GPS sensor, and GSM module along with an Arduino. This proposed system purely depends on the psychology of a woman which deals with measuring of brainwaves and heartbeat rate within a given limit. And on further exceeding, the data will be detected depending on the level of waves emitted. These values would trigger the GSM to receive the data and thereby it sends an automatic popup alert in the form of message along with the location. The entire system will be activated to safeguard woman in all aspects.

1. INTRODUCTION

This chapter gives a detailed introduction about the domain and also discusses the a variety of existing solutions available along with their disadvantages. It also emphasizes the necessity of the suggested society, in which women uphold the title of Goddess. Every cultural setting is revered because it represents the planet. Despite the fact that they are revered as the embodiment of wealth, wisdom, and power, there are grave concerns about the safety of women in India. This discrimination beside women has mainly increased in recent decades. As the threat against the women increases rapidly, to reduce that and to enhance here we propose a system in order to provide a security precaution, so that women never feel helpless while facing social challenges. It has always been a concern for many people and committees around the world that women's security plays an important role today and that they are developed by security measures.

It is obvious when we shop for products that some people have a misunderstanding of what it means to be a woman, so every effort must be made to ensure that this doesn't lower their social standing. This paper discusses a lot of issues related to wireless technology for women and kids, specifically Arduino

DOI: 10.4018/978-1-6684-9819-4.ch010

Copyright © 2023, IGI Global. Copying or distributing in print or electronic forms without written permission of IGI Global is prohibited.

as an embedded device. This model can be designed in such a way that it detects the brainwaves with the heart beat such that we can be able to check their state of emotion also the location of victim allows the rescue system to take an action accordingly, based on the electronic gadgets like GSM and GPS.

2. LITERATURE SURVEY

In (Ansari et al., 2017) SM(sutra mega) system will help women in tough situations to inform about attacks and also in giving their exact location to a nearby police station for necessary action. In this, the author designed a device. The LPC controller,ARM controller, GPS, shock circuit and GSM make up the system. Women who are in danger may press the emergency button. Women under a dangerous situation can press the emergency button. So that, the location along with data is being taken the helping message can be transferred to the emergency number. (Bhardwaj & Aggarwal, 2014) Anup CJ created a device that looks like a typical belt to protect women from these circumstances. This design includes a buzzer, zapper, GPS, and GSM circuit. Women can use the device with the emergency button when they feel uneasy. The weapon's Zapper circuit produces shocks with a high voltage to stun the attacker non-lethally. Additionally, a buzzer is activated, continuing to make a loud yelling noise to attract nearby people's assistance. As a result, they are no longer in danger.

In (Chand et al., 2015) IoT (Internet of things) based women safety device designed by Shreyasha Choudary uses sensors to connect devices to the internet using a suitable platform. It cannot be connected with monitors and is meant to be carried around for a long time as an additional package. This makes it unique in terms of safety precautions for quick data recovery. In (Campbell et al., 2001) Sneha consist of android application,main device, and portable camera. Android application uses Phone GPS or GPS of the main device to locate the victim in the critical situation they were pressing the emergency button needs a separate device to locate along with android application. The retrieving of data location from the devices to be slow at some areas. As it cannot predict the exact location and may lead to danger to anyone.

In (Descamps et al., 2000) Junaid Mohammed an android app based on their mobile application for the health monitoring. They developed an application called ECG android app, which provides the user with to interact with the woman data through the app. waves and the data logging functionality in the background, the sorted data can be upload in the Wi-Fi cloud, which future analysis for monitors the records of the person, it will by this idea will develop an internet of things. A system has been developed that uses an IOIO-Microcontroller monitoring ECG to keep track of patients and their ECG waves from anywhere in the world.

In (Fan, 2018) D. G., M., Pavithra, G., & Subhashini presented ladies device and application-FEMME is a large product made utilizing in the wired form, and it can be purchased or the application may be put in smart phones and used in emergencies. The quickest and easiest approach to help someone in need is through FEMME. The product's use is increased by making it smaller so that it can be worn as a pendant or even as a watch. Voice keyword recognition might also be included, which would trigger the device to notify the contacts of an emergency. The product's use is increased by making it smaller so that it can be worn as a pendant or even as a watch. Voice keyword recognition might also be included, which would trigger the device to notify the contacts of an emergency.

Chougula, Naik, Patil, and Das describe a smart girl security system that is portable and resembles a belt in (Gowri & Anandha Mala, 2015).

In (Gowri & Anandha Mala, 2015) Smart girl security system that is portable and looks like a belt is described in (Gowri & Anandha Mala, 2015) by Chougula, Naik, Patil, and Das. This device was created after seeing a few devices and applications, such as an app that was introduced by the well-known TV show Gumrah that was broadcast on a channel. The founder of ILA Security and SHE (Society Harnessing Equipment) developed a shock-and-disorientation alarm that generates current to help victims flee. The System has a lot of advantages, but it also has a lot of disadvantages. Due to the high battery consumption, having a battery backup option will be more practical.

According to (George et al., 2014) Bhavale, M. D., et. al. (2015) based on Wireless Security and GPS Unified move toward for Women and Children. The suggested system comprises of a wearable safety gadget with a camera for capturing the attacker's image and an emergency button for delivering alarms. The victim's location and any accompanying evidence are tracked using GPS, and the attacker's image is captured. To all required contacts, an emergency message with an image link will be sent. This solution can utilize Arduino and a microcontroller built on a Linux board more effectively. Haversine and trilateration algorithms are used. The model's main drawback is its cost; putting all of the ideas into one system would hurt your wallet and make the system bigger, which turns it into a non-portal gadget. being costly and difficult to transport.

In (Hussain et al., 2016) Menasinkai K., et., al. The Internet of Things (IOT)-based smart security solution for women uses a device that can shock victims with 3800KV and ILA security. As the victim will likely be panicked and more likely to run away, which will increase the likelihood that they will go missing, the device sending the victim's instant location in real time can be shared with the appropriate people. As a result, it will be simple to identify and locate the path to success. (Koss, et al., 1996) Paradkar, A., & Sharma, D. all-in-one Intelligent Safety System for Security divides the technologies used in the papers into three categories and lists out their drawbacks. They then propose a model that incorporates the benefits of all the technologies—including Arduino, AI, and machine learning—and adds a few more to them in order to address their drawbacks and offer a better security system for women. The model has some novel ideas that were not previously considered, but the main drawback is that it is expensive to implement all of the ideas into a single system, which also makes the system non-portable.

Figure 1. Smart security architecture

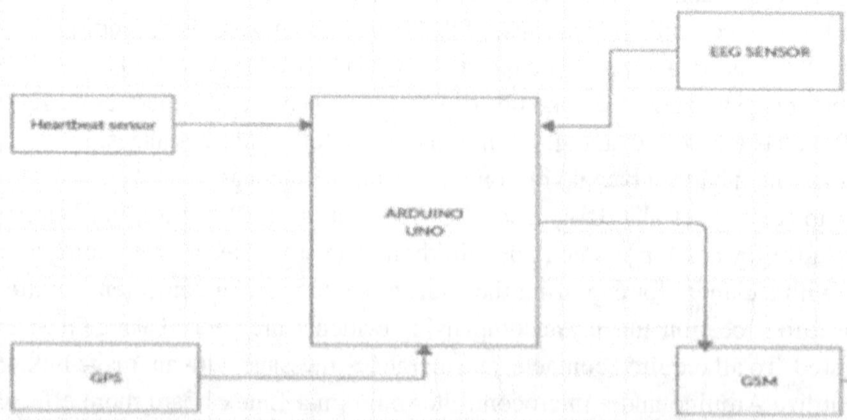

3.PROPOSED SYSTEM

This system involves an EEG sensor, Heartbeat sensor, GPS sensor and GSM module along with an Arduino. This proposed system purely depends on the psychology of a woman which deals with measuring of brainwaves and heartbeat rate within a given limit. From the brain the alpha, beta and gamma waves are detected. On further exceeding, the data's will be detected depending on the level of waves emitted. These values would trigger the GSM and thereby it sends an automatic popup message along with the location to the nearby police station. On further moves we connect this with an app. This proposed system will prove to be very useful in saving lives as well as protecting women.

Figure 1 depicts the systematic view of the proposed design which sends the waves from the sensor and that data is processed based on various conditions and it would function accordingly. We propose different modules and sensors for the system to ease the work fast.

3.1 Arduino UNO

Arduino is a piece of open-source physical preparation equipment that requires a microcontroller board and a secure upgrade environment in order to be modified. The fundamentals of Arduino are simple enough for children to learn. Unlike other microcontrollers, which only function in the Windows operating system, Arduino can run on any operating system that includes Windows, Linux, and Macintosh. The Arduino can be used to control various devices, such as lights and various mechanical assemblies, as well as to get commitments from various switches and furthermore sensors. The core of Arduino is a physical figuring stage, which needs to be altered using a language that is executed through wiring. An Arduino uno microcontroller that can carry out the following multiple actions is shown in Fig. 3.2.

Figure 2. Arduino uno

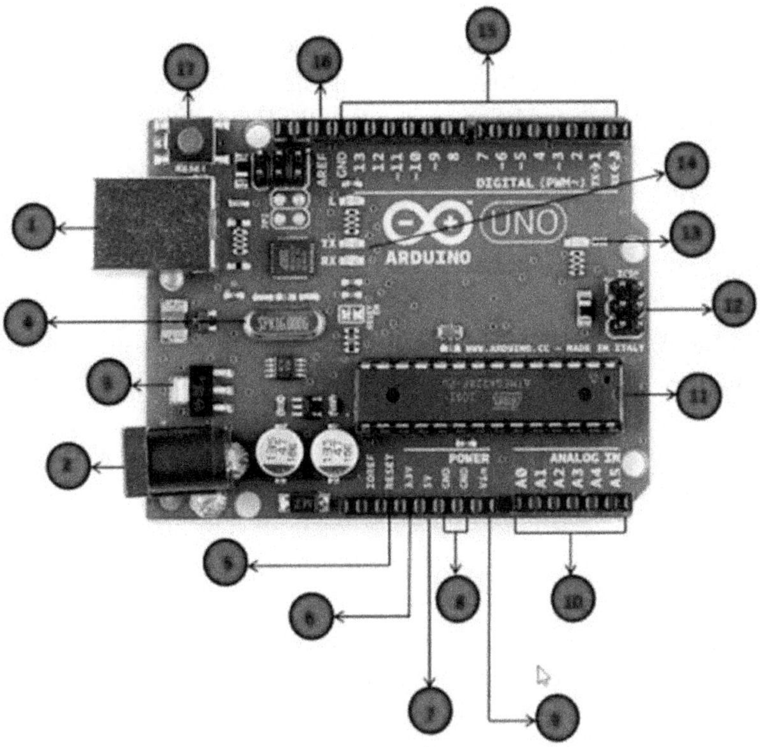

Power USB

An Arduino board can be powered by any USB cable connected to your computer. Only the USB connection and the USB cable need to be connected.

Power (Barrel Jack)

By attaching an Arduino board to the Barrel Jack, it is possible to power it directly from the AC mains power supply.

Voltage Regulator

The voltage regulator's job is to stabilise the DC voltages used by the processor and other components in addition to managing the voltage supplied to the Arduino board.

Crystal Oscillator

The Arduino is assisted by the crystal oscillator in resolving some time-related issues. how the Arduino clock works. The crystal oscillator, is the answer. 16.000H9H is the printed number that has been placed atop the Arduino crystal. It indicates that the frequency range is 16 MHZ, or 16,000,000 Hertz.

Arduino Reset

We can start our Arduino programme from scratch by performing a reset on the board. There are two ways that the UNO board can be reset. first, by pressing the board's reset button (17). Second, we can attach an external reset button that can be created to the RESET Arduino pin

Pins (3.3, 5, GND, Vin)

- 3.3V (6) – Supply 3.3 output volt
- 5V (7) – Supply 5 output volt
- The majority of the Arduino board's components operate without issue at 3.3 and 5 volts.
- .GND (8)(Ground) – The Arduino board has a number of GND pins that can be connected to a circuit.
- Vin (9) – This pin can also be used to power the Arduino board using an AC power supply or other external power source.

Analog pins

Five analogue input pins (A0 through A5) are present on the Arduino UNO board. It can read the signal from a temperature or humidity sensor, for example, and convert it into a digital value that the microprocessor can read. The Arduino UNO board has five analogue input pins (A0 through A5). For instance, it is capable of reading the signal from a temperature or humidity sensor and converting it into a digital value that the microprocessor can read.

3.2 Heartbeat Sensor

Figure 3. Heartbeat sensor

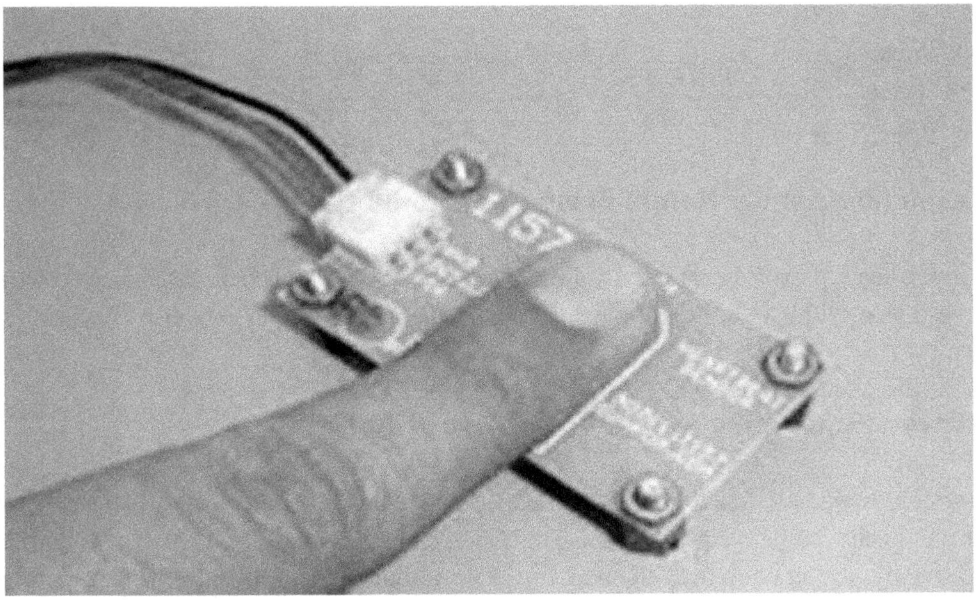

A digital output of the heartbeat is supposed to be generated when a finger is kept on a heartbeat sensor. When the heartbeat detector is in operation, the beat LED flashes in time with each heartbeat. The Beats Per Minute (BPM) rate can be measured by directly connecting the microcontroller to this digital output. Its fundamental premise is that every pulse results in a change in the amount of light. The sensor is composed of a light detector and an extremely bright LED. The LEDs must produce the greatest amount of light, move uniformly across the finger, and be detected by a detector.

3.3 Electroencephalogram (EEG)

The EEG (electroencephalogram) is connected to the users head to detect the brain waves. This technique focuses on brain waves which are produced by synchronized electrical pulses communicating with each other. This electrical activity is generated by cortical layers of the brain. These electrical signals are picked up from the gray matter regions which have pyramidal cells that communicate with each other. The EEG measures the electricity that is produced whenever these pyramidal cells fire in large clusters in a coordinated pattern and radiates to the scalp surface. A cap with electrode sensors is fitted on to the head and it's connected to the EEG scanner and the activity from the person's brain can be recorded and the emotions and the thoughts are recorded. This leads to 98% accuracy rate. EEG is the most versatile and cost-efficient solution. It provides temporal resolution. Can be used with minimal training and allows low spatial resolution.

The two electrodes are placed behind the ear and another electrode will be placed inside the ear. The typical electrodes which are used for this acquisition are made up of Silver- Silver Chloride (Ag-AgCl).

Since these electrodes have low impedance, low offset voltage, low noise, and high stability, they are used in the acquisition of EEG signals.

The 4 types of brainwaves

- Beta Waves
- Alpha Waves
- Theta Waves
- Delta Waves

Beta Waves (Frequency Range From 14 Hz to 30 Hz)

Additionally, when a person is conscious, awake, attentive, and alert, beta waves are most frequently detected. Beta waves with a low amplitude are associated with active connection, as well as being busy or anxious.

Alpha Waves (Frequency Range From 7 Hz to 13 Hz)

Alpha waves are frequently linked to a calm, relaxed, and clear state of mind. Alpha waves are rarely present during intense cognitive processes like thinking, mental calculus, and problem-solving, but they can be induced by closing one's eyes and relaxing. Young adults frequently have theta rhythms detected in EEG measurements (frequency range: 4 to about 7 Hz). Theta activity with an amplitude greater than 30 millivolts (mV) is less common in older people, with the exception of when they are sleepy.

Delta Waves (Frequency Range up to 4 Hz)

The majority of delta activity is found in newborns. In older people, deep sleep stages are related to delta waves. Patients with absence seizures, which cause brief, unexpected lapses in attention, have been shown to experience delta waves incidentally.

Figure 4. EEG

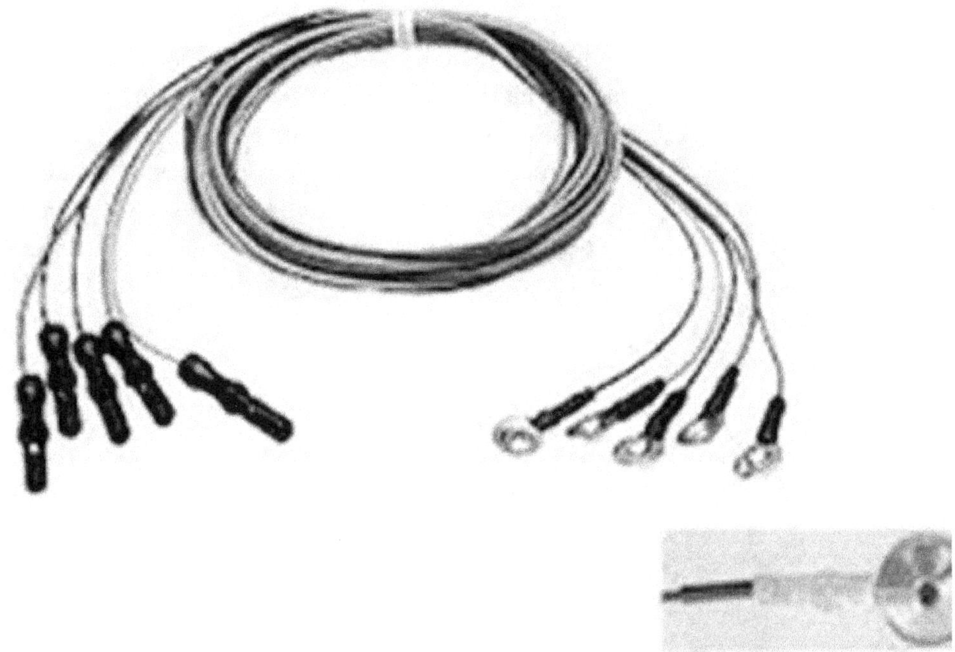

Figure 5. GSM

3.4 GPS

A location tracker is the Global Positioning System. It keeps track of the location by recording the longitude and latitude. This data will be used by the GPS Coder Module to locate the exact address of that place, including the street name, nearby intersection, etc. In all weather conditions, day or night, anywhere on or near the Earth, reliable positioning, navigation, and timing services are continuously provided to users worldwide thanks to a microcontroller that is directly connected to the USART. In the event that GPS is turned off, the system only sends the longitude and latitude via SMS. Internet usage is thus required. The physical principles underlying GPS may appear to be quite straightforward, and they are not difficult to comprehend. However, as is often the case, there is a long way from theory to practice.

Figure 6. GPS

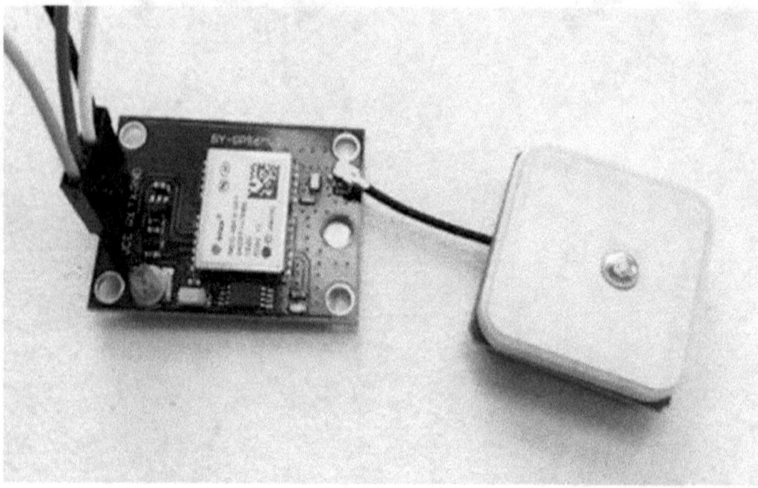

3.5 Bread Board

A bread board's function is to establish electrical connections between the parts in order to stabilize the system. Breadboards have a lot of tiny sockets, some of which are electrically connected to one another. There are numerous metal strips that physically connect to sockets on the board's underside. Additionally, it enables even electricity to flow over them. The metal strips can't be seen. Typically, a bread board has four sections. sections, two outer and two inners.

Power supply is in the outer sections. A common power supply here connected to Vcc at 5 volts is provided for all the connections of the components. Bluetooth Bluetooth is a wireless technology standard used to create personal area networks and exchange data over short distances between fixed and mobile devices using UHF radio waves in the industrial, scientific, and medical radio bands, from 2.40 GHz to 2.48 GHz.

4. IMPLEMENTATON RESULTS

Figure 7. Heartrate detection

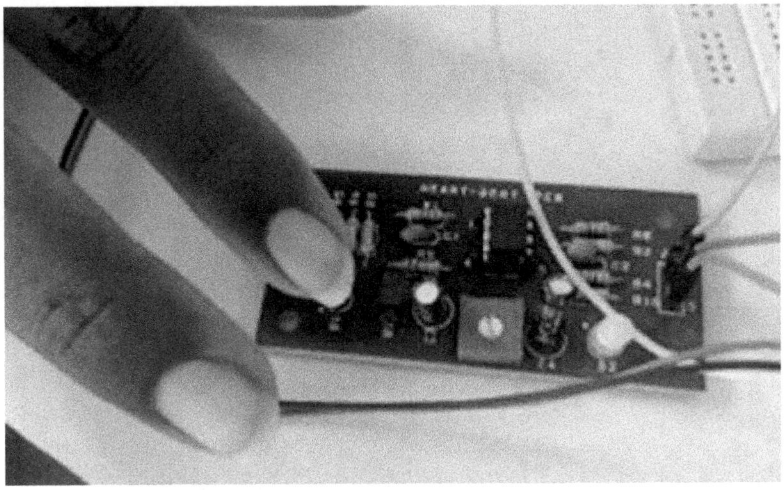

Based on changes in the optical power of an LED as it travels through the blood, the heartbeat can be measured. The perfect measurement of the heart rate is revealed by the light's intensity. The pulses begin to receive data from the body by staying in touch with the sensor pin, and it eventually returns values.

Figure 8. Implementation with component

To determine the proper values of the waves and pulses, the heartbeat sensor and the EEG sensor's two pins are connected. According to the specified pins, the entire circuit is connected to the breadboard. To plot and track the data, an Arduino UNO is connected to a computer

Figure 9. Values detection

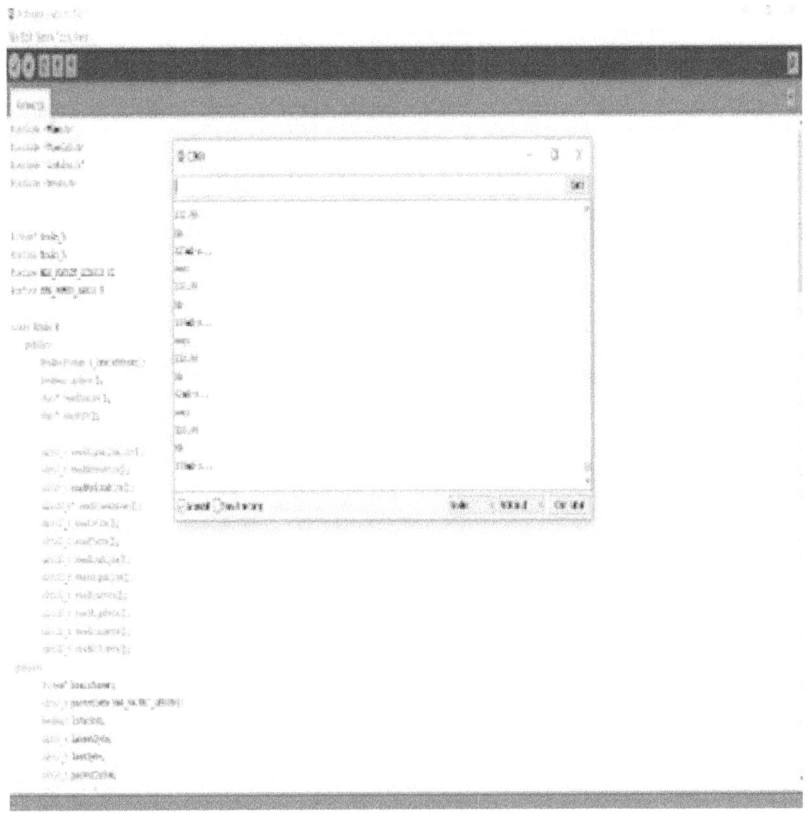

The EEG rate and heartbeat rate are output after the four waves and pulse rate have been measured and combined. The serial monitor displays accurate readings from both sensors following a successful compilation.

Figure 10. Wave detection

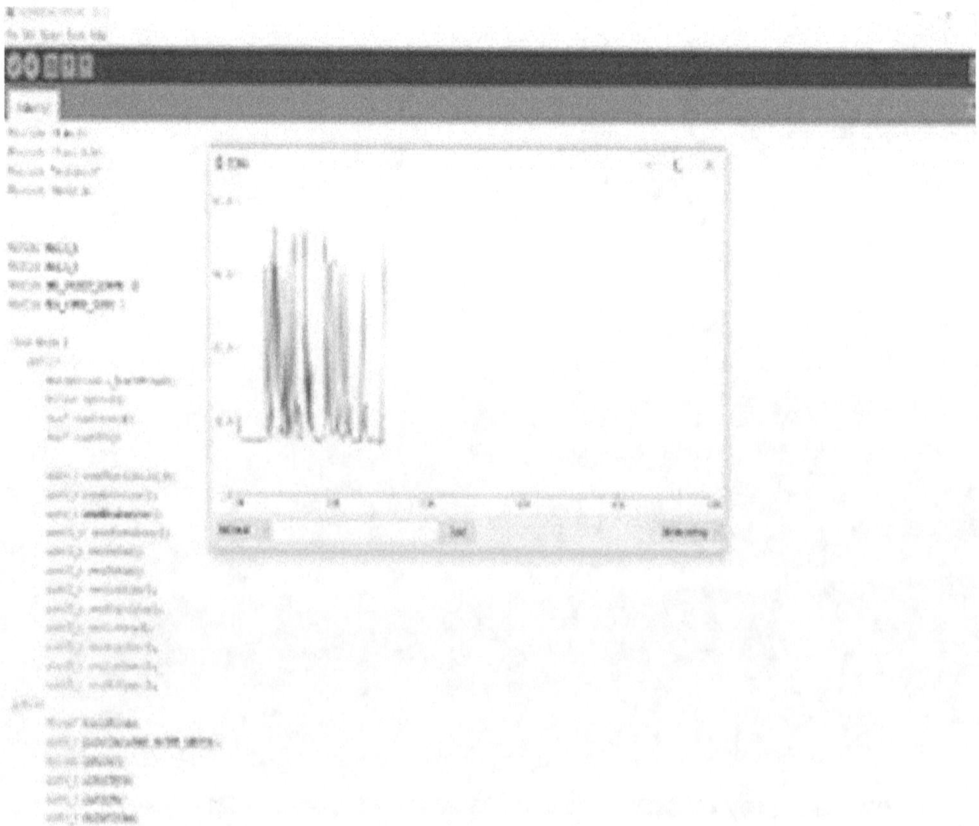

With the help of blue lines, the Serial Plotter displays the variations in waves and pulses as a graph so that it is simple to identify the variations and obtain precise results. The best way to predict readings might also be to use.

5. CONCLUSION

The goal of this proposed system is to design and create a small device that can benefit from an emergency response system and personal security system, both of which are beneficial to women in criminal situations. It is a low-cost system that can be used to transmit user information in a specific location and send out an immediate alert in the event of a crime against women. The system is effective because it has a cutting-edge sensor built in that emits waves in the form of informational data. So, even when women lose consciousness, the victim can still be found easily. The fear of harassment among women can also be dispelled by this system. This proposed system can be used as a tiny wearable device, such as a watch or a pendant, with additional research and innovation. Women are given security by this. In the modern world, people demand to feel safe and secure.

REFERENCES

Ansari, A., Pratiksha, B., Tejal, M., & Yelmame, M. (2017). Women Security System using GSM & GPS. *IJIRSET, 6*(3).

Bhardwaj, N., & Aggarwal, N. (2014). Design development of "Suraksha". A women safety device. *Int J Inf Computer Technology 4*(8), 787–792.

Campbell, R., Ahrens, C. E., Sefi, T., Wasco, S. M., & Barner, H. E. (2001). Social reactions to rape victims: Healing and hurtfull effects on psychological and physical health outcomes. *Violence and Victims, 16*(3), 310–367. doi:10.1891/0886-6708.16.3.287 PMID:11437118

Chand, D., Nayak, S., Bhat, K. S., & Parikh, S. (2015). A mobile application for Women's Safety: WoS App. *IEEE Region Conference TENCON*. IEEE.

Descamps, M. J., Rothblum, E., Bradford, J., & Ryan, C. (2000). Mental health impact of child sexual abuse, rape, intimate partner violence, and hate crimes. National lesbian health care survey, 9(5).

Fan, D., Ruiz, L., & Gong, J. (2018). An Energy Harvesting Modeling and Profiling Platform for Sensor Networks. *IEEE Journal of biomedical conference in health informatics, 22*(1).

George, R., Anjaly Cherian, V., & Antony, A. (2014, April). An intelligent security system for violence against women in public places. *IJEAT, 3*(4), 64–68.

Gowri, S., & Anandha Mala, G. S. (2015, June). IR system for investigation in textual data. *Indian Journal of Science and Technology, 8*(12), 1–7.

Hussain, S. M., Nizamuddin, S. A., Asunico, R., Ramaiah, C., & Vikram Singh, A. (2016) Prototype of an intelligent system based on RFID and GPS technologies for women safety. In: *IEEE 5th international conference on reliability, infocom technologies and optimization (ICRITO), 5*. 10.1109/ICRITO.2016.7784986

Koss, M.P., Figuerdo, A.J., Bell, I., Tharan, S. (1996). Traumatic memory characteristics: a cross validate mediational model of response to rape among employed women. *Abnormal Psychol, 105*(3), 421–432.

Chapter 11
Women and Millets:
Historical Perspective

Shanmuga Sundari
https://orcid.org/0000-0001-5755-474X
BVRIT HYDERABAD College of Engineering for Women, India

Harshini Reddy Penthala
BVRIT HYDERABAD College of Engineering for Women, India

Akshita Mogullapalli
BVRIT HYDERABAD College of Engineering for Women, India

Vidyullatha Sukhavasi
BVRIT HYDERABAD College of Engineering for Women, India

Anand Nayyar
https://orcid.org/0000-0002-9821-6146
Duy Tan University, Vietnam

ABSTRACT

Millets are a group of small-seeded grasses that have been cultivated for thousands of years in different parts of the world. In many cultures, millets have been considered a staple food, and they continue to be a vital source of nutrition for numerous people today. In the context of women's history, millets have played a significant role in many societies. In India, millets have been a traditional food for centuries, and women have played a critical role in their cultivation and consumption. Historically, millets have also been an important source of income for women in many societies. Millets have also played a significant role in women's health. Millets are highly nutritious and excellent sources of protein, fiber, and vitamins. They are also gluten-free, making them a valuable food source for people with celiac disease or gluten intolerance. Millets are a nutritious and healthy food that offers a variety of health benefits. Incorporating millets into your diet can help improve overall health and well-being.

DOI: 10.4018/978-1-6684-9819-4.ch011

Copyright © 2023, IGI Global. Copying or distributing in print or electronic forms without written permission of IGI Global is prohibited.

1. INTRODUCTION

1.1 Definition of Millets

Millets (Gowda et al., 2022) are a group of small-seeded grasses that have been cultivated for thousands of years in various parts of the world, including India, China, Africa, and South America. These grains are highly nutritious and have been a vital source of food for people living in regions with poor soil and limited water resources. Millets are a rich source of nutrients, including fiber, protein, vitamins, and minerals. They are also highly drought-resistant and can grow in poor soil conditions, making them ideal for cultivation in arid and semi-arid regions.

Millets can be broadly classified into two categories: major millets and minor millets. Major millets include sorghum, pearl millet, finger millet, and foxtail millet, while minor millets include kodo millet, little millet, barnyard millet, and proso millet. Each of these millet varieties has its own unique nutritional properties and cultivation requirements.

Figure 1. Definition of millets

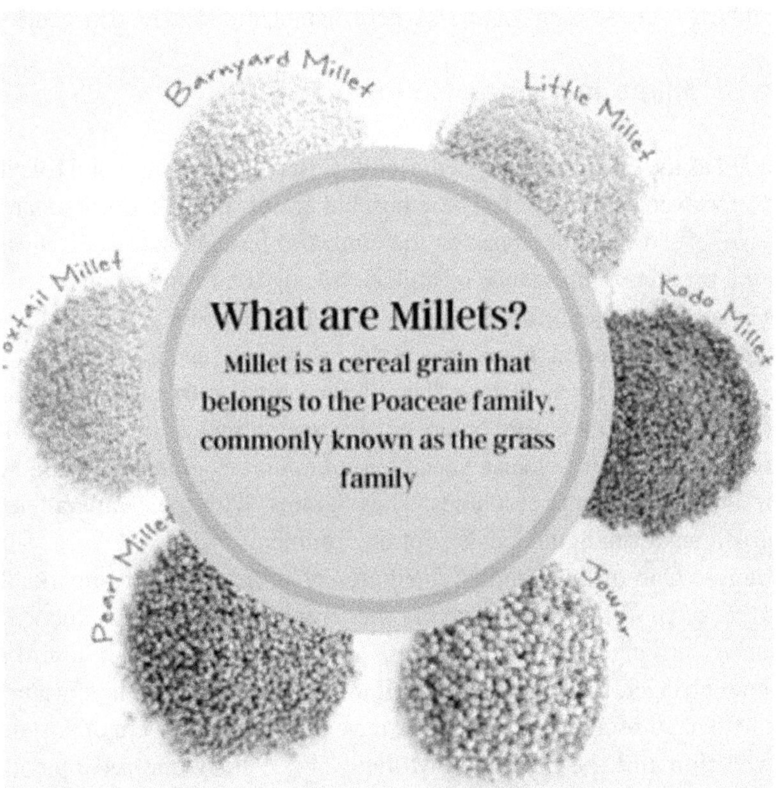

Major millets, such as sorghum and pearl millet, are some of the most widely grown and consumed millets in the world. Sorghum is an important crop in many African countries and is used to make porridge, bread, and beer. Pearl millet is a staple food in many parts of India and is used to make roti, a type of flatbread. Finger millet is another major millet that is commonly grown in India and is used to make porridge and flour for baking.

Minor millets, such as kodo millet (Yousaf et al., 2021), little millet, barnyard millet, and proso millet, are less commonly grown and consumed but are still an major source of nutrition (Samtiya, Aluko, Dhaka, Dhewa, & Puniya, 2023) for many people. Kodo millet, for example, is a staple food in some parts of India and is rich in protein and fiber. Little millet is another minor millet that is commonly grown in India and is used to make porridge and snacks.

Despite their nutritional value and resilience, the popularity of milo has declined in recent years due to the rise of modern agricultural practices and the introduction of other grains. The Green Revolution, which began in the 1960s, led to the widespread adoption of high-yielding varieties of rice and wheat, which are now the primary crops grown in many parts of the world. This shift away from traditional crops like millets has led to a decrease in millet cultivation and intake, which has had negative consequences for both human health and the environment.

Recently, there has been a renewed popularity in millets as a sustainable and nutritious food source. Millets are gaining popularity as a healthy alternative to refined grains like wheat and rice. They are also being recognized for their potential to promote food security and sustainable agriculture in regions with poor soil and limited water resources. As such, efforts are being made to promote millet cultivation and consumption, particularly in developing countries where malnutrition and food insecurity are major issues.

1.2 Importance of Millets in Human History

Millets, often regarded as the unsung heroes of agriculture, have played a pivotal role in human civilization (Li et al., 2020) for over 10,000 years. These humble grains, initially domesticated in the cradle of Asia and Africa, have left an indelible mark on the annals of human history. This essay embarks on a journey to unravel the profound importance of millets, tracing their roots from the dawn of agriculture to their resurgence as modern-day nutritional and environmental saviors.

An Ancient Staple: Millets found themselves at the heart of ancient civilizations, gracing the tables of societies such as the Indus Valley and China. Their widespread cultivation and consumption were not confined to these regions alone; they were embraced across continents. In Africa, millets emerged as a lifeline, serving as a primary source of sustenance for communities grappling with the twin challenges of poor soil quality and limited water resources. These grains, with their unparalleled resilience, held the promise of nourishment even in the harshest of environments.

Nature's Resilience: One of the standout attributes of millets is their remarkable adaptability to adverse conditions that often stymie the growth of other crops. Millets are the embodiment of resilience and drought resistance, thriving where others falter. Their ability to flourish in arid and semi-arid regions has been a game-changer, reducing the agricultural burden in areas where water is scarce and soil quality is wanting. Millets, through their hardiness, have heralded a new era of sustainable agriculture.

The Green Revolution and the Decline of Millets: The 1960s witnessed a paradigm shift in agriculture with the advent of the Green Revolution. While this agricultural transformation brought about a surge in productivity, it also marked the beginning of a decline in millet cultivation, particularly in India. High-yielding varieties of rice and wheat took center stage, luring farmers with promises of greater profitability. As a consequence, the once-ubiquitous millets saw their popularity wane.

The Resurgence of Millets: However, the story of millets does not conclude with their decline. In recent years, millets have been experiencing a renaissance, propelled by a renewed appreciation for their nutritional prowess and environmental sustainability. These grains, often touted as "nutri-cereals," have emerged as a treasure trove of protein, fiber, vitamins, and minerals. They stand as a beacon of hope for individuals with dietary restrictions, being naturally low in fat and gluten-free.

Environmental Stewards: Beyond their nutritional virtues, millets champion environmental stewardship. Their ecological footprint is significantly lighter compared to other crops, demanding less water and fewer inputs. In an era where environmental concerns loom large, millets offer a sustainable agricultural alternative that resonates with farmers and consumers alike.

Recently, there has been a renewed growth in millets due to their nutritional benefits and environmental sustainability. Millets are a rich source of protein, fiber, vitamins, and minerals. They are also low in fat and gluten-free, making them a popular for people with dietary restrictions.

Moreover, millets have a lower carbon footprint than other crops, as they require less water and inputs to grow. As a result, millets are considered to be an environmentally sustainable crop, making them an attractive option for farmers and consumers concerned about the impact of agriculture on the environment.

Figure 2. Agriculture and nutritional benefits of grains

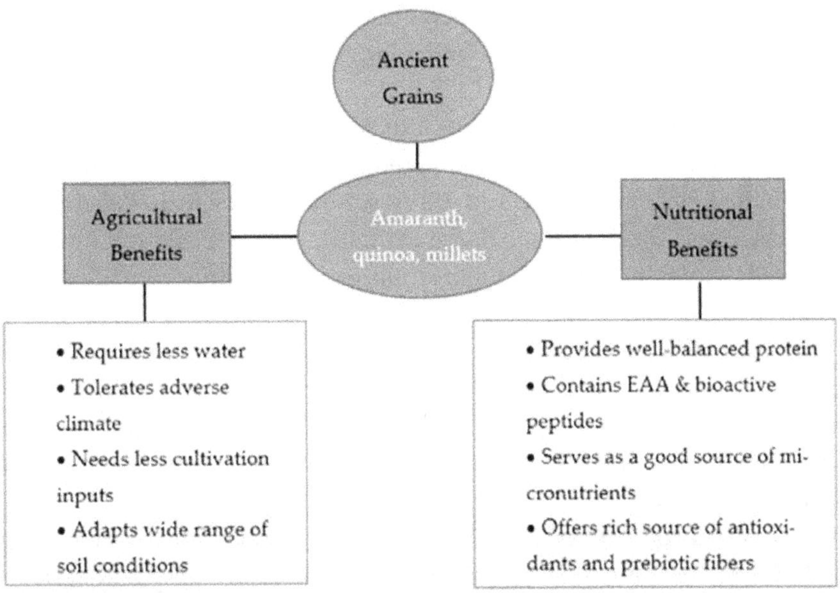

1.3 Women's Role in Millet Cultivation and Processing

The role of women in millet cultivation and processing has been crucial for centuries. In many societies, millet farming and processing (He et al., 2015) have been considered women's work. Women are responsible for planting, harvesting, processing, and storing millets. They have also played a vital role in conserving the diverse varieties of millets.

Women's role in millet cultivation and processing is often undervalued and overlooked, leading to their exclusion from policy-making and development programs. This exclusion limits women's reach to utilities, such as credit, land, and technology, and hinders their ability to produce and sell millets.

Consequently, women farmers face a disproportionate impact of climate change, food insecurity, and malnutrition.

Despite these challenges, women farmers have been at the forefront of promoting sustainable agriculture and preserving traditional farming practices. Women have developed innovative and sustainable farming techniques that have helped to increase millet yields, improve clay vitamins, and decreased the negative effects of climate swings. Their knowledge and skills have been critical in promoting the consumption of millets as a healthy and nutritious food.

In many societies, women are also responsible for processing millets. Women use traditional techniques to de-hull and grind millets, making them into flour or other food products. They have a deep understanding of the various types of millets and their nutritional value. Women's knowledge of millet processing is passed down from generation to generation, preserving traditional techniques and knowledge.

2. MILLET CULTIVATION

Millet cultivation stands as an enduring chapter in the epic saga of human agriculture. Its roots dig deep into the past, with origins stretching back to the Neolithic era, a remarkable 10,000 years ago. In this narrative, we delve into the fascinating tale of millet farming, tracing its journey from humble beginnings to global significance.

Agricultural Antiquity: Millet's journey as a cultivated crop commenced during the Neolithic era, marking a pivotal moment in human history. It was among the first crops to be tamed by human hands, its seeds sown in different corners of the world. Early agriculturalists recognized the unparalleled resilience of millets, harnessing their ability to thrive in regions where fertile soil was a rarity, and water resources were limited. The robustness of millets transformed them into an indispensable source of sustenance for our forebears.

The Global Spread of Millets: As humans traversed the globe, they carried with them the gift of millet cultivation. The craft of growing millets became a global phenomenon, with distinct varieties adapting to the unique needs of diverse regions. From the sun-soaked fields of India to the terraced landscapes of China, from the arid stretches of Africa to the fertile soils of South America, millets found a home in the hearts and soils of farmers across the world.

The Advent of Modern Agriculture: The 1960s marked a transformative moment in agriculture with the onset of the Green Revolution. This revolutionary era brought high-yielding varieties of rice and wheat to the forefront, promising unprecedented agricultural productivity. Millets, once paramount, began to recede into the background as these new cereals took center stage. The decline of millet cultivation came at a cost, erasing generations of traditional knowledge and practices linked to these resilient grains.

Millet Farming: A Regional Symphony: The techniques employed in millet farming are a harmonious blend of tradition and region-specific adaptations. Traditional millet farming, a practice rooted in sustainability, often involves minimal tillage and the use of organic fertilizers. Crop rotation, a time-honored technique, sees millets paired with legumes to enhance soil fertility. Rain-fed agriculture, prevalent in some regions, relies on nature's blessings, while others employ irrigation to nurture these hardy grains. Traditional millet farming practices epitomize the essence of sustainable agriculture, working in harmony with nature rather than against it.

Millet farming, an ancient practice that has weathered the storms of time, reminds us of the profound resilience of both the crop and the human spirit. Its resurgence in the modern era, fueled by an understanding of its nutritional and ecological value, heralds a promising future where tradition and innovation coexist in harmony.

2.1 Origins of Millet Cultivation

Millet is one of the oldest cereal grains cultivated by humans. Its origins are believed to be in Asia and Africa, where it was first domesticated over 7000 years ago. Millet has been a staple crop in many cultures throughout history, and it continues to be a vital food source for millions people today.

The cultivation of millets began during the Neolithic period, which marked the beginning of human civilization. During this period, people were transitioning from hunting and gathering to farming. Millets were among the first crops to be domesticated in dry regions of Asia and Africa, where they were grown in areas with limited rainfall and poor soil quality. Millets were the perfect crop for these conditions, as they were hardy and could survive in regions where other crops failed to thrive.

One of the earliest known millet crops was found in a village in northern China, where scientists discovered evidence of millet cultivation dating back to 7000 BCE. The crop was likely grown for food, fuel, and building materials. Millets were also grown by early civilizations in India, Egypt, Greece, and Rome. In ancient India, millets were a principle food, particularly in the arid regions of Rajasthan and Gujarat. In Egypt, millets were used to make bread, while in Greece and Rome, millets were primarily used as animal feed.

The spread of millet cultivation can be attributed to several factors, including trade and migration. Millets were introduced to Europe during the Bronze Age, and from there, they spread to the rest of the continent. The spread of millets was likely due to the movement of people and the exchange of goods along trade routes. Millets were also introduced to South America by Spanish colonizers during the 16th century.

Millets have played in development of cultivation and human civilization. They were the first crops to be domesticated in dry regions, and their hardiness allowed them to be cultivated in areas where other crops could not grow. Millets were an essential source of food, fuel, and building materials, and they played a vital role in the development of early civilizations. The widespread cultivation of millets is a testament to their adaptability and hardiness in diverse growing conditions.

2.2 Spread of Millet Cultivation

Millet, a versatile and hardy cereal grain, has made an extraordinary journey across the world, adapting to diverse climates and establishing itself as a vital crop for many communities. Its global spread is a testament to its remarkable resilience and significance in addressing food security challenges.

2.2.1 Adaptability: Millets in Varied Climates

One of the most remarkable features of millet cultivation is its adaptability to a wide range of environmental conditions. Millets can thrive in regions characterized by arid, hot, and even clayey soils, where many other crops struggle to survive. This unique trait has made millet a lifeline for farmers in some of the world's most challenging agricultural landscapes.

2.2.2 India: The Millet Giant

India, with its rich agricultural heritage, is a prominent contributor to global millet production. The nation's millet output accounts for approximately 40% of the world's total production. Among the various millet varieties cultivated in India, pearl millet, finger millet, and foxtail millet stand out as the most commonly grown. These millets play a pivotal role in the dietary fabric of the country.

Pearl millet, in particular, has found a niche in arid regions such as Rajasthan, Gujarat, and Haryana. Its capacity to thrive in water-scarce conditions and withstand high temperatures has made it a staple crop for millions in these regions. Beyond human consumption, millets are also utilized as fodder for livestock, contributing to the overall sustenance of rural communities.

2.2.3 China: A Millet Heritage

In China, millets have enjoyed a revered status in culinary traditions for millennia. Foxtail millet and proso millet are two of the most commonly cultivated varieties in the country. These grains serve as the foundation for a range of traditional dishes, including porridge and noodles. The deep-rooted cultural significance of millets in Chinese cuisine highlights their enduring role in the nation's food culture.

2.2.4 Millets Across Asia

The influence of millets extends beyond India and China into other parts of Asia. Nations like Nepal, Thailand, and Vietnam also engage in millet cultivation. These countries appreciate the crop's adaptability and nutritional value, contributing to regional food security.

2.2.5 African Millet Belt: Sahel Region

The Sahel region in Africa, comprising countries such as Niger, Nigeria, and Mali, depends heavily on millets as an essential food crop. Varieties like pearl millet and sorghum play a fundamental role in the creation of staple dishes like porridge, couscous, and even beer. In East Africa, finger millet has gained prominence, used to prepare porridge, bread, and various other food items.

2.2.6 Millets in South America: A Colonial Legacy

The introduction of millet to South America can be traced back to the colonial era when Spanish colonizers brought these grains to the continent. Today, countries like Brazil and Argentina continue to cultivate millets, although their scale of production remains smaller compared to Asian and African nations.

Figure 3. Types of millets

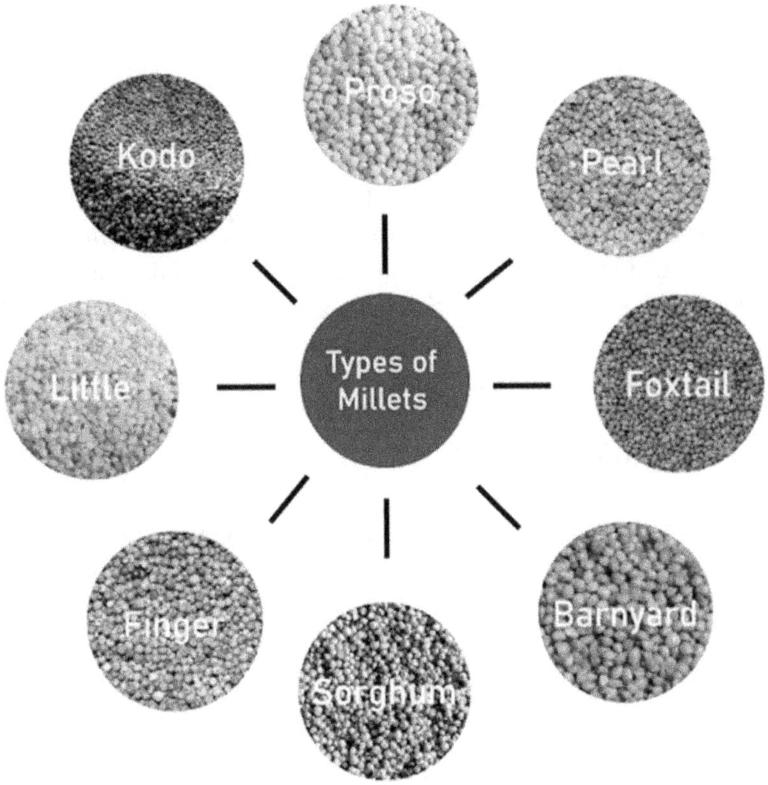

2.3 Millet Varieties and Their Nutritional Value

Millet varieties are important for their unique nutritional content, which can provide a variety of health benefits. Here are some of the major and minor millets and their nutritional values:

- Sorghum: Sorghum is a major millet that is known for its rich antioxidant content. It is also a fount of fiber, protein, and minerals such as iron, magnesium, and phosphorus. Sorghum is commonly used to make flour, porridge, and beer.
- Pearl millet: Pearl millet is a major millet that is rich in iron, calcium, and protein. It is also a good source of fiber and has a low glycemic index, which means it can help regulate blood sugar levels. Pearl millet is used to make various food products such as bread, porridge, and couscous.
- Finger millet: Finger millet is a major millet that is rich in protein, fiber, and minerals such as iron and calcium. It is also gluten-free and has a low glycemic index, which makes it a good food for people with diabetes. Finger millet is used to make various food products such as porridge, bread, and cookies.
- Foxtail millet: Foxtail millet is a major millet that is a good source of carbohydrates, protein, and fiber. It is also rich in minerals such as iron, magnesium, and phosphorus. Foxtail millet is used to make porridge, noodles, and other food products.

- Kodo millet: Kodo millet is a minor millet that is a good source of protein, fiber, and minerals such as iron and calcium. It is also gluten-free and has a low glycemic index. Kodo millet is used to make various food products such as porridge, bread, and dosa.
- Little millet: Little millet is a minor millet that is a good source of protein, fiber, and minerals such as iron and calcium. It is also gluten-free and has a low glycemic index. Little millet is used to make various food products such as porridge, bread, and dosa.
- Barnyard millet: Barnyard millet is a minor millet that is a good source of protein, fiber, and minerals such as iron and calcium. It is also gluten-free and has a low glycemic index. Barnyard millet is used to make various food products such as porridge, bread, and dosa.
- Proso millet: Proso millet is a minor millet that is a fount of protein, fiber, and iron and magnesium. It is also gluten-free and has a low glycemic index. Proso millet is used to make various food products such as porridge and flatbread.

2.4 Millet Farming Techniques

Millet, known for its adaptability and rich nutritional content, has been cultivated for thousands of years across diverse regions of the world. Millet farming techniques have evolved in response to local conditions, climates, and the specific type of millet being grown. This adaptability has contributed to millets' status as essential crops for communities globally. In this comprehensive exploration, we will delve into various millet farming techniques, from intercropping to conservation agriculture, highlighting their importance and impacts on food security and sustainable agriculture.

2.4.1 Intercropping: Enhancing Biodiversity and Yields

Intercropping, a widely adopted technique in millet farming, involves planting two or more crops in the same field simultaneously. This strategy offers several benefits to millet farmers. One of the most common intercropping practices is combining millets with leguminous crops such as beans, peas, or lentils.

The symbiotic relationship between millets and legumes is advantageous for several reasons. Legumes are nitrogen-fixing plants (Bangar et al., 2021) that enhance soil fertility by converting atmospheric nitrogen into a form that can be used by plants. This natural nitrogen fixation reduces the need for synthetic fertilizers in millet cultivation. Moreover, legumes provide an additional source of protein, both for human consumption and as livestock fodder.

Intercropping also contributes to biodiversity and pest management. The diversity of plant species within a field can deter specific pests that target a single crop. In essence, intercropping minimizes the risk of total crop failure due to pest infestations.

2.4.2 Crop Rotation: Nurturing Soil Health

Crop rotation is another crucial millet farming technique that promotes soil health and fertility (Li et al., 2023). It involves the systematic alternation of different crops in the same field during successive growing seasons. The practice of crop rotation offers several advantages in millet cultivation.

Firstly, crop rotation breaks the cycle of pests and diseases. Different crops have distinct nutrient requirements and vulnerabilities, which can disrupt the life cycles of pests and pathogens that target a particular crop. This minimizes the need for chemical pesticides and fosters more sustainable pest management.

Secondly, crop rotation enhances soil fertility by optimizing nutrient utilization. Certain crops are known as nutrient fixers; for instance, legumes like soybeans fix nitrogen in the soil. When these crops are rotated with millets, they enrich the soil with essential nutrients, reducing the dependence on synthetic fertilizers.

2.4.3 Mixed Cropping: Maximizing Yield and Minimizing Risk

Mixed cropping, a technique where different crops are planted together in the same field, contributes to improved millet yields and reduced risk. The concept behind mixed cropping is to maximize resource utilization by planting crops with complementary growth patterns and nutrient requirements.

This technique diversifies the food production system and minimizes the vulnerability of crops to adverse weather conditions or pest infestations. If one crop suffers due to unfavorable conditions, the other may thrive, ensuring a more reliable overall yield. Furthermore, mixed cropping minimizes soil erosion by maintaining ground cover throughout the growing season.

Figure 4. Farmers seed system

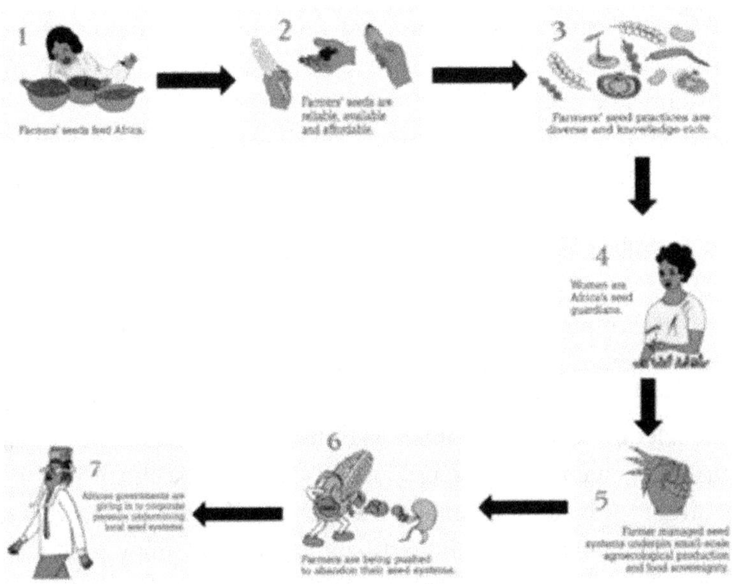

2.4.4 Mulching: Conserving Moisture and Soil Health

Mulching is a vital millet farming technique employed to conserve moisture, reduce soil erosion, and enhance soil health. It involves covering the soil's surface with organic or inorganic materials, such as straw, leaves, or plastic.

Mulch acts as a protective barrier against soil moisture loss through evaporation, especially in arid or semi-arid regions where water is scarce. Conserving soil moisture is crucial for millet cultivation, as these grains are renowned for their ability to thrive in dry conditions.

In addition to moisture conservation, mulch provides multiple benefits for soil health. It reduces temperature fluctuations in the soil, promoting microbial activity and nutrient cycling. Organic mulch materials decompose over time, adding valuable organic matter to the soil, which enhances its structure and fertility.

2.4.5 Conservation Agriculture: Preserving Soil Integrity

Conservation agriculture is a comprehensive approach aimed at minimizing soil disturbance and maximizing soil (Devi et al., 2023) cover. This technique seeks to promote soil health, reduce erosion, and optimize crop yields. Conservation agriculture is particularly relevant to millet farming, given the grain's preference for semi-arid and arid regions where soil conservation is crucial.

The core principles of conservation agriculture include minimal or no tillage, maintenance of crop residues as soil cover, and crop rotation or diversification. These practices collectively foster a conducive environment for millet cultivation.

Minimum Tillage or No-Till Farming: Minimal soil disturbance reduces erosion risk, minimizes nutrient loss, and maintains soil structure. By not plowing or tilling the soil extensively, conservation agriculture helps preserve its integrity.

Maintenance of Crop Residues: Retaining crop residues on the field's surface as mulch or cover crop materials contributes to soil protection. These residues shield the soil from erosive forces, maintain moisture, and improve organic matter content.

Crop Diversification: Integrating a variety of crops into the rotation or simultaneously cultivating different species in the same field enhances overall biodiversity, soil health, and pest management.

2.4.6 Innovations in Millet Farming: Modern Approaches

Beyond traditional millet farming techniques, contemporary agricultural practices and technologies are emerging to further optimize millet production. These innovations encompass improved seed varieties, precision agriculture, irrigation systems, and integrated pest management strategies.

Improved Seed Varieties: The development and dissemination of high-yielding and pest-resistant millet varieties contribute to increased crop productivity and food security.

Precision Agriculture: Advanced technologies such as GPS-guided tractors and drones are being leveraged for precise planting, irrigation, and fertilizer application, improving resource efficiency.

Irrigation Systems: In regions with access to water resources, irrigation systems, both traditional and modern, are employed to ensure consistent crop growth, especially during dry spells.

Integrated Pest Management: Combining biological, chemical, and cultural pest control methods allows farmers to manage pests effectively while minimizing environmental impact.

In conclusion, millet farming techniques have evolved over time, showcasing the adaptability and resilience of these grains. Intercropping, crop rotation, mixed cropping, mulching, and conservation agriculture are fundamental approaches employed by farmers worldwide. These techniques not only enhance millet yields but also promote sustainable and environmentally friendly agriculture. Moreover, innovations in millet farming are paving the way for increased productivity and food security, ensuring that millets continue to be a valuable source of nutrition for communities across the globe.

Figure 5. Ancient millets processing

Figure 6.

Figure 7.

3. WOMEN AND MILLET CULTIVATION

Across the globe, millet cultivation stands as a vital thread woven into the cultural fabric of numerous societies. However, the true strength of this agricultural practice lies not only in the crops themselves but in the nurturing hands of women who have been the unsung heroes of millet farming for generations. In this exploration, we shed light on the pivotal role women play in the cultivation, processing, and preservation of millets, underscoring their profound knowledge, skills, and contributions.

In diverse corners of the world, women have borne the responsibility of sowing the seeds of millet farming. This multifaceted role begins with the selection of quality millet seeds, a task that requires a keen understanding of varietal traits and local climate conditions. Women bring their deep knowledge of soil, weather patterns, and traditional agricultural practices to ensure a bountiful harvest.

The journey of millet cultivation traverses the entire agricultural cycle (Singh et al., 2023), with women at the forefront of each stage. The planting and nurturing of millet crops demand diligence and care, as women meticulously tend to their fields. As the millet matures, women are responsible for the vital task of harvesting. Armed with sharp sickles and an intimate knowledge of when each crop is at its peak, they ensure that not a single grain is lost to premature harvesting or overripeness.

Millet's journey from field to plate is a testament to the skill and expertise of women. Post-harvest, women are entrusted with the intricate task of threshing, winnowing, and processing millets. Threshing involves the separation of grains from stalks, a process that requires precision to avoid damage. Winnowing, the art of separating chaff from grains, showcases women's dexterity as they toss the mixture into the air, letting the wind carry away unwanted debris. Their contribution continues as they meticulously clean, grind, and store millets, ensuring the family's sustenance throughout the year.

Beyond mere agricultural endeavors, women play a pivotal role in preserving the rich cultural heritage intertwined with millet farming. Many employ traditional processing techniques that have been passed down through generations. These time-honored practices are not merely a means to an end but a celebration of cultural continuity, forging a deep connection with the past.

Women's involvement in millet seed preservation extends to the crucial practice of seed exchange within and between communities. Seed exchange has been a cornerstone of maintaining and promoting local crop diversity for centuries. Women, often the primary actors in this practice, facilitate the resilience of their communities by preserving traditional seed varieties and sharing them with others.

In this section, we will explore the role of women in millet cultivation in different parts of sphere, their knowledge and skills in millet farming, and their contribution to millet seed preservation and exchange.

3.1 Women's Involvement in Millet Cultivation in Different Regions of the World

Millet is a principal crop. It is an essential fount of food and income for millions of people in these regions, and women have played a significant role in millet cultivation. In many communities, women are responsible for all aspects of millet farming, from land preparation and planting to harvesting and processing.

In Africa, women's involvement in millet cultivation is widespread. They are responsible for preparing the land by clearing the bush, digging holes for planting, and weeding the fields. Women also plant millet seeds, which they have either saved from the previous harvest or obtained from other sources. They use traditional techniques such as intercropping and crop rotation to maximize yields (Kumari et al., 2024) and improve soil fertility. Additionally, women are responsible for harvesting the crop, which involves cutting the stalks and separating the grain from the chaff.

After harvesting, women process the millet into flour, which is used to make porridge, bread, and other food products. Millet flour is a staple food in many African communities, and women's knowledge of traditional processing techniques such as threshing and winnowing is essential in separating the grain from the chaff. Women also have knowledge of the appropriate millet processing techniques that help to maintain the nutritional value of the crop. For example, in some communities, women remove the breakfast food and germ layers of the milos to produce white flour, which has a longer shelf life but is less nutritious than whole-grain flour.

In Asia, women have also played a significant role in millet cultivation, particularly in India. Women in India are involved in all aspects of millet farming, from land preparation and planting to harvesting and processing. They use traditional techniques such as intercropping and crop rotation to maximize yields and improve soil fertility. Women in India also process millets into flour, which is involved to make various foods namely roti, dosa, and idli.

The involvement of women in millet cultivation is not limited to traditional methods. Women in many regions have also adopted modern technologies that have helped to improve their yields and reduce post-harvest losses. For example, in some communities, women use improved storage methods such as hermetic bags to preserve millet grains and reduce pest damage.

Despite their significant contributions to millet farming, women face various challenges that limit their participation in the sector. These challenges include limited access to credit, land, and technology. In many communities, women do not own land and are therefore unable to make decisions about what crops to grow or how to use the land.

Figure 8. Women in millets cultivation

To communicate on these provocations and support gentle sex's involvement in millet cultivation, various initiatives have been implemented. For example, the International Crops Research Institute for the Semi-Arid Tropics (ICRISAT) has made sex-inclusive seed systems, which promote women's contribution in seed proffering, processing, and marketing. The initiative also provides training and capacity building for women in millet farming techniques, entrepreneurship, and marketing.

In India, the Self-Employed Women's Association (SEWA) has been instrumental in empowering women in millet farming. The organization provides training on sustainable farming practices, seed production, and processing. SEWA also offers financial and technical support for women to start their own millet-based businesses, such as flour mills and food processing units. Through these initiatives, women have been able to increase their income and improve their livelihoods.

3.2 Women's Knowledge and Skills in Millet Farming

Millet is a resilient crop that is well-suited for growing and farming in arid and semi-arid regions. It is a foremost nutrition in many parts of Africa and Asia. Millet is also becoming increasingly popular in Western countries, where it is recognized as a gluten-free and nutrient-dense alternative to wheat and other grains (Samtiya, Aluko, Dhaka, Dhewa, & Puniya, 2023).

Women have played a vital role in the cultivation of millet for centuries. They have developed a wealth of knowledge and skills that are passed down from generation to generation, ensuring the continued success of millet farming. This knowledge and skills encompass all aspects of millet cultivation, from land preparation to processing and marketing.

Land Preparation: Women have a deep understanding of the land and the best practices for preparing it for millet cultivation. They know which crops to plant alongside millet to improve soil fertility, such as legumes and root crops. They also know which weeds to remove and which to leave in place, as some weeds can actually help to enrich the soil. Women are also skilled in soil conservation techniques, such as terracing and contour plowing, which can help to prevent erosion and preserve soil moisture.

Seed Selection: Choosing the right seeds is essential for successful millet cultivation, and women have a vast knowledge of the best millet seed varieties for different soil types and weather conditions. They also know how to select seeds that are disease-free and have high germination rates. Women also know how to store seeds properly, ensuring that they remain viable for future planting.

Planting: Planting millet is a delicate process that requires careful attention to detail. Women know the ideal seed depth, spacing, and planting technique for different millet varieties. They also know how to protect the seedlings from pests and diseases, such as by applying natural pesticides and fertilizers. Women also understand the importance of timing, planting millet at the optimal time to ensure maximum growth and yield.

Weeding: Weeds can has a footprint on millet cultivation, and women are skilled in the best weed control techniques. They know how to hoe and hand-weed effectively, removing weeds without damaging the millet plants. Women also understand the timing of weeding, as weeds need to be removed before they compete with the millet plants for nutrients and water.

Harvesting: Harvesting millet is a labor-intensive process that requires skill and precision (Karpagapandi et al., 2023). Women know the ideal time for harvesting millets, which is essential in ensuring high yields. They also understand the best harvesting techniques, such as cutting the stalks or pulling them out of the ground. Women are also skilled in threshing and winnowing, separating the grain from the chaff to prepare it for processing.

Processing: Millet grains have a hard outer layer that needs to be removed before consumption, and women are skilled in traditional techniques such as pounding or grinding the grains. They use a mortar and pestle or a mechanical grinder to unfasten the air exposed layer, making the grains worthy for rustle up and consumption. After the grains are ground, they need to be stored properly to avoid spoilage, and women have a wealth of knowledge on traditional storage techniques such as using woven baskets, clay pots, or other containers that can protect the grains from pests and moisture.

Marketing and Selling: In many regions of the world, women are responsible for selling the millet crops in local markets or to traders. They often have specialized knowledge of market prices and trends, as well as the preferences of local consumers. Women also have knowledge of the nutritional value of different millet varieties, which can help them to promotemillet consumption as a healthy and nutritious food choice.

Millet is an dominant assortment for many farmers, especially in Africa and Asia, where it is a staple food. Millet farming is primarily done by women, who have a significant amount of knowledge and skills in millet cultivation. In addition to their role in farming, women are into the marketing and selling of millet crops. Women are responsible for selling the millet crops in local markets or to traders, which requires specialized knowledge of market prices and trends, as well as the preferences of local consumers.

Women farmers often face numerous challenges when selling their crops. These challenges include lack of access to markets, inadequate storage facilities, and limited bargaining power. To overcome these challenges, women have developed various strategies to sell their crops. One of the strategies is to form cooperatives or groups that can pool their resources and sell their crops collectively. This strategy not only increases their bargaining power but also allows them to access markets that they would not have been able to access individually.

Women also use their specialized knowledge of market prices and trends to negotiate better prices for their crops. They often keep up-to-date with the latest market prices and can quickly respond to changes in demand. For instance, if there is a sudden increase in demand for millet, women farmers can quickly increase their production to take advantage of the high prices. They can also negotiate with traders to get a fair price for their crops.

In addition to their knowledge of market prices and trends, women also have knowledge of the dietary practicality of different juar varieties. This knowledge is essential in promoting millet consumption as a healthy and nutritious food choice. Millet is a rich in protein, fiber, and essential nutrients. It is also gluten-free, building it an excellent choice for homo sapiens with gluten intolerance. Women can use their knowledge of the nutritional value of millet to promote its consumption and increase demand for their crops.

To promote millet consumption, women often use various marketing strategies such as product branding, packaging, and labeling. They also participate in trade fairs and exhibitions to showcase their products and attract potential buyers. These marketing strategies help to increase the visibility of millet and make it more appealing to consumers.

3.3 Women's Role in Millet Seed Preservation and Exchange

Millet is a staple food crop that is widely grown and consumed in numerous regions, particularly in Africa and Asia. Millet farming is a complex process that requires a significant amount of knowledge and skills, ranging from land preparation and seed selection to harvesting and processing. Women play a vital role in millet cultivation, as they have a significant amount of knowledge and skills that have been passed down through generations. Women's involvement in millet farming is particularly important in regions where milo is a principle dietary crop and where traditional seedling options are localized to municipal growing conditions and climate (Sanodiya et al., 2023). In this essay, we will discuss the role of women in millet farming, focusing on their knowledge and skills in land preparation, seed selection, planting, weeding, harvesting, processing, and seed preservation and exchange.

Figure 9. Millets preservation by tribal women

Women's Knowledge and Skills in Millet Farming: Women have a significant amount of knowledge and skills in millet farming, which has been passed down through generations. In land preparation, women have knowledge of traditional techniques such as intercropping and crop rotation, which can improve soil fertility and maximize yields. Intercropping involves planting two or more crops in the same field, while crop rotation involves alternating the crops grown in the same field to reduce soil erosion and maintain soil fertility. Women also have knowledge of the ideal planting season and weather conditions for millet cultivation, which is essential in ensuring successful crop growth.

In seed selection, women have knowledge of the best millet seed varieties for different soil types and weather conditions. They also have knowledge of seed storage techniques and how to preserve seeds for future planting. Millet seed selection is critical in ensuring high yields and crop resilience. Women often have specialized knowledge of local seed varieties, such as their nutritional value, growth characteristics, and resistance to pests and diseases. They are responsible for selecting and saving seeds from the best-performing plants, which can help to maintain the genetic diversity of local millet varieties.

Once the outer layer of the millet grain is removed, the women proceed to clean the grains. Cleaning involves removing any foreign matter that may be mixed with the grain, such as stones, leaves, or dirt. Women use different methods to clean millet grains, depending on the region and the availability of resources. In some areas, they use a traditional winnowing basket to separate the grain from the chaff, while in other areas, they use a sieve or a mesh to remove any impurities. In some cases, they use water to wash the grains, which can help to remove any debris and improve the grain's quality.

After the grains are cleaned, they are ready for grinding. Pulverize is the citation of crushing the millet seedlings into a fine powder that can be utilized for cooking or baking. Women use different tools and methods for grinding millet, depending on the availability of resources and the amount of millet they need to process. In some regions, they use a mortar and pestle to grind the grains, while in other areas, they use a mechanical grinder or a mill. The use of mechanical grinders and mills can significantly reduce the time and labor required for grinding, making it easier for women to process larger quantities of millet.

Once the millet is ground, it is ready for consumption or storage. Women have knowledge of traditional storage techniques, such as using woven baskets, clay pots, or other containers that can protect the grains from pests and moisture. Proper storage is essential to avoid spoilage and maintain the quality of the millet. Women also have knowledge of the appropriate conditions for storing millet, such as storing it in a cool, dry place, away from direct sunlight or moisture.

Women also play an important role in exchanging millet seeds within and between communities. Seed exchange is a practice that has been used for centuries as a way to maintain and promote local crop diversity. Women are often the primary actors in seed exchange networks, as they have established relationships with other women in their communities and can share seeds through informal channels. This allows for the preservation and exchange of traditional seed varieties, which are adapted to local growing conditions and climate, and are often more resilient to pests, diseases, and weather variability than modern varieties.

Seed exchange is particularly important for millet cultivation because it is a crop that is mainly grown by smallholder farmers who may not have access to formal seed markets or improved seed varieties. In addition, millet is a crop that is often used for household consumption rather than for sale, which means that seed exchange is an essential means of ensuring food security in times of crop failure or other emergencies.

Women's role in millet seed exchange is not only important for promoting local crop diversity but also for empowering women themselves. Seed exchange networks provide women with opportunities to connect with other women in their communities, share knowledge and skills, and develop social and economic relationships. In some cases, seed exchange networks have also led to the formation of women's cooperatives, which provide women with a collective voice and enable them to negotiate better prices for their crops and other products.

However, the role of women in seed exchange is often overlooked or undervalued by policymakers and development organizations. In many cases, seed exchange is seen as a traditional practice that is not compatible with modern agricultural practices or formal seed markets. This has led to policies and programs that promote the use of modern seed varieties and discourage traditional seed exchange practices. Such policies and programs often fail to recognize the importance of traditional seed varieties and the role of women in seed exchange, and can have negative impacts on local crop diversity and food security.

To address these issues, it is important to recognize the importance of women in millet seed exchange and to support the development of seed exchange networks and women's cooperatives. This can be done through the provision of training and capacity-building programs that promote the storage and avail oneself of conventional seed varieties, as well as the establishment of policies and programs that recognize the importance of traditional seed exchange practices and support the development of women's cooperatives.

One of the key challenges faced by women are discrimination and unequal treatment in labor markets and other economic activities also limit women's participation in millet cultivation. Women often face lower wages and less favourable working conditions than men, which makes it difficult for them to earn a decent income from their labor. In addition, women are often responsible for unpaid care work such as cooking, cleaning, and caring for children and the elderly, which can limit their ability to participate in paid labor and other economic activities. Limited access to education and training is another constraint that limits women's participation in millet cultivation.

This is a crucial challenge that needs to be addressed to ensure that women farmers have access to the latest technologies and innovations that can help increase their productivity and income.To address the challenges facing women in millet production, there is a need for a comprehensive and integrated approach that involves all stakeholders. This approach should prioritize the following actions:

1. Investing in women's education and training in millet production: Governments and development organizations should prioritize providing women farmers with education and training opportunities in millet production, including access to new technologies, best practices, and market information.
2. Improving access to agricultural inputs and resources: Women farmers need access to high-quality seeds, fertilizers, and other agricultural inputs to improve their productivity and income. Governments and development organizations should invest in improving access to these resources for women farmers.
3. Strengthening women's thoughts about managerial processes: Women's participation in ruling issues related to agriculture and rural development should be promoted and strengthened. This will help to ensure that their voices and perspectives are taken into account in policies and programs that affect their lives.
4. Promoting market access and opportunities: Governments and development organizations should invest in improving market infrastructure, providing market information and training, and promoting and all related to markets.
5. Supporting climate change: Efforts to adapt to and alleviate the impacts of climate change on millet production should be prioritized, with a focus on building resilience and promoting sustainable agriculture practices.

4. MILLET PROCESSING

Millet is a highly nutritious grain that has been consumed for centuries across the world. Its popularity is due to its versatility in preparation, taste, and numerous health benefits. Millet processing involves several stages, starting with harvesting the grain and removing the husk before it can be eaten or processed into other forms. However, with the advent of modern technology, millet processing has undergone significant changes.

One modern method of millet processing involves the use of machines, such as threshers, winnowers, and milling machines. These machines make the processing of millet faster and more efficient. The harvested millet is threshed to separate the grains from the stalks, cleaned to remove impurities, and milled to remove the husk. The resulting flour can then be used to make various dishes, such as porridge, bread, and cakes.

Traditional methods of millet processing are still prevalent in many communities, modern methods have revolutionized the processing of millet. These methods are faster and more nutritious. Incorporating modern millet processing methods into traditional practices can help preserve the nutritional content of millet grains while promoting diversity in the diet.

Households are a common place where millet processing takes place. In many communities, millet is a women's crop, and women are primarily responsible for its processing. The harvested millet is threshed, cleaned, and milled in households, with women using traditional methods, such as pounding with a mortar and pestle or grinding with a stone mill. Households may also engage in collective millet processing activities, where women come together to process their millet harvests collectively.

Processing centers are another common place where millet processing takes place. These centers are usually located in rural areas and serve as collection points for millet farmers. At these centers, the millet is cleaned, sorted, and milled using modern processing equipment. The processed millet flour is then packaged and sold to wholesalers or retailers.

Mills are also a common place where millet processing takes place. These are large-scale processing facilities that use modern equipment to process millet. Mills may be located in urban or rural areas and serve both domestic and industrial markets. The processed millet flour may be used to produce various products, such as breakfast cereals, snacks, and beverages.

In many communities, millet processing is also done on a small scale, with individuals using portable processing equipment, such as hand-cranked mills or motorized mills. These portable mills are particularly useful in remote areas where access to large-scale processing facilities is limited.

4.1 Traditional Methods of Millet Processing

Millet is one among of the ancient cultivated crop, with a history that dates back over 10,000 years. The origins of millet cultivation are not precisely known, but it is believed to have originated in the African Sahel region, which extends from the Atlantic coast to the Red Sea. From Africa, millet cultivation spread to India, China, and other parts of the world.

The traditional methods of millet processing have evolved over time and vary across different regions. Millet processing methods have been shaped by factors such as local climate, soil, and available technology. Millet processing has been a vital aspect of many cultures and communities, with each region developing its unique methods and techniques.

In Africa, traditional millet processing methods involve harvesting the crop by hand and threshing the grains with wooden sticks or by trampling them with cattle. The threshed grains are then winnowed by throwing them in the air and allowing the wind to blow away the chaff. The cleaned grains are then pounded with a mortar and pestle to remove the outer layer and reduce them to a coarse flour. In many cultures, millet is a staple food, and the traditional millet processing methods have been passed down from generation to generation.

In India, traditional millet processing methods involve soaking the millet grains in water for several hours before draining them and spreading them on a mat or cloth to dry. Once the grains are dry, they are roasted on a flat griddle until they turn golden brown. The roasted grains are then ground into a fine flour using a stone mill or mortar and pestle.

In the US, millet processing methods vary depending on the intended use of the millet. For instance, millet used for birdseed is often harvested mechanically and cleaned with machines. However, millet used for human consumption may be processed using traditional methods.

Figure 10. Millets processing by women

Some US-based organizations are promoting traditional millet processing methods as a way to promote sustainable agriculture and preserve cultural heritage. For instance, the Manna Project International, a nonprofit organization that works in Latin America, has been promoting the use of traditional millet processing methods in Nicaragua as a way to promote food security and cultural heritage.

While millet is not a traditional crop in the United States, there is growing interest in using traditional millet processing methods to promote sustainable agriculture and cultural heritage.

All above, traditional millet processing methods have a rich history that reflects the diversity of cultures and regions where millet is cultivated. These methods have helped preserve the nutritional content of the millet grains and promote the cultural heritage of many communities. While modern processing techniques have replaced some traditional methods, many communities still use traditional methods to process millet, and these methods continue to are a crucial bit part in promoting nutrients security and cultural diversity.

Figure 11. Millets cleaning process

The following are the steps involved in traditional millet processing:

1. Harvesting: Millet is typically harvested when the grains are fully mature and the stalks have turned yellow or brown. The harvest is usually done by hand, and the millet stalks are cut close to the ground using sickles or other sharp tools.
2. Threshing: After harvesting, the next step in traditional millet processing is threshing, which involves separating the millet grains from the stalks. Traditional threshing methods involve spreading the harvested millet on the ground and beating it with sticks or threshing flails to remove the grains from the stalks.
3. Winnowing: Once the millet is threshed, it is then winnowed to remove the chaff and other impurities. Traditional winnowing involves throwing the threshed millet into the air, allowing the wind to carry away the lighter chaff and debris, leaving only the heavier millet grains.
4. Cleaning: After winnowing, the millet grains are cleaned to remove any remaining impurities. This is often done by hand, using sieves or other tools to sift through the grains and remove any debris.

5. Milling: The final step in traditional millet processing is milling, where the millet grains are ground into flour. In many communities, traditional milling involves pounding the seeds with a mortar and grinding them using a stone mill. The millet flour obtained can be used to make various dishes, such as porridge, bread, and cakes.

Ancient barley cultivation methods can be laborious and expensive to perform, but they help preserve the nutritional content of the millet grains. Millet is a rich source of vitamins, minerals, and dietary fiber, and consuming it in its whole grain form can help promote good health and prevent chronic diseases. Traditional millet processing methods also help preserve the cultural heritage of many communities.

Women often play a crucial role in traditional millet processing. In many communities, millet is considered a women's crop, and women are primarily responsible for its processing. Women have a wealth of knowledge and experience in traditional millet processing methods, and they pass down these skills from one generation to another. Women often engage in collective millet processing activities, where they come together to process their millet harvests collectively, and this helps promote social cohesion and unity.

4.2 Women's Role in Millet Processing

In many traditional millet cultivating communities, women stand as the unsung heroines of millet processing. This often-overlooked facet of agricultural life carries profound significance in ensuring food security, preserving cultural heritage, and empowering women. Within the pages that follow, we embark on a journey to uncover the critical role that women play in millet processing, shedding light on the laborious yet vital tasks they undertake.

The Harvesting Heroines: Across the diverse millet cultivating landscapes of Africa, women emerge as the primary harvesters. Armed with sharp knives or sickles, they deftly cut millet stalks, gathering them into bundles for the threshing process. The art of threshing, honed over generations, requires women to use sticks or employ methods like trampling to separate the grains from the chaff. This delicate and skillful chore lies at the heart of ensuring that the harvest is bountiful and that the grains are ready for further processing.

Winnowing Wisdom: Winnowing, another crucial step in millet processing, is a task that predominantly falls within the purview of women. With a rhythmic grace born of tradition, they toss the threshed grains into the air, allowing the wind to perform its part by carrying away the unwanted chaff. This age-old technique exemplifies the seamless harmony between nature and human effort, a synergy that women have mastered through time and practice.

The Grinding Guardians: In the Indian subcontinent, women take center stage in many aspects of millet processing. Beyond harvesting, they are entrusted with the arduous journey from grain to flour. This multifaceted process begins with soaking the millet grains in water, followed by the delicate art of drying and roasting. Finally, women are the keepers of the grinding stones or the mechanical grinders, transforming the roasted grains into the fine millet flour that will form the basis of countless meals.

Cultural Stewards: Beyond their indispensable role in food security, women's involvement in millet processing plays a pivotal role in cultural preservation. The techniques they employ have been passed down through generations, and by continuing these traditions, women become stewards of their cultural heritage. In a rapidly changing world, this bridge to the past becomes increasingly important in preserving the rich tapestry of cultural diversity.

Undervalued Contributions: Regrettably, the contribution of women in millet processing often remains undervalued and underrecognized. Their labor-intensive efforts, the culmination of years of experience, may not receive the acknowledgment it deserves. Moreover, many women lack access to resources or training opportunities that could enhance their millet processing techniques and overall productivity.

Empowerment Through Recognition: To harness the full potential of women's involvement in millet processing, recognition and support are paramount. Recognizing the pivotal role they play, not only in food security but also in preserving cultural heritage, is the first step toward empowerment. Access to resources, education, and training can enhance their skills, improve productivity, and elevate their status within their communities.

In this exploration of women's role in millet processing, we uncover a profound and often underappreciated dimension of agricultural life. Women, with their unwavering commitment and expertise, are the driving force behind the transformation of millet from field to table. Their empowerment is not just a matter of gender equality but a vital component in the sustenance of traditions, food security, and the continued flourishing of millet cultivation.

4.3 Nutritional Benefits of Traditional Millet Processing Techniques

Traditional millet processing techniques offer several nutritional benefits that are lost when modern processing methods are used. Millet is a worthwhile way to obtain minerals and vitamins, and dietary fiber, and traditional processing methods help preserve these nutrients.

One significant nutritional benefit of traditional millet processing methods is that they help retain the natural dietary fiber present in the grain. Dietary fiber is essential for digestive health and helps regulate blood sugar levels. Traditional processing methods such as pounding and grinding help preserve the fiber-rich outer layer of the grain, which is often removed in modern processing methods.

Traditional millet processing methods also help retain the natural vitamins and minerals present in the grain. Milo is a healthy rise of several essential B vitamins and vitamin E. These vitamins are vital for maintaining healthy skin, nerve function, and cardiovascular health. Millet is also rich in minerals such as iron, magnesium, and phosphorus, which are essential for bone health, muscle function, and energy production.

Figure 12. Millets processing methods and their applications

Benefits of millets:

1. Rich in nutrients: Millets are a good source of essential nutrients like vitamins, minerals, fiber, and protein. They are rich in calcium, iron, phosphorus, magnesium, and potassium.
2. Gluten-free: Millets are gluten-free, making them a good option for people with celiac disease or gluten intolerance.
3. Low glycemic index: Millets have a low glycemic index, which means they release glucose into the bloodstream slowly, keeping blood sugar levels stable. This makes millets a good option for people with diabetes.
4. Helps in weight loss: Millets are high in fiber, which helps to keep you feeling full for longer periods and reduces overeating. This can help in weight loss and weight management.
5. Good for heart health: Millets are rich in fiber, which can help reduce cholesterol levels and lower the risk of heart disease.
6. Helps in digestion: Millets are high in fiber, which promotes healthy digestion, prevents constipation, and reduces the risk of colon cancer.
7. Good for bone health: Millets are a good source of calcium, magnesium, and phosphorus, which are essential for strong bones and teeth.

5. MILLETS IN HISTORY

The history of millets is a journey through time, tracing the footsteps of these remarkable grains from their ancient origins to their pivotal role in nourishing some of the world's earliest civilizations. Millets, often overlooked in modern diets, have an enduring legacy deeply embedded in the annals of human history. This essay embarks on an exploration of millets' historical significance, highlighting their cultivation, utilization, and contributions to the development of early human societies.

- **Millet's Early Beginnings:** Millets, among the world's oldest cultivated grains, have left their indelible mark on the pages of history. These resilient grains were some of the first to be cultivated in prehistoric times, signifying their fundamental importance to early agricultural practices. As humans transitioned from nomadic hunting and gathering to settled farming communities, millets emerged as vital staples that sustained these burgeoning civilizations.
- **India's Indus Valley Civilization:** Millets played a pivotal role in the development of agriculture and civilization in the Indian subcontinent. Dating back to around 3300 BC, the Indus Valley Civilization is renowned for its agricultural advancements, and millets were at the forefront of this agrarian revolution. Grown alongside wheat and barley, millets emerged as a critical source of sustenance for the people of this ancient civilization.
- **Beyond Borders: Millets' Global Influence:** As history unfolded, millets transcended geographical boundaries, spreading their influence across regions and cultures. Their adaptability and ability to thrive in diverse climates made them invaluable crops in various parts of the world, including Africa, Asia, Europe, and South America.

In this exploration of millets' historical significance, we embark on a journey that spans millennia. From their humble beginnings as some of the earliest cultivated grains to their roles in nourishing the great civilizations of antiquity, millets have left an indelible imprint on human history. Beyond their historical legacy, millets continue to hold promise as resilient, nutritious grains with the potential to address contemporary challenges in agriculture and food security. In the pages that follow, we delve deeper into the multifaceted world of millets, unearthing their nutritional prowess, ecological resilience, and modern resurgence.

5.1. Millets in Ancient Civilizations

Millets are small-seeded grasses that have been used as a staple food in many ancient civilizations. These ancient civilizations recognized the nutritional value of millets and incorporated them into their daily diets. Millets are a source of protein, fiber, vitamins, and minerals, making them a healthy alternative to other grains. In this chapter, we will explore the historical use of millets in various ancient civilizations.

Millets in Ancient India: India is considered to be the birthplace of millets, where they have been cultivated for over 5000 years. Millets were used as a staple food in ancient India, particularly in the southern regions. The ancient Indian texts, such as the Vedas, mention the use of millets in various dishes. Millets were considered to be the food of the poor and were consumed by people of all castes. The most commonly cultivated millets in ancient India were pearl millet, finger millet, and foxtail millet.

In Ancient China: Millets were also used as a staple food in ancient China, where they were known as "the grain of the people." The most commonly cultivated millet in ancient China was foxtail millet.

Foxtail millet was used to make porridge, which was consumed by people of all classes. Millets were also used to make alcoholic beverages in ancient China.

In ancient Egypt: Millets were used as a staple food for both humans and animals. Millets were used to make bread and porridge, and they were also fed to animals such as camels and donkeys. The most commonly cultivated millet in ancient Egypt was pearl millet.

In Ancient Greece: Millets were also used in ancient Greece, where they were known as "millet bread." Millets were considered to be a food of the poor and were consumed by people of all classes. Millets were used to make a type of bread called millet bread, which was made by mixing millet flour with water and baking it in an oven.

In ancient Rome: Millets were known as "panicum" and were used to make a type of porridge called puls. Puls was consumed by both the rich and the poor and was often served with vegetables or meat. Millets were also used to make a type of bread called panis quadratus, which was made by mixing millet flour with water and baking it in an oven.

Figure 13. Nutritional benefits

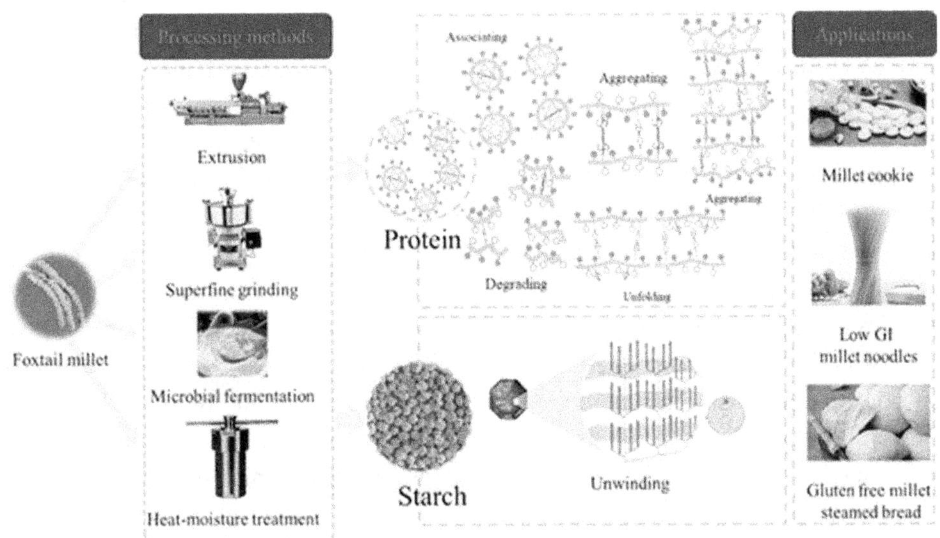

These civilizations recognized the nutritional value of millets and incorporated them into their daily diets. Millets were consumed by people of all classes and were used to make a variety of dishes, including bread, porridge, and alcoholic beverages. Today, millets are still consumed in many parts of the world and are considered to be a healthy alternative to other grains.

5.2. Millets in Traditional Diets

Millets, the resilient grains of antiquity, have held an esteemed position on the world's plates, particularly in regions where environmental adversities challenge the cultivation of more conventional grains. Amidst this global backdrop, India emerges as a testament to the enduring legacy of millets in traditional diets. For thousands of years, these unassuming grains have been woven into the culinary fabric of India, offering sustenance and nutrition to communities grappling with limited rainfall and poor soil conditions. The culinary repertoire of millets spans an impressive range, encompassing their versatile use in porridges, flatbreads, savory dishes, and even the creation of alcoholic beverages like beer and spirits.

The Cultural Tapestry of Millets: In India, millets have been an integral part of traditional diets, reflecting not only their adaptability but also their cultural significance. These grains have been more than just a source of sustenance; they represent a way of life, a connection to the land, and a celebration of culinary diversity. From the pearl millet rotis of Rajasthan to the ragi porridge of Karnataka, millets have left an indelible mark on the culinary tapestry of the nation.

Nutritional Powerhouses: Millets are not just culinary marvels; they are also nutritional powerhouses. Rich in protein, fiber, and an array of essential nutrients such as iron, calcium, and B vitamins, millets offer a holistic nutritional package. In regions where access to diverse foods may be limited, millets play a pivotal role in addressing malnutrition and dietary deficiencies. Their ability to provide a wide spectrum of nutrients makes them invaluable for promoting health and well-being.

Gluten-Free Marvels: In an era where gluten sensitivity and celiac disease affect a substantial portion of the population, millets rise as champions of dietary inclusivity. Their inherent gluten-free nature ensures that they can be enjoyed by individuals with gluten sensitivities without compromising on taste or nutrition. This attribute not only caters to dietary restrictions but also fosters culinary creativity, enabling individuals to explore a diverse range of gluten-free recipes.

Balancing Blood Sugar: The low glycemic index of millets is a hidden gem in the quest for balanced blood sugar levels. Unlike high-glycemic grains that trigger rapid spikes in blood sugar, millets offer a more gradual and sustained release of energy. This unique property is of paramount importance in the context of managing and preventing conditions like diabetes. By providing a steady source of energy and helping to control blood sugar levels, millets emerge as allies in the global battle against metabolic disorders.

Medicinal Treasures: Beyond their culinary and nutritional prowess, millets have earned a place in traditional systems of medicine, such as Ayurveda, as medicinal treasures. They are believed to possess cooling properties, making them invaluable in alleviating a range of conditions, including indigestion, ulcers, and diarrhea. Furthermore, millets are believed to harbor anti-inflammatory and antioxidant properties, opening doors to potential preventive measures against chronic diseases such as cancer, heart disease, and diabetes.

A Nutritional and Culinary Odyssey: The pages that follow delve deeper into the multifaceted role of millets in traditional diets. We will explore the diverse culinary traditions that revolve around millets, their exceptional nutritional content, their gluten-free nature, and their potential to contribute to better blood sugar management and overall health. This journey into the world of millets reveals not just their nutritional richness but also their cultural and medicinal significance, making them more than just grains but a way of life.

5.3. Millets and Food Security

Food security is a major concern for many countries around the world, particularly in developing nations. The high cost of staple foods and the instability of their prices has made it difficult for many people to access the food they need to survive. Millets have the potential to play a significant role in addressing this problem, as they are relatively inexpensive, easy to grow, and can thrive in harsh environmental conditions. This section will discuss how millets can contribute to food security and the challenges that need to be addressed to maximize their potential.

Millets are traditionally grown in regions with low rainfall and poor soil quality, making them ideal for cultivation in areas where other crops would struggle to grow. They are also able to mature quickly, with some varieties able to be harvested within 60 days of planting. This rapid growth rate means that millets can be used to provide food in times of crisis, such as during droughts or other natural disasters. The low cost of millets also means that they are accessible to people living in poverty, making them an important source of nutrition for many.

In addition to being a source of nutrition for humans, millets are also used as animal feed. This can help to support livestock farming, which is an important source of income for many farmers in developing countries. Millet straw can also be used for fuel or as building material, providing additional benefits to farmers.

Despite their potential benefits, there are several challenges that need to be addressed to maximize the potential of millets in addressing food security. One of the main challenges is the lack of investment in research and development of millets. This has resulted in limited knowledge of the best practices for cultivating millets and developing new varieties with improved yields and nutritional content.

There is also a need to address the perception of millets as a “poor person’s food”. This perception has contributed to a decline in the consumption of millets in many regions, as people increasingly turn to more expensive and processed foods. Promoting the nutritional benefits of millets and their versatility in cooking can help to change this perception and increase their consumption.

Figure 14. Approach towards agriculture and healthcare

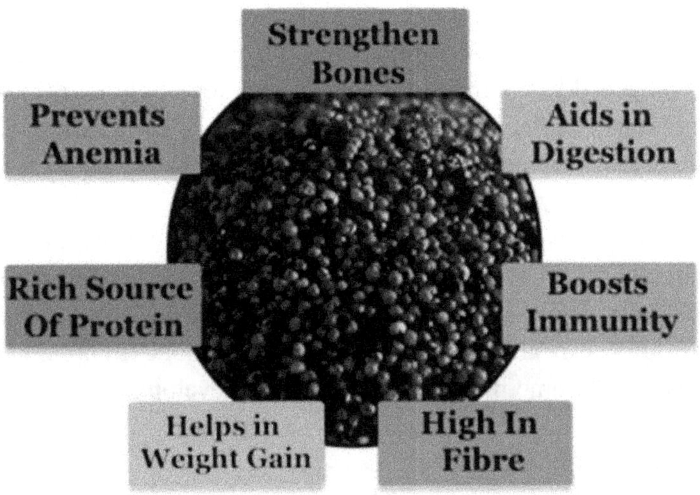

To address these challenges, there is a need for increased investment in research and development of millets, including the development of improved varieties with better yields and nutritional content. Governments and development organizations can also provide support for the development of millet processing and marketing infrastructure, to help farmers access markets and increase the economic benefits of millet cultivation.

In traditional diets, millets continued to play an important role as a source of nutrition and sustenance. In many cultures, millets were seen as a food for the poor, but this perception is changing as people become more aware of the nutritional benefits of millets. Millets are also becoming popular among health-conscious consumers, as they are gluten-free and have a low glycemic index, making them a good option for people with celiac disease or diabetes.

6. CONCLUSION

Millet is an ancient grain that has been a comestibles crop in countless regions of the sphere for thousands of years. It is highly nutritious, gluten-free, and adaptable to a range of growing conditions. Millets are lavish in essential enriched such as protein, fibre, supplements, and casting, making them an important food source for many communities, particularly in regions where access to other sources of protein and micronutrients is limited.

The role of women in millet cultivation and processing has been overlooked for too long. Women have always been involved in the cultivation and processing of millets, but their contribution has often been invisible or undervalued. In many regions, women are responsible for the majority of agricultural work and play a vital role in maintaining household food security. Women's knowledge and skills in millet farming are essential for sustainable and equitable development in rural areas, and their contributions should be recognized and supported by policymakers and development organizations.

6.1 Importance of Millets for Human Health and Food Security

Millets are an important crop for millions of people around the world, especially in regions where other crops struggle to grow due to poor soil conditions or drought. Juar are a species of wildflowers with tiny embryos that are ploughed for their edible seeds, which are highly nutritious and have been used as a staple food for a lifetime. Millets are grown an assortment of quinoa, including Africa, Asia, and Latin America, and there are many different including oyster millet, are found around the world, finger millet, and sorghum.

Research has shown that daily squandering of millets can have a range of health benefits. For example, studies have found that millets can help to reduce the ongoing anxiety possibility of illness. This is because millets are high in antioxidants, which guide to safe guard the humans against the harm done by reactive substances. Millets are also an excellent source of plant-based protein, which is essential for maintaining muscle mass and promoting healthy growth and development.

6.1.1 Nutritional Richness of Millets

Millets are not only hardy but also nutrient-dense. They are rich in essential vitamins and minerals, including iron, calcium, magnesium, and B vitamins. These nutrients are vital for various bodily functions, including bone health, energy metabolism, and overall well-being. Additionally, millets are a valuable source of dietary fiber, which aids in digestion, regulates blood sugar levels, and helps maintain a healthy weight.

6.1.2 Health Benefits of Millet Consumption

Regular consumption of millets has been associated with a range of health benefits. Research has shown that millets can play a significant role in reducing the risk of chronic diseases such as diabetes, cardiovascular disease, and certain types of cancer. Their high fiber content contributes to better glycemic control and helps manage diabetes. Furthermore, millets' low glycemic index makes them an ideal choice for individuals aiming to stabilize blood sugar levels.

Millets also contribute to heart health due to their low-fat content and the presence of heart-healthy nutrients like potassium and magnesium. These minerals help regulate blood pressure and reduce the risk of hypertension, a major risk factor for heart disease.

Moreover, millets contain phytochemicals and antioxidants, which have been linked to a reduced risk of cancer. These compounds help combat oxidative stress and inflammation, processes implicated in cancer development.

In an era where plant-based diets are gaining popularity, millets shine as an excellent source of plant-based protein. Their protein content, coupled with a balanced amino acid profile, makes them a valuable addition to vegetarian and vegan diets, aiding muscle maintenance and growth.

6.1.3 Decline in Millet Cultivation and Consumption

Despite their numerous health benefits, millets have seen a decline in both cultivation and consumption in recent years. The reasons for this decline are multifaceted, including the rising popularity of other grains like rice and wheat. These grains often receive more attention and research funding, leaving millets in the background.

Additionally, the loss of traditional knowledge and practices related to millet cultivation and processing has contributed to the decline. As societies modernize and transition to monoculture farming practices, traditional crops like millets face neglect. Economic factors also play a role, as millet cultivation may be perceived as less profitable than other cash crops.

Despite the many health benefits of millets, their cultivation and consumption have declined in recent years. This is due to a range of factors, such as the increasing popularity of other crops like rice and wheat, as well as the loss of traditional knowledge and practices related to millet cultivation and processing. In many regions, millets are considered to be a poor people's food and are often associated with rural poverty and backwardness. As a result, there is a lack of investment in millet production and marketing, and many farmers have shifted to other crops that are considered to be more profitable.

There are also opportunities to promote millets as a sustainable food source that can help to address issues such as food security and climate change. Milo are dry spell strong and can cultivate in nutrition less soil situations, which makes them a resilient crop that can withstand the effects of climate change. Additionally, promoting the cultivation and intake of juar can help to diversify nourishment systems and reduce dependence on a fewer units of pasture, which can improve food security and resilience.

6.2 Recognition of Women's Contribution to Millet Cultivation and Processing

Women's contribution to millet cultivation and processing is a critical aspect of sustainable and equitable development in rural areas. Women have been cultivating millets for thousands of years, passing down traditional knowledge and skills from one generation to the next. In many regions of the world, women are responsible for the majority of agricultural work and play a central role in household food security. However, despite their vital contributions, women's role in millet farming has often been undervalued and overlooked.

To address these challenges, there is a need for policies and programs that promote gender equality in agriculture. Initiatives that ameliorate colleen's reach to assets such as land, credit, and technology can help to reduce gender disparities in agricultural productivity and income. Moreover, programs that provide training and support for women in millet cultivation and processing can help to build their capacity and enhance their skills. For example, women can be trained in the use of improved millet processing technologies such as mechanical threshers, which can significantly reduce the time and effort required for processing.

The recognition and support of women's contributions to millet cultivation and processing can have significant impacts on food security and human health. Women's traditional knowledge of millet cultivation and processing techniques can help to preserve local crop diversity, maintain soil health, and promote sustainable agricultural practices. Sorghum are extremely alimental and have been shown to reduce the danger of developing recurring diseases equivalent to diabetes, cardiovascular disease, and types of carcinomas. By promoting women's involvement in millet farming, policymakers and development organizations can help to increase the availability of healthy and nutritious food for communities and contribute to the attain of the Global Development Standards.

In conclusion, the recognition of women's contributions to millet cultivation and processing is essential for sustainable and equitable development in backcountry. Women play a trenchant role in millet farming, processing, and seed preservation, and their contributions should be supported and valued by policymakers and development organizations. Promoting gender equality in agriculture through custom and preparations can help to enhance womenfolk's dispensation, improve their livelihoods, and increase their contributions to agricultural production. By recognizing and supporting women's role in millet farming, we can help to ensure that this essential crop is cultivated and consumed for generations to come.

REFERENCES

Bangar, S. P., Ashogbon, A. O., Dhull, S. B., Thirumdas, R., Kumar, M., Hasan, M., Chaudhary, V., & Pathem, S. (2021). Proso-millet starch: Properties, functionality, and applications. *International Journal of Biological Macromolecules, 190*, 960–968. doi:10.1016/j.ijbiomac.2021.09.064 PMID:34536472

Devi, K., Kumar, V., Kumar, V., Mahajan, N., Kaur, J., Sharma, S., Kumar, A., Khan, R., Bishnoi, M., & Kondepudi, K. K. (2023). Modified cereal bran (MCB) from finger millet, kodo millet, and rice bran prevents high-fat diet-induced metabolic derangements. *Food & Function, 14*(3), 1459–1475. doi:10.1039/D2FO02095E PMID:36648164

Gowda, N. N., Siliveru, K., Prasad, P. V., Bhatt, Y., Netravati, B. P., & Gurikar, C. (2022). Modern processing of Indian millets: A perspective on changes in nutritional properties. *Foods, 11*(4), 499. doi:10.3390/foods11040499 PMID:35205975

He, L., Zhang, B., Wang, X., Li, H., & Han, Y. (2015). Foxtail millet: Nutritional and eating quality, and prospects for genetic improvement. *Frontiers of Agricultural Science and Engineering, 2*(2), 124–133. doi:10.15302/J-FASE-2015054

Karpagapandi, L., Kalaiselvan, A., & Baskaran, V. (2023). Nutraceutical and Functional Attributes of Foxtail Millet (Setaria italica). In Nutri-Cereals (pp. 141-177). CRC Press.

Kumari, P., Pareek, V., Kajla, P., & Khurana, S. (2024). Composition, structure and functionality of starch isolated from Kodo millet. In *Non-Conventional Starch Sources* (pp. 253–278). Academic Press. doi:10.1016/B978-0-443-18981-4.00009-4

Li, J., Li, W., Hu, J., Li, C., & Cui, X. (2023). Proso Millet Peroxidase-mediated Degradation and Detoxification of Rhodamine B in Water. *Environmental Technology*, 1-33.

Li, S., Xian, F., Guan, X., Huang, K., Yu, W., & Liu, D. (2020). Neural protective effects of millet and millet polyphenols on high-fat diet-induced oxidative stress in the brain. *Plant Foods for Human Nutrition (Dordrecht, Netherlands)*, *75*(2), 208–214. doi:10.100711130-020-00802-6 PMID:32078112

Samtiya, M., Aluko, R. E., Dhaka, N., Dhewa, T., & Puniya, A. K. (2023). Nutritional and health-promoting attributes of millet: Current and future perspectives. *Nutrition Reviews*, *81*(6), 684–704. doi:10.1093/nutrit/nuac081 PMID:36219789

Sanodiya, L. K., Sanodiya, P. K., Prajapati, A., Kumawat, P., & Singh, A. (2023). *The sustainability of millets and the advantages they bring for human health*. Research Gate.

Singh, N., Kumari, P., & Bassi, J. G. N. (2023). Nutrient rich Kodo millet, importance and value addition: An overview. *The Pharma Innovation Journal.*

Yousaf, L., Hou, D., Liaqat, H., & Shen, Q. (2021). Millet: A review of its nutritional and functional changes during processing. *Food Research International*, *142*, 110197. doi:10.1016/j.foodres.2021.110197 PMID:33773674

Chapter 12
Economic Empowerment of Women, Millet Farming, and Sustainable Development

Anita Medhekar
https://orcid.org/0000-0002-6791-4056
Central Queensland University, Australia

ABSTRACT

The Indian government, with the support of 72 nations, proposed to the United Nations Food and Agriculture Organisation to declare 2023 as the "International Year of the Millets." Millets have numerous benefits-are gluten free with high nutritional properties, grown in dry-arid regions, drought and disease resistance, environmental sustainability, and empowering women for socio-economic development. The rural women in developing countries are helping to sustain the millet economy, which has high nutritional value and very tolerant to extreme semi-arid and drought weather conditions experienced due to climate change. This chapter will examine the Indian government's policy towards developing a millet economy amidst the impacts of polycrisis situation experienced around the world; to empower rural women in the agricultural sector, focusing on the economic dimensions; to bring about social transformation and economic significance of millet farming, value adding, improving livelihood, and social transformation for sustainable development through climate smart millet economy in India.

BACKGROUND

In the 21st century, the government of India has implemented many programs to empower women in every sector of the economy through education and providing employment opportunities. The Indian government's policy is focussed on the economic dimensions to bring about social transformation, economics progress, and prosperity through programs linked with women's empowerment in general, and for rural women in particular, through inclusive agriculture transformation. In 2012, Indian government policy initiative for *'Nutritional Security'* provided incentives for rural women to take charge of their food and nutritional security through intensive millet cultivation and production. Millets the ancient grains have a

DOI: 10.4018/978-1-6684-9819-4.ch012

Copyright © 2023, IGI Global. Copying or distributing in print or electronic forms without written permission of IGI Global is prohibited.

5000-year-old rich intangible cultural heritage from Indus-Valley civilization to 3rd millennium BC with a large potential to achieve social, economic, and environmental benefits for sustainable development (Dal Corso et al., 2022; Das & Ismail, 2023; Morrison, 2016). Millets have significant health benefits, packed with essential nutrients, are gluten-free, high in fibre, and low glycaemic-index, providing long-term immunity to the consumer. Millets have multiple uses as human food, animal fodder and for brewing beer (Dal Corso et al., 2022).

Nealy 97% of variety of Millets are cultivated in dry lands and consumed as staple food in developing countries with vulnerable population, to overcome food scarcity, and climate resilient crop in extreme drought or high salinity conditions in India, Sub-Saharan Africa, and Asia (Azare et al., 2020; Meena et al., 2021; Meldrum et al., 2020; Rana & Sahoo, 2023; Saxena, 2018 & 2019; Shah et al., 2023; Singh et al., 2022). With the current climate crisis, the importance of millets in sustainable agriculture has gained renewed attention in developing countries such as India and Niger. Millets are climate-resilient sustainable crops and can survive and thrive in drought to extreme high temperatures in arid and semi-arid regions of developing countries, with poor soil without any expensive fertilizers (Gujral, 2017). Millets are resistant to crop disease and pests and help to improve soil health and reduce soil erosion. India is the largest producer of 26.6% of world millets, with 83% of its land under Millet cultivation in whole of Asia (Sood et al., 2019). Millets are popular with consumers who are vegetarians and value traditional Ayurvedic sattvic food, indigenous, and organic food grains (Sarkar et al., 2015), with high nutritional energy and health benefits (Kumar et al., 2023).

Before the 'Green Revolution' of 1960s in India, millets were 40% of total grain production, which dropped to 19% of global share, and replaced by oilseeds and large-grain cereals like wheat and rice for consumption. Given the background the Indian government declared 2018 to be celebrated as the *"National Year of Millets"* to drive food security and introduced various schemes and minimum support price, to provide incentives and support the cultivation of superfoods "Millets" by small scale subsistence farmers. Women in rural India are actively participating in growing millets to sustain their families food and income. The President of India awarded *Nari Shakti Award* (Women Power Award) for women empowerment on behalf of the *All-India Millet Sisters (AIMS) Network*, to a widowed woman named *Mogulamma from Zaheerabad village in Telangana state*, in 2018, for being a successful tribal rural women farmer for growing millets on her two-acre land, end hunger and ensure food security for their families and encourage millet bio-diverse ecological farming practises to her AIMS sister (Down to Earth, 2023; Muthamilarasan & Prasad, 2021; Rana & Sahoo, 2023). Later *Mogulamma* also received the UNESCO Equator Prize in New York for environmental conservation and sustainable farming practices and reduce gender discrimination in the society.

The Indian government took the lead with the support of 72 countries, United Nations General Assembly, Food and Agriculture organisation (FAO) in their 75th session declared in March 2021, the year 2023 as the *"International Year of the Millets" (2023-IYM)* (United Nations, 2023). The aim of the 2023-IYM is to raise awareness by informing people about the nutritional and health benefits of consuming millets, creating domestic and global demand, and also their suitability for sustainable cultivation and production under adverse and changing climatic conditions being climate and drought resilient and developing markets and efficient supply-chain for millets (FAO, 2023; Government of India Budget, 2023-24).

Figure 1. International year of millets
(FAO, 2023)

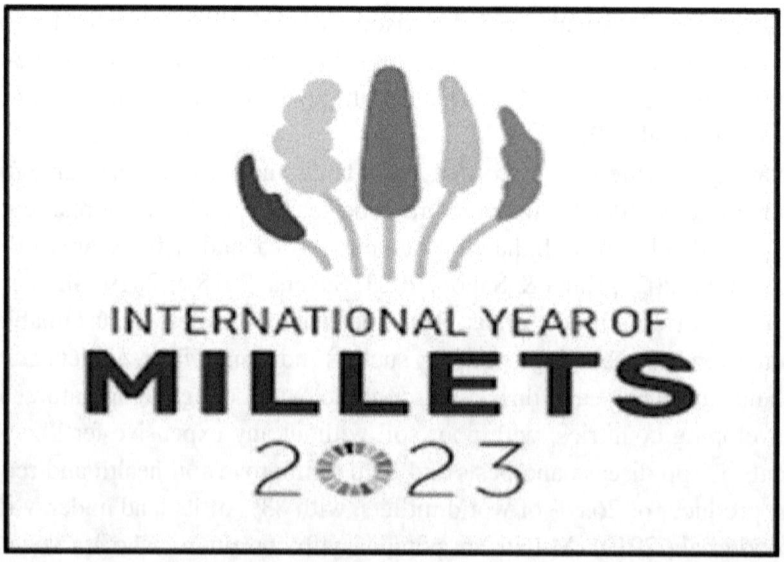

This chapter aims to explore the role of women in millet farming in adverse-dry arid and harsh climatic conditions being an arid regions resilient crop, and rooted in ancient Indian cultures and civilisations, develop policies to support women's economic empowerment, and their contribution to value adding, improving livelihood, nutrition, economic development, and social transformation for sustainable development of rural farming societies.

The purpose of this chapter is: (i) To examine the Indian government's policy to empower women in the agricultural sector and growing economic importance and significance of the role of women in growing millets, managing millets small and medium business enterprises, for sustainable development. (ii) To explore the economic dimensions of the role of women in farming millets, adding value, provide income, and enhance wellbeing for economic progress and prosperity of their communities. (iii) To identify and recommend a public-private partnership (PPP) framework to support women in agriculture, as a cooperative such as forming self-help-groups, given the increasing feminisation of agriculture and female headed households in rural India, due to men migrating to cities in search of employment, climate change and supply chain shocks/disruptions experienced during the pandemic and the war.

This chapter is structured as follows. The first introductory section of the chapter will explore the growing economic importance and significance of millets as a crop and the role of women in farming millets, providing livelihood, and creating sustainable societies. Next section provides the literature review millet farming, its nutritional properties and sustainable development in case of India in particular and global context in general. Section- 3 will examine and discusses the case study examples in relation to role of women as agents of change in Millet cultivation and empowerment. Section-4 discusses the challenges faced by women in millet farming in India and other developing countries for their economic empowerment and creating sustainable societies. The concluding section presents discussions for managerial and policy implications, conclusion and future research directions to advance the knowledge regarding the economic benefits from women's participation in Millet cultivation and empowerment for solving the deep rooting problem of gender linked poverty, malnutrition, infant and maternal mortality, and stereotype gender biases, and thus ensuring food security, economic prosperity, employment opportunities, and sustainable societies.

LITERATURE REVIEW

Millet Economy and Food Security

Millets are hardy small-seeded grass such as foxtail, that can grow on arid regions of Asia and Africa with minimal input of water, fertilizers and pesticides and are resilient to challenging changing climate conditions and giving good yield, and therefore are an ideal solution for countries to increase self-sufficiency for farmers, rural women farmers empowerment, ensuring national food security by diversifying crop rotation, increasing genetic diversity and rural poverty alleviation (Gujral, 2017; Meldrum et al., 2020; Muthamilarasan & Prasad, 2021; Nagaraj et al., 2013; Padulosi et al., 2015; Padulosi et al., 2022; Rana & Sahoo, 2023; Sun et al., 2023; Verma et al., 2023; Yapi et al., 2000). Further, given the impact of supply-chain disruptions due to COVID-19 pandemic (Al Saidi, 2023) and shortage of wheat and other cereal grains on food security, due to Russia-Ukraine war (Hassen & ElBilali, 2022), millets has been brought to the forefront for our plate by the Indian government scientist working on millet farming and its nutritional value. Therefore, cultivating 'Millets' locally, will reduce reliance on imported grains like wheat, rice, corn, and oats. Millets are adaptable to adverse climatic conditions, supply shortages caused due to the global health crisis, war, and natural disasters.

All types of Millets are gluten free, having a high nutritional value being rich in protein, fibre, minerals, and vitamins. Further, millets described by Meek (2022, p.387) as "marginalised foods" consumed by non-elite lower class, indigenous population but now "rediscovered and revalued as branded miracle superfoods", has been an integral part of traditional cuisines in different communities in India in particular and sub-Saharan Africa. Although their prominence in the dietary composition has declined over the past few decades but rediscovered as niche power food. Thus, building awareness regarding millet farming and encouraging consumption of millet will have substantial impacts on improved health, environment, and income in communities around the world, as well as revive the diminishing cultural values which are all core to the realization of the SDGs (Das & Ismail, 2023).

Meek's (2020), research on re-emergence of Millet economy as a local-regional food in India is based on the bio-regional vision logic of place and authenticity of agricultural land practices, and food to be locally produced, precured and consumed. Further, Meek (2022) has clearly identified three key points in the rise of Millets food politics and branding of millets as super foods: **(i)** emergence of Millets as 'super-food' rich in nutrients to solve global hunger and nutritional problems replacing 'inferior foods', **(ii)** depoliticizing hunger through 'millets as miracle food narrative', and adopting interventionist western solutions in development ideology to solve malnutrition, and **(iii)** to grow local agricultural foods, and procure and consume locally produced goods.

The economic significance of the *'miracle small-grain millets'* are numerous for achieving sustainable deelopment goals. **(i)** Growing millet is ecologically sustainable farm practice in arid/drought prone regions with less dependence on chemical fertilizers and pesticides, **(ii)** creates employment opportunities in rural areas for farmers with small holdings, **(iii)** empowers rural women in growing nutritious millets and achieving food security for their families, **(iv)** well established millet cooperatives provides access to local, national and global markets in a millet economy, **(v)** millets have high nutritional value containing calcium, essential minerals protein, natural fibres and is gluten free, **(vi)** producing and consuming millets provide sustainable, nutritional security, and holistic wellbeing (Babele et al., 2022; Bandyopadhyay et al., 2017; Mees, 2022; Meldrum et al., 2020; Muthamilarasan & Prasad, 2021; Rana & Sahoo, 2023; Rao & Basavaraj, 2015; Tiwari et al., 2022). The world's current top 10 producers of millets shown in Table-1 (FAO, 2023).

Table 1. World's top 15 leading millet producers

Rank	Country	Production (Kilo Tonnes 1KT =1000)
1	India	13,750 KT
2	China	2,300
3	Niger	2,100
4	Nigeria	1,900
5	Sudan	1,750
6	Mali	1,750
7	Senegal	1100
8	Ethiopia	1000
9	Burkina Faso	800
10	Chad	700
11	Russia	600
12	USA	400
13	Tanzania	350
14	Nepal	300
15	Myanmar	200

Source: FAO statistics (update 9-July 2023)
https://www.helgilibrary.com/indicators/millet-production/

The Indian export of super-food millets has increases from US$ 28.5 million (2019-20), to US$ 26.97 million (2020-21), and in (2021-22) US$ 43.32 million. India exports millets to major countries of Egypt, Libya, Nepal, Oman Saudi Arabia, Tunisia, UAE, UK, US, and Yemen. Major millet importing countries are Belgium, Brazil, Germany, Mexico, Netherlands, Italy, UK, USA, and Japan (FAO, 2023).

Types of Millets and Nutritional Benefits

Millet is highly nutritious super food, and a good source of antioxidants, essential nutrients, protein, fibre, and vitamins and good for the consumers. Millets are often grown and consumed as a staple food in parts of the world's dry/arid regions of India (Foxtail millets (*Kangni*), Browntop Millets (*Andu korralu*), Little millets (*Samalu*), Kodo millets (*Arikalu*) and Barnyard millets (*Jhangora* Oodalu) are the five positive millets, coined with respect as "*Siridhanyalu (Millets)*" by the 'Millet Man of India- Dr Khadar Valli, the ultimate solution to good health and sustainable agriculture (Valli, 2018). Millets are easily grown in extreme weather conditions, are drought, diseases, and pest resistant, thus reducing the need for inputs such as irrigation, fertilizers and chemical pesticides. Therefore, making it a reliable food security for farmers where other grains are not well suited to grow in the local climate and ensures nutritional food security with many nutritional health benefits (Babele et al., 2022; Bhat et al., 2018; Fischer et al., 2016; Meldrum et al., 2020; Rana & Sahoo, 2023; Shah, et al., 2023; Singh et al., 2022; Sun et al., 2023; Verma et al., 2023).

All types of Millets have high nutritional value and are an important source of income and food for many small-scale women farmers in developing countries (Azare et al., 2020; Bandyopadhyay et al., 2017; Basavaraj et al., 2010; Bunkar et al., 2021; Di Stefano et al., 2017; Gujral, 2017; Meek, 2022; Meldrum et al., 2020; Rana & Sahoo, 2023; Sabuz et al., 2023; Singh, 2022). Millets are high in anti-oxidant, aids in sleep, weight loss, help in slowing muscle degradation, controls diabetes, reduce high blood pressure, reduce the risk of colon cancer, celiac disease, and coronary artery disease. According to Valli (2018) Different types of millets should not be mixed together. They should be cooked and eaten one type of grain at a time, to obtain the most nutritional benefits. Nine key types of Millets with their nutritional value is explained: https://krishijagran.com/agripedia/know-about-different-types-of-millet/

1- **Finger Mill**et also known as *'Ragi'* is native to India and the eastern parts of Africa and widely cultivated in African, Asian, and South American countries. It is an important source of dietary fibre and used to make variety of dishes, has a low glycemic index and medicinal properties, with rich source of proteins, vitamins, and minerals, particularly iron, calcium, and B vitamins, making it suitable for people with diabetes, amnesia, and high blood pressure.

2- **Foxtail Millet** known as *'Kangni'* is native to Asia and commonly grown in India, China, and Italy. It is consumed as a substitute for rice, in making bread, pilaf, polenta, porridge, and salad. Foxtail millet is highly nutritious in protein, fibre, and minerals such as iron, potassium, and magnesium. It is also a good source of B vitamins, thiamine, and niacin, which are important for energy metabolism and brain function.

3- **Proso mill**et also known as *'Proso'* in India, is high in nutritional source of fibre, proteins, minerals, iron, vitamin-B, and zinc. It is low in fat, cholesterol-free and a staple food in many countries, particularly in India, China, and Eastern Europe. It can also be used as flour for cooking flatbreads, baked dishes, soups, stews, and porridges.

4- ***Pearl mill***et also known as *'Bajra'* in India, is a staple food grown in arid-dry regions with limited water resources in Africa and India, with a huge potential to address food security and malnutrition problems in developing countries. Bajra is an important source of affordable nutritional food, and its flour is used to make a variety of dishes, including bread, porridge, and flatbread. The grain can also be used to feed livestock and make biofuel. Pearl millet is high in nutrients, essential amino acids, minerals, protein, fibre, and a good source of iron, and zinc.

5- **Barnyard mill**et also known as *'Jhangora'* is native to India and has been cultivated for thousands of years for its nutritious seeds. It has many health benefits and is rich in fibre, protein, and essential nutrients such as iron, calcium, and B vitamins. It is ground into flour and used to make dosas, rotis, and other types of flat-breads. Full grain can be used in soups, porridge, and other grain-based dishes. It is believed to have anti-inflammatory properties, which may help to reduce the risk of chronic diseases such as heart disease and diabetes. It is also high in antioxidants, which can help to protect against oxidative stress and improve overall health.

6- **Kodo millet** is native to India and parts of Southeast Asia, grown by small holding farmers being environmentally sustainable and healthy. It is a highly nutritious and been grown for thousands of years in these regions. Kodo has a nutty flavour and can be used in a variety of dishes, including bread, porridge, and is a substitute for rice or wheat. It is a rich source of protein, dietary fibre, and essential vitamins, iron, calcium, and magnesium. Kodo is low in fat and calories, therefore, making it a healthy choice for those wanting to maintain a healthy weight.

7- **Amaranth millet** family includes *quinoa*. It is native to South America and has been cultivated for thousands of years by Indigenous people, being highly nutritious. Amaranth millet is highly nutritious, contains high protein levels, fibre, essential amino acids, and minerals such as calcium, iron, and magnesium. One of the most notable benefits of amaranth millet is its high level of antioxidants, which help to protect the body against free radicals and reduce the risk of cancer and heart disease. It is also a good source of lysine, an amino acid that is important for the growth and repair of tissues in the body. The flour is used in a variety of dishes, including bread, porridge, and beer.

8- **Sorghum millet** known as *'Jowar'* is native to and staple food in India and Africa. It is now grown in many parts of the world, including Asia, Europe, and North America. Sorghum millet can be white, red, and brown in colour, and is considered one of the world's most important food crops. Sorghum is a nutritious grain that is high in protein, fibre, and important minerals such as iron, zinc, and magnesium. It is also rich in antioxidants, which can help to reduce inflammation and improve overall health. Sorghum is a substitute for rice for humans and also livestock-feed. The flour, is used to make bread, noodles, porridge, and other dishes.

9- **Broomcorn millet** is brown red, or yellow in colour. It is native to Africa and Asia, and is grown in many countries around the world, including China, India, Russia, and the USA. It is primarily used for feeding livestock and as human food source in arid regions of Asia and Africa to make bread, pasta, and cereal. It is a good source of protein, fibre, and various essential vitamins and minerals. It is also low in calories and has a relatively low glycemic index, making it a suitable choice for people with diabetes or other metabolic conditions. It is also used to make a traditional Chinese alcoholic beverage called baijiu, and beer.

The next section will examine, as to how millet economy is linked with women's empowerment and sustainable development.

Millets and Sustainable Development Goals

In 2016 United Nations announced 17 -Sustainable Development Goals to be achieved by 2030. These 17-SDG goals are not binding and builds on the United Nations Millennium Development Goals for 2000. The countries should endeavour to take responsibility and aim to develop programs and projects to meet these 17-SDG. These goals aim is to irradicate all forms of poverty from the world and make a conscious effort to build an "inclusive, sustainable, and resilient future for the people and the planet" (UN-SDG, 2023). To achieve sustainable development and enhance income, empowerment, and nutrition in rural India via millet farming (Padulosi et al., 2015; Rana & Sahoo, 2023) it is important to simultaneously achieve interconnected goals of equitable economic growth, environmental protection and social inclusion of all vulnerable groups and women in the development process, which is crucial for the wellbeing of the individuals, society, local economy, and the world as whole.

Millets are sustainable ancient crops in the fight against climate change as it has a low carbon footprint and can help to reduce greenhouse gas emissions (FAO, 2023). Millets are *climate smart and environmentally friendly* crop grown in a sustainable manner in arid regions of India and other parts of east-Asia such as Thailand, to ensure food security for the future (Bopp, 2020; Ceasar & Maharajan, 2022;

Fischer et al., 2015). It is also *farmer/cultivator friendly* crop being drought, disease, and pest resistant, needs fewer inputs to cultivate in dry-arid regions of developing countries in Asia, Africa, and the world. Millets are *consumer friendly* crop with very high nutritional value containing high antioxidants, low glycemic index and medicinal properties, and a rich source of fibre, proteins, vitamins, minerals, iron, and calcium as discussed in the section above for each type of millet.

Therefore, given the 'polycrisis' situation with multiple catastrophise that the world has and is experiencing since the pandemic (World Economic Forum, 2023), *'The International Year of Millets-2023'* will help not only in creating greater awareness of millet production, but also contribute to ensuring incomes for local farmers, local food security, mitigate climate change impact, poverty eradication, having nutritional value, focus on consuming local bio-regional food produced within 161 Kilometres/100 miles (Delind et al., 2006; Meek, 2022). This will help to achieve various SDGs, particularly in the drought prone regions of developing countries that are threatened by climate change. Millets can be a farmers choice for 'climate smart' drought resistance cultivation, which does not require irrigation, chemical fertilizers, and pesticides, ensuring food security, including the smart-food consumption choice of the consumers, for sustainable consumption and production (Bandyopadhyay et al., 2017).

As a result of increase in population and socio-economic development there is a growing demand for food, water, and energy resources along with an increase in the prices and cost of living. There is a gap between the demand and supply for these resources. Further, global farming, food production and retail sector is facing many challenges, of hunger, malnutrition of children and women in developing countries, limited resources, increasing population demand and climate change impact. To overcome these challenges, it is important to adopt environmentally sustainable crop cultivation/production practices, resilient value chains, and consumers access to affordable, nutritious, and healthy diet. Varity of Millets can be cultivated as they are environmentally sustainable and can be easily grown in adverse dry climate and arid regions, with minimal inputs, and is an affordable source of nutrition for healthy diet.

Millet-smart production, millet-smart consumption, and millet education awareness can help to achieve the following SDGs (Goals-1, 2, 3, 5, 8, 12, 13, 15 and 17) for women's empowerment and capacity development for addressing climate change via sustainable agriculture practises, food security and nutritional security.

- **Goal 1- No Poverty:** End poverty in all forms especially in the rural areas for women who engage in millet farming on 1-2 hectare of land, by empowering them with market access to millet cooperatives, education to sell their surplus, and be self-sustaining.
- **Goal 2- Zero Hunger**: Important to support local farmers to grow Millets, as sustainable agriculture practices provides solution for zero hunger with access to affordable food, poverty eradication, and sustainable economic development. Millets are climate smart small grains, consumed as staple food in developing countries with vulnerable population, to overcome food scarcity in difficult drought times and ensures food security in arid regions of Asia and Africa.
- **Goal 3- Good Health and Well-Being:** For sustainable development it is essential to consume variety of millets with high nutritional value, rich in protein, fibre, minerals, vitamins, various essential minerals, and gluten free, ensures good health and promotes well-being for all age groups.
- **Goal 5 - Gender equality** can be achieved by empowering women and girls involved in Millet farming, cooperatives, millet-based restaurants, and value-added processing of millet-based products, by having government support and incentive programs to provide market access, additional

inputs like seeds, required to improve yield per hectare, education and training in value added millet products.

- **Goal 8 - Decent Work and Economic Growth:** Creating millet friendly economic conditions, by supporting farmers for millet cultivation though education, incentives, support price, market access, millet processing and value adding, and resilient millet supply chain will provide work for women farmers and other jobs related to the Millet industry such as millet related retail shops, specialist millet chefs, and promote sustainable employment and economic growth.
- **Goal 12 - Responsible consumption and production:** Consumption and production Millets of Millets is good for the consumers, producers, environment, economy, and planet earth.
- **Goal 13 - Climate Action:** Cultivating Millet is beneficial being climate resilient as it is highly efficient to convert more carbon-dioxide to oxygen by C4 Photosynthesis method and mitigates climate change impacts. The carbon footprint (CO2) of Millets, which is vegetarian protein is 25% less than wheat and lower than rice being climate resilient crop, with great potential to promote food security and combat climate change with sustainable farm practices (Gujral, 2017; Tiwari et al., 2022).
- **Goal 15- Life on Land:** Cultivate 'Millets' to protect the environment and save biodiversity. Millets being antient grains and are found naturally in the arid regions specially in Asia and Africa. Millets are climate friendly and are sustainable crop, combat land degradation and soil erosion, halt biodiversity loss, drought, disease, and pest resilience, and needs fewer inputs to cultivate in dry-arid regions.
- **Goal 17-Partnerships:** Local, national, and global partnerships are essential for promoting and providing support to local farmers growing Millets from all stakeholders, government agriculture department, private sector, food processing industry to add value to millets, millets being climate resilient form a basis for sustainable food supply chain education to raise awareness about various benefits from cultivating and consuming millets, researchers- scientists, nutritionists, and economist.

CASE STUDY: INDIAN GOVERNMENT POLICY ON WOMEN'S ECONOMIC EMPOWERMENT AND MILLET FARMING

The rural women in India have been growing millet in 1-2 hectare of land holdings in the arid/drought prone regions, and in the 21st century are empowering themselves and helping to sustain the Millet economy, society, and their families (Hazaresingh, 2020; Padulosi et al., 2015; Saxena, 2019;Tiwari et al., 2022). Millet is an ancient grain cultivated in the Indus-Valley civilisation, grown in arid to semi-arid regions of Africa and Asia. India is the largest producer of millets followed by Niger, Sudan, and Nigeria. October the 15th, is celebrated by the United Nations as *"International Rural Women's Day"* which was celebrated in 2021 with the theme of " *Rural Women Cultivating Good Food for all*" and October 16th is also celebrated as World Food Day. The Indian Government celebrates the day as *"Rashtriya Mahila Kisan Diwas*" translated as, "*National Women Farmer's Day*" recognising rural women and women farmers roles in sustaining rural societies by cultivating superfoods Millets.

The Indian Government's 2022-23 Budget Speech, Finance Minister announced 2023 – an empowered and inclusive economy for all Indians. The Government's philosophy of *"Sabka Saath Sabka*

Vikas"(means- "Everyone's Support and Everyone's Progress*)* has facilitated inclusive development specially covering women farmers to produce and export Millets- 'Shree Anna' in the world.

According to *Government of* India Budget (2023-24)...***"Economic Empowerment of Women:*** *Deendayal Antyodaya Yojana National Rural Livelihood Mission has achieved remarkable success by mobilizing rural women into 81 lakh Self Help Groups. We will enable these groups to reach the next stage of economic empowerment through formation of large producer enterprises or collectives with each having several thousand members and managed professionally. They will be helped with supply of raw materials and for better design, quality, branding and marketing of their products. Through supporting policies, they will be enabled to scale up their operations to serve the large consumer markets, as has been the case with several start-ups growing into 'Unicorns'"*

Union Budget (2022), also announced 2023 as the International Year of Millets to provide support for post-harvest value addition, enhancing domestic consumption, production and for branding Indian millet products nationally and internationally. Millets also known as "Shree Anna" foods production and consumption helps to provide nutrition, food security and welfare for farmers. The aim of the Indian Government is to establish in an International Centre for Millets in India with excellence in best practices, research, and technologies at the Indian Institute of Millet Research, Hyderabad.

Opportunities from Growing Millets, the Miracle Food

According to "Millet Man of India - Dr Khadar Valli" an independent scientist, environmentalist and food expert received one of the highest Indian awards of '*Padmashri*' in 2022, for promoting the consumption and production of Millets on small holdings by women, at a subsistence farm level as a superfood, with high nutritional benefits, climate resilient, but also a crop for the future with global warning related raising temperatures, which requires fewer inputs as fertilizers, irrigation or pesticides.

According to Valli (2023) Women with small land holdings of 2-3 acres, or they lease small holdings from large farmers, produce millets at home, ensure food security and also sell the surplus in the market, thus providing rural women with employment and source of economic independence. Millet farming ensures nutritional food security for rural women and improves the health of the soil and livestock. Millets can be farmed in low fertile, arid, waste land, or saline soil, fulling nutritional needs of the family. Women are empowered and also sell the surplus in the millet cooperative and buy livestock. Millets production and consumption therefore must be made affordable, accessible, available with its nutritious properties and providing employment and income opportunity for small women farmers in developing countries. Millet education to raise consumer awareness and to bring about a behavioural change by nudging the consumer to make an informed choice is essential, through institutionalised marketing and promotion of health benefits of millets because they are cheaper

to cultivate, suited to the Indian climatic conditions and soil, drought resistance and given its health benefits to reduce blood pressure, diabetes, hypertensions, improve bone health, slowing ageing, colon cancer, gastric ulcers, and reduce cholesterol (Annor et al., 2017).

Verma et al. (2023), in their resent paper compare the traditional millet crops considered as orphaned crops (Babele et al. 2022; Padulosi et al., 2022), which are unnoticed, but perform better in terms of environmental sustainability are drought and pest resistance and require less water and no chemical fertilizer compared to the major crops- of rice, wheat, and maize in terms of their external resource use and environmental sustainability. That is millets have a short life cycle of 10-12 weeks of harvest time, require less irrigation/water and rainfall of 20 cm, best sown in semiarid and arid drought prone regions, with C4 photosynthesis category and can survive with high temperature and low soil moisture. Whereas in comparison, the three major commercial crops rice, wheat, and maize require intensive external irrigation resource use of high water, more than 120-140 cm rainfall, harvest life cycle of 20-24 weeks, loam texture soil, moderate water holding capacity and warm temperature for growing.

Further, Shah et al.(2023) recent qualitative study of interviews with 25 millet entrepreneurs found that producers, retailers, and consumers are becoming more aware of the ecological, economic, and nutritional health benefits of millets production and consumption. This popularity of millets in India is gaining acceptance in the market since COVID-19 pandemic amongst the urban-metro population as a trendy health food to build immunity against corona-virus and has resulted in increasing demand by consumers and interest by millet producers and entrepreneurs to invest in millet-based value added ready-to-eat and ready-to-cook food products, via retail shops with cooking instruction recipes, fast food, local restaurants serving millet-based cuisine, bakery products, and gastronomic millet tourism in India.

Since declaring 2018, as the 'India's National Year of Millets'. In 2021-22, leading states in India with land area in hundred thousand hectares under millets production in ascending order are Rajasthan, Maharashtra, Uttar Pradesh, Karnataka, Madhya Pradesh, Gujarat, Haryana, and Tamil Nadu - see Figure -1 (Government of India: Millets Portal, 2023).

Figure 2. Millet producing Indian state 2021-22
(Government of India: Millets Portal, 2023)

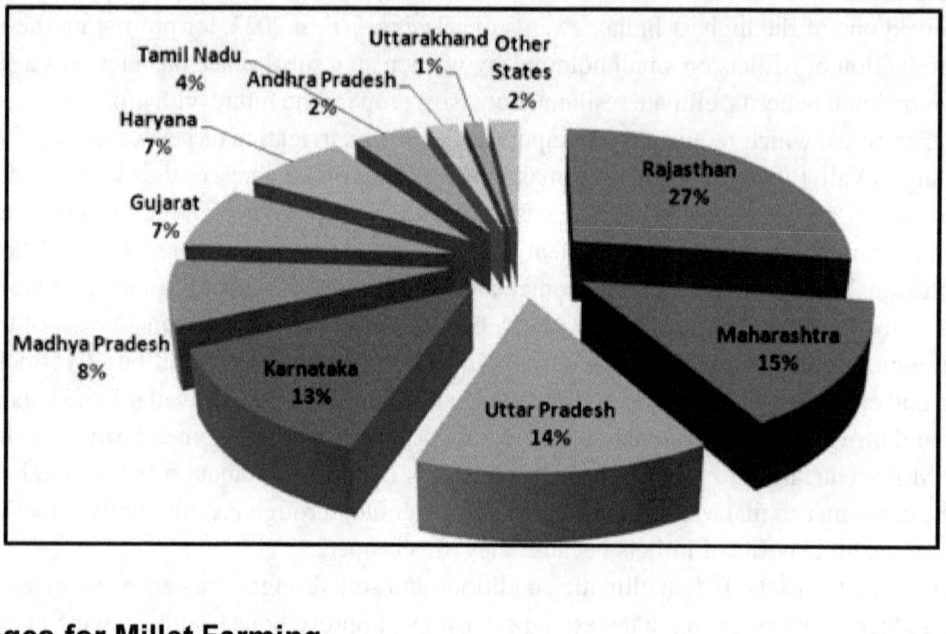

Challenges for Millet Farming

There are numerous challenges faced by women engaged in millet farmers in arid regions with small holdings.

1- ***Promotion of millets nutritional properties:*** Millets constitutes 10% -15% of world food consumption in rural areas, it is not known for consumption by the middle class. However, in the last few years in the western countries and urban areas very few varieties of whole millets and millet flour is available in the health food stores demanded by the vegetarians and those with gluten free diet. Millets are an alternative to wheat and rise consumption and production.

2- ***Millet Information workshops***: It is important to promote cultivation, production, value adding and consumption of millets through workshops and seminars highlighting types of millets nutritional properties, along with climate resistant, good for the land for sustainable agricultural practices, and pest resistant benefits.

3- ***Education:*** Millet education and training for the rural women about its nutritional benefits, better farming practices for sustainable land management, to improve yield per hectare, market information, market development, wholesale, and retail price transparency to get better price, and training in value added millet products.

4- ***Trade bottleneck*** – Information services and market access needs to be improved for the small holding women farmers through the millet cooperatives, price transparency, market development to provide confidence to the rural women to farm, market, integrate into millet grain value chain and gain the benefits of economic independence and market access with confidence.

5- ***Climate Resistant***: Sustainable agriculture practises are required, due to an increase in dry lands because of global warming, fall in yield due to vagaries of the rainfalls causing drought or floods, and poor farm management practices (Azare et al., 2020). Therefore, cultivating millets have a superior role to play being climate and pest resistant small- seed crops (Taraz, 2017). According to United Nations (2023) estimates, 90 million people in Asia and Africa depend on millets diet. More than 55 percent of global production comes from Africa, 40 percent from Asia, and Europe represents around three percent of the world market.

6- ***Conflict between export of millets and domestic consumption***: Indian government wants to boost exports of millets and millet-based value-added products. The boosting exports to high-income countries and branding millets as an organic super-nutritious food, sold at a high price to developed countries, purchased by affluent people in health shops, may make it unaffordable to the rural poor and low-income people of India and other developing countries in Africa and Asia where it is the staple diet for the rural population. The challenge is to find a balance to make millets accessible by the rationing methods at an affordable support price ceiling, for the rural population.

7- ***Lack of consumers awareness of nutritional benefits and familiarity:*** Urban consumers in particular lack awareness and familiarity about nutritional benefits of various types of millets and how to cook and use them in different recipes. They have an inferior image about millets which are usually consumed by people living in the rural areas. Hence, it is important to raise consumer awareness through health, wellness and marketing campaigns and cooking shows, so that they can make an informed choice and make a behavioural change by buying and consuming millets.

8- ***Startup programs for women headed small and medium business enterprises:*** it is challenging to educate women and women entrepreneurs and to motivate them by the government millet promotion programs and non-government organisations focussing on nutritional health of women and children in rural areas and urban poor, to get into millet framing, and value adding by producing millet based ready-to-eat and ready-to-cook meals, and provide financial support for starting millet based health food stores and vegetarian restaurants serving sweets and savory meals made from millets.

Therefore, millets have the potential to solve future global nutritional security challenges, as a sustainable alternative to major cereals of rice and wheat, which are expensive to grow and has less nutritional value. With climate-resilient qualities, Millets are a smart food, and are therefore considered good for health, good for the environment as they survive with less water and have a low-carbon footprint, and good for the farmer as they are more resilient and climate-smart and provides socio-economic empowerment to rural women in decision making (Government of India: Millet Portal, 2023; Rana & Sahoo, 2023; Saxena, 2019; United Nations, 2023). In the search for nutritional value, and climate/drought resilient solutions, millets could be the crucial link in the future sustainable food supply chain.

DISCUSSION

Given the interconnected risk and catastrophise experienced in the past 5 years, or *Polycrisis* (World Economic Forum, 2023*)*, caused due to multiple emergencies faced by the world, of global health emergency, pandemic linked border closures and supply chain disruptions, followed by a Russia-Ukraine armed conflict, inflation, rising price of fuel, gas and cost-of-living, and failure to mitigate climate change related future impacts have shown that food security, nutritional security, food availability and food accessibility is still a concern for all, especially developing countries. Therefore, to declare 2023 as the *"International Year of the Millets"* is at the right time to raise awareness of Millets by government, institutions, farmers, climate scientists and individuals, to bring about agricultural, bio-security, and dietary diversity by cultivating the traditional miraculous 'Millet' varieties. Government has provided incentives for Millet based Start-ups programs by providing subsidies for seeds and involving food processing companies such as Britannia, in producing ready-to-eat biscuits and other millet-based health foods and promoting millets as health food.

As summed up by Gujral *"Millets are one-stop solution in the wake of climate change, water scarcity, and drought conditions along with high nutritive value to provide sustainable food security"* (Gujral, 2017, p.666).

The supply chain of millets has five key stages from farmers/ producers, cooperative societies or wholesalers, millet processors with their processing mills, retails (wholesale, local markets, and health shops) to end use consumers. Despite its high costs and unaffordability in the recent year, due to its nutritional awareness amongst the elite high income and health-conscious consumers and high demand in India and by overseas countries, the price of millets and its processed products have increased, making it unaffordable for lower- and middle-income groups. Economic benefits of millet crops to the farmer and consumers are numerous. They are cheaper to grow, highly nutritious super-food, are drought-resistant, require very less water, and easy to grow in dry-arid regions without any expensive external inputs as chemical fertilizers and pesticides. In comparison rice, and wheat needs far more water to grow. Millets are also good for the consumer with its balanced nutritional properties providing a holistic staple diet and ensures nutritional and food security. It is good for the environment, as it is drought and pest resistant crop, helping to flight climate change by reducing the greenhouse effect that is helps in converting carbon-dioxide to oxygen being plant C-4 cereal category (Singh et al., 2022).

With rising input costs, falling ground water levels, falling profits for the farmers, and erratic climate change impacts causing shortage of rice due to drought and wheat due to Russia-Ukraine war; millets are the best crops which are cheaper to grow and are drought resistant. Millets should be a part of national food security programs and governments should encourage minimum support price for the farmers for millet production, just as for rice and wheat.

Indian government wants to boost exports of millets and millet-based value-added products and be one of the top three exporting countries. Currently Canada, Russia, Ukraine, China, and US holding the top positions. These are developed countries (with Ukraine currently facing devastation by war). India should not forget that it has large population living in rural areas depending on subsistence farming agriculture and consume millets. With branding millets as nutrition miraculous super-food, the demand for it will increase by the affluent and health-conscious consumers and thus prices will rise and could become unaffordable for small farmers and poor people in rural areas, as they may sell all the produce for income, and thus impacting their own nutritional security. For example, recently in June 2023, India banned export of non-basmati white rice to the world, to help lower the rising domestic price of rise due to increases cost of fertilizers and El-Nino impact of drought and secure its availability first to the Indian citizens (The Economic Times, 2023). A similar step may have to be taken for millets if the price rises due to increased demand from other countries.

Figure 3. Public-private partnerships for millet production and consumption

Public-Private Partnership (PPP) for Millets Production and Consumption

Public: Government agriculture department policy, financial support, programs, subsidies, millet benefits education/awareness, government cooperative, Fair Price shops, program implementation support, rural women's empowerment, market access, branding, and capacity building

Private Industry: Millet Farmers, Millet Industry, Investment, agriculture millet harvesting technology, education, supply chain distribution networks, millet cooperatives, retail entrepreneurs, value adding millet products, millet retail, millet-based restaurants

National Research Institutes filed research and development by government scientists about sustainable power-food millet production, climate resistant crop, nutritional benefits, population growth and demand , food, and nutritional security goals, encourage millet farming to combat climate change in the future

University: Academia, Research centers, university faculties of agriculture economics, environment, and nutrition for sustainable development of the economy

Start-Up initiatives: Public and private support for start-ups for millet product technologies, retail, branding as organic health food, advertising, packaging, digital services, millet awareness of its nutritional health benefits, millet festivals, and millet

Through public-private partnerships with the key stakeholders government, private sector researchers, food industries like Britannia Biscuits and marketing promotional agencies can help to transform 'Millets' from orphaned grain to mainstream super food for sustainable nutritional food security Padulosi et al.(2022) (Figure-2). Indian government at national and global levels want to popularise, promote, and raise awareness of prioritising millets production, cultivation, value-added products, and consumption of millets through various media channels, cooking shows, recipe books, nutritional education of millets, research seminar, workshops, conferences, and publications. The government's aim is to highlight the various benefits of millets: nutritional, promote peace, pest resistant, climate resistant, bio-diversity, and investment in sustainable production, and management of millet values chain by providing farming equipment and technology to the small and medium farmers, access to millet cooperatives, and education for farmers in millet cultivation, harvesting and value adding.

CONCLUSION

Millet accounts for only 3% of the global grain trade and gaining popularity in India and abroad amongst the health-conscious consumers, millet farmers, retailers, and restaurant industry. The nutritional, economic, and environmental potential of ancient Millet grains with its ayurvedic health properties are infinite and priceless. This chapter provides an overview of the role of women in millet farming in India, its nutritional benefits and economic dimension of social, economic, and environmental sustainability to combat climate change being drought resistant crop requiring few external resource inputs. With increase in global consumption of wheat, rice, and maize, which requires lot of water, good soil and fertilizers, demand for millets has declined in recent decades also in India. However, with the declaration of 2023, as international year of the Millets it is hoped that consumers and the producers recognise nutritional, economic, and environmental health benefits of millets in their daily life and will make a conscious choice to change their behaviour and consume millets.

The Indian government has a major role by declaring 2023-International Year of the Millets, and encouraged women farmers and millet entrepreneurs in the rural and urban areas and provided startup support with seeds, technical expertise and education to inform them of the benefits of millet farming: (i) suitability for sustainable cultivation and production under arid to semi-arid adverse and changing climatic conditions, (ii) millets being pest and drought resilient crop (iii) requires minimal or no external inputs of chemical fertilizers, water and pesticides, (iv) cultivating Millet is beneficial being climate resilient as it is highly efficient to convert more carbon-dioxide to oxygen by C-4 Photosynthesis method and mitigates climate change impacts, (v) and developing markets and efficient global value supply chains for millets and, finally (vi) nudge the consumers and make them aware of the significant health and nutritional benefits of consuming millets and proving the same through the public distribution rationing system, holding talks by millet researchers about nutritional benefits and how to cook millet dishes which are packed with essential nutrients, are gluten-free, high in fibre, low glycaemic-index, and providing long-term immunity to the consumer, creating domestic and global demand (Government of India Budget, 2023-24).

Likewise, researchers also have a role to play such as agricultural scientists, plant science researchers, soil and climate change scientists, agronomist, nutritionists, millet chefs, and food processing industries. They need to innovate harvesting technologies and strategies for sustainable millet farming, increasing output per acre of land, reduce loss pre/post-harvest, add value without losing its nutritional properties.

Further, food processing industries that add value and make food items which are ready-to-cook with recipes and ready-to-eat fast food such as small and medium business enterprise headed by women entrepreneurs can make millet-based biscuits, flakes, and chips, and motivate the consumers to make an informed decision, through ethically branding and marketing their products as millet based organic health foods.

Given the dietary globalisation by multinationals companies using genetic seeds and chemical fertilizers and pesticides, which makes the farmers depend on monopoly multinationals companies such as Monsanto for seeds. There is an urgent need to recognise and promote the local cultural and culinary knowledge of millets, it is nutritional value, create bio-diverse agricultural practices, and ecological benefit to consumers, producers, and decision-makers, in order to improve production efficiencies, investment in research and development and food sector supply-chain linkages. Therefore, importance of millets as solving the problems of going global as a nutritious food for solving problems related to hunger, poverty, mal-nutrition, food security, supply-demand imbalance based on vagaries of monsoon, sustainable farming, and climate change, has gone global, since the announcement of *'2023-Inetrnational Year of the Millets'*.

Future research can explore qualitatively with in-depth interviews with women entrepreneurs in millet industry, and rural women farmers producing millets covering factors of empowerment, food security, millet incentives, social, economic, and environmental sustainability about millets. In-depth interviews can also be conducted with key millet sector stakeholders such as government agriculture department officials, university researchers, millet educators, economist, entrepreneurs engaged in processing millet based ready-to-eat products and ready-to-cook meals, organic health grocery stores, millet cafés and restaurants, value added millet products, climate change experts, nutritionist, millet chefs, consumers, and Millett tourism agencies organising cooking class.

Further, millets nutritional quality and food varieties can also be measure through a quantitative survey applying quality factors/construct to measure customer satisfaction, loyalty to millets, nutritional benefits, health benefits, quality of the millet products, and consumer experience. Further, market demand for millets from domestic and foreign consumers and the supply-side stakeholders can identify factors for developing new value-added, ready to eat millet products such as chips and biscuits.

REFERENCES

Al Saidi, M. (2023). Caught off guard and beaten: Ukraine war and food security in the Middle East. *Frontiers in Nutrition*, *10*, 983346. doi:10.3389/fnut.2023.983346 PMID:36895270

Annor, G. A., Tyl, C., Marcone, M., Ragaee, S., & Marti, A. (2017). Why do millets have slower starch and protein digestibility than other cereals? *Trends in Food Science & Technology*, *66*, 73–83. doi:10.1016/j.tifs.2017.05.012

Azare, I.M., Dantata, I.J., Abdullahi, M.S., Adebayo, A.A. & Aliyu, M. (2020). Effects of Climate Change on Pearl Millet (Pennisetum glaucum [L. R. Br.]) Production in Nigeria. *Journal of Applied Sciences and Environmental Management*, 157-162. doi:10.4314/jasem.v24i1.23

Babele, P. K., Kudapa, H., Singh, Y., Varshney, R. K., & Kumar, A. (2022). Mainstreaming orphan millets for advancing climate smart agriculture to secure nutrition and health: Systematic Review. *Frontiers in Plant Science*, *13*, 902536. doi:10.3389/fpls.2022.902536 PMID:36035707

Bandyopadhyay, T., Muthamilarasan, M., & Prasad, M. (2017). Millets for next generation climate-smart agriculture. *Frontiers in Plant Science*, *8*, 1266. doi:10.3389/fpls.2017.01266 PMID:28769966

Basavaraj, G., Rao, P., Bhagavatula, S., & Ahmed, W. (2010). Availability and utilization of pearl millet in India. *Journal of SAT Agricultural Research*, 8.

Bhat, B. V., Rao, B. D., & Tonapi, V. A. (2018). *The Story of Millets*. ICAR-Indian Institute of Millets Research, Hyderabad. https://www.millets.res.in/pub/2018/The_Story_of_Millets.pdf

Bopp, J. (2020). Local notions of alternative practices: Organic food movements in Bangkok, Thailand and Chennai, India. *Sustainability (Basel)*, *12*(5), 1952. doi:10.3390u12051952

Bunkar, D. S., Goyal, S. K., Meena, K. K., & Kamalvanshi, V. (2021). Nutritional, Functional Role of Kodo Millet and its Processing: A Review. *International Journal of Current Microbiology and Applied Sciences*, *10*(1), 1972–1985. doi:10.20546/ijcmas.2021.1001.229

Ceasar, S. A., & Maharajan, T. (2022). The role of millets in attaining United Nation's sustainable developmental goals. *Plants, People, Planet*, *345-349*(4), 345–349. Advance online publication. wileyonlinelibrary.com/journal/ppp3345. doi:10.1002/ppp3.10254

Dal Corso, M., Pashkevych, G., Filipovic, D., Liu, X., & Matuzeviciute, G. M. (2022). Between Cereal Agriculture and Animal Husbandry: Millet in the Early Economy of the North Pontic Region. *Journal of World Prehistory*, *35*(3-4), 321–374. doi:10.100710963-022-09171-1

Das, K., & Ismail, S. (2023). Millet Mania: The Resurgence of Ancient Grains in Indian Agriculture. *Just Agriculture, Multidisciplinary e- Newsletter*, *3*(6).

Delind, L. B. (2006). Of bodies, place, and culture: Re-situating local food. *Journal of Agricultural & Environmental Ethics*, *19*(2), 121–146. doi:10.100710806-005-1803-z

Di Stefano, E., White, J., Seney, S., Heckmat, S., Mc Dowell, T., Sumarah, M., & Reid, G. (2017). A Novel Millet-Based Probiotic Fermented Food for the Developing World. *Nutrients*, *9*(5), 529. doi:10.3390/nu9050529 PMID:28531168

Down To Earth. (2020). *Daughters of the soil: Meet the fiery Dalit women working to preserve millet biodiversity*. Video. [Youtube]. https://www.youtube.com/watch?v=E_HHEF1guH8

Fischer, H. W., Reddy, N. L., & Rao, M. L. S. (2016). Can more drought resistant crops promote more climate secure agriculture? Prospects and challenges of millet cultivation in Ananthapur, Andhra Pradesh. *World Development Perspectives*, *2*, 5–10. doi:10.1016/j.wdp.2016.06.005

Food and Agriculture Organisation. (2023). *International Year of Millets*. FAO. https://www.fao.org/millets-2023/en

Government of India. (2023). *Millet Portal*. Indian Millets Production, APEDA. https://apeda.gov.in/milletportal/Production.html

Government of India Budget. (2023-24). *Government of India Budget Speech 2023-24*. Ministry of Finance. https://www.indiabudget.gov.in/doc/budget_speech.pdf

Gujral, S. K. (2017). Millets-the Climate Resilient Crop. *Journal of Emerging Technologies and Innovative Research*, *4*(6), 663–667.

Hassen, T. B., & ElBilali, H. (2022). Impacts of the Russia-Ukraine War on Global Food Security: Towards More Sustainable and Resilient Food Systems? *Foods*, *11*(15), 2301. doi:10.3390/foods11152301 PMID:35954068

Hazaresingh, H. (2020). 'Our Grandmother Used to Sing Whilst Weeding': Oral histories, millet food culture, and farming rituals among women smallholders in Ramanagaram district, Karnataka. *Modern Asian Studies*, 1–35.

Kumar, S., Sridhar, R., Monika, S., Kumar, A., Bhayankar, Raghavan, M., Tiwari, H., Kumar, A., Singh, S., & Yadav, R. (2023). A comprehensive review on Millets: A potential source of energy and nutrients for health. *International. Journal of Environment and Climate Change*, *13*(9), 2531–2538. doi:10.9734/ijecc/2023/v13i92487

Meek, D. (2022). From marginalized to miracle: Critical bioregionalism, jungle farming and the move to millets in Karnataka, India. *Agriculture and Human Values*, *39*(3), 871–883. doi:10.100710460-021-10287-x

Meena, R.J., Joshi, D.S., Bisht, J.K., & Kant, L. (2021). Global scenario of Millets Cultivation. *Millets and Millet Technology*, (pp. 35-50). Springer. doi:10.1007/978-981-16-0676-2_2

Meldrum, G., Roy, S., Lauridsen, N., & King, O. (2020). *India impact brief: Promoting kodo and kutki millets for improved incomes, climate resilience and nutrition in Madhya Pradesh*. Bioversity International. cgiar.org/bitstream/handle/10568/109361/India%20Impact%20Brief%20A4.pdf

Morrison, K. (2016). From millets to rice (and back again?): Cuisine, cultivation, and health in southern India. In *A companion to South Asia in the past* (pp. 358–373). Wiley Blackwell. doi:10.1002/9781119055280.ch23

Muthamilarasan, M., & Prasad, M. (2021). Small Millets for Enduring Food Security Amidst Pandemics. *Trends in Plant Science*, *26*(1), 33–40. doi:10.1016/j.tplants.2020.08.008 PMID:32900620

Nagaraj, N., Basavaraj, G., Rao, P., Bantilan, C., & Haldar, S. (2013). Sorghum and Pearl Millet Economy in India: Future Outlook and Options. *Economic and Political Weekly*, *48*(52), 74–81. https://www.jstor.org/stable/24477899

Padulosi, S., King, I. S., Hunter, D., & Swaminathan, M. S. (Eds.). (2022). *Orphan Crops for Sustainable Food and Nutrition Security: Promoting neglected and Underutilized Species.: Issues in Agricultura Biodiversity*. Routledge.

Padulosi, S., Mal, B., King, O. I., & Gotor, E. (2015). Minor Millets as a Central Element for Sustainably Enhanced Incomes, Empowerment, and Nutrition in Rural India. *Sustainability (Basel)*, *7*(7), 8904–8933. doi:10.3390u7078904

Rana, K., & Sahoo, M. (2023). Millet-based Farming Operations and Women: A Micro-Analysis in Koraput District, Odisha. *Science and Culture*, *89*(9-10), 304–310. doi:10.36094c.v89.2023.Millet_Based_Farming_Operations_and_Women.Rana_Sahoo.304

Rao, P. P., & Basavaraj, G. (2015). *Status and prospects of millet utilization in India and global scenario. Millets: Promotion for food, feed, fodder, nutritional and environment security, proceedings of global consultation on millets promotion for health & nutritional security. Society for Millets Research, ICAR*. Indian Institute of Millets Research.

Sabuz, A. A., Rana, M. R., Ahmad, T., & Molla, M. M. (2023). Health-Promoting Potential of Millet: A Review. *Separations*, *10*(80), 80. doi:10.3390eparations10020080

Sarkar, P., Kumar, L., Dhumal, C., Panigrahi, S., & Choudhary, R. (2015). Traditional and ayurvedic foods of Indian origin. *Journal of Ethnic Foods*, *2*(3), 97–109. doi:10.1016/j.jef.2015.08.003

Saxena, R. (2019). Feminization of Indian Agriculture. *International Journal of Advanced Scientific Research and Management*, *4*(7), 79–83.

Saxena, R., Vanga, S., Wang, J., Orsat, V., & Raghavan, V. (2018). Millets for Food Security in the Context of Climate Change: A Review. *Sustainability (Basel)*, *10*(7), 2228. doi:10.3390u10072228

Shah, P., Dhir, A., Joshi, R., & Tripathy, N. (2023). Opportunities and challenges in food entrepreneurship: In-depth qualitative investigation of millet entrepreneurs. *Journal of Business Research*, *155*(B), 113372. . doi:10.1016/j.jbusres.2022.113372

Singh, S. (2022). Know about different types of millets. *Krishijagran* https://krishijagran.com/agripedia/know-about-different-types-of-millet/

Singh, S., Suri, S., & Singh, R. (2022). Potential and unrealized future possibilities of Browntop millet in the food sector. *Frontiers in Sustainable Food Systems*, *6*, 974126. doi:10.3389/fsufs.2022.974126

Sood, S., Joshi, D. C., Chandra, A. K., & Kumar, A. (2019). Phenomics and genomics of finger millet: Current status and future prospects. *Planta*, *250*(3), 731–751. doi:10.100700425-019-03159-6 PMID:30968267

Sun, M., Yan, H., Zhang, A., Jin, Y., Lin, C., Luo, L., Wu, B., Fan, Y., Tian, S., Cao, X., Wang, Z., Luo, J., Yang, Y., Jia, J., Zhou, P., Tang, Q., Jones, C. S., Varshney, R. K., Srivastava, R. K., & Huang, L. (2023). Milletdb: A multi-omics database to accelerate the research of functional genomics and molecular breeding of millets. *Plant Biotechnology Journal*, *21*(11), 1–20. doi:10.1111/pbi.14136 PMID:37530223

Taraz, V. (2017). Adaptation to climate change: Historical evidence from the Indian monsoon. *Environment and Development Economics*, *22*(5), 517–545. doi:10.1017/S1355770X17000195

The Economic Times. (2023, July 21). India bans export of non-basmati white rice. *The Economic Times*. https://economictimes.indiatimes.com/news/economy/agriculture/govt-bans-export-of-non-basmati-white-rice/articleshow/101989078.cms?from=mdr

Tiwari, H., Naresh, R. K., Kumar, L., Kataria, S. K., Tewari, S., Saini, A., Yadav, R. K., & Asati, R. (2022). Millets for Food and Nutritional Security for Small and Marginal Farmers of North-West India in the Context of Climate Change: A Review. *International Journal of Plant and Soil Science*, 1694–1705. doi:10.9734/ijpss/2022/v34i232594

Union Budget. (2022-23). *Government of India: Summary of the Union Budget 2022-23*. Government of India. https://static.pib.gov.in/WriteReadData/userfiles/SUMMARY2022.pdf

United Nations. (2023). *Millets – good for people, the environment, and farmers*. UN. https://unric.org/en/millets-good-for-people-the-environment-and-farmers/

United Nations Sustainable Development Goals {UN-SDG}. (2023). *The Sustainable Development Agenda*. UN. https://www.un.org/sustainabledevelopment/development-agenda/

Valli, K. (2018). "Siri Dhanya (Millets)"is the Ultimate Solution to Good Health- Dr. Khadar Valli presented at Tech Mahindra campus https://www.youtube.com/watch?v=HtOBAulpR24

Verma, P. R. K., Lakhavat, S., & Yadav, V. K. (2023). Millets: Sustainable treasure house of bioactive components. *International Journal of Food Properties*, *26*(1), 1822–1840. doi:10.1080/10942912.2023.2236317

World Economic Forum. (2023). *We're on a brink of a 'polycrisis'- how worried should we be?* WEF. https://www.weforum.org/agenda/2023/01/polycrisis-global-risks-report-cost-

Yapi, A. M., Kergna, A. O., Debrah, S. K., Sidibe, A., & Sanogo, O. (2000). *Analysis of the Economic Impact of Sorghum and Millet Research in Mali*. International Crops Research Institute for the Semi-Arid Tropics, Patancheru, Hyderabad, Andhra Pradesh. http://oar.icrisat.org/1084/

ADDITIONAL REFERENCES

FAO. (2023). *Opening Ceremony of the International Year of Millets 2023*. Video. [Youtube]. https://www.fao.org/webcast/home/en/item/6085/icode/

Gupta, S. M., Arora, S., Mirza, N., Pande, A., Lata, C., Puranik, S., Kumar, J., & Kumar, A. (2017). Finger millet: A "certain" crop for an "uncertain" future and a solution to food insecurity and hidden hunger under stressful environments. *Frontiers in Plant Science*, *8*, 643. doi:10.3389/fpls.2017.00643 PMID:28487720

INDIASTAT. (2020). *Statistical information about India*. INDIASTAT. https://www.indiastat.com/agriculture data/2/stats.aspx

Kothari, S. L., Kumar, S., Vishnoi, R. K., Kothari, A., & Watanabe, K. N. (2005). Applications of biotechnology for improvement of millet crops: Review of progress and future prospects. *Plant Biotechnology (Tsukuba)*, *22*(2), 81–86. doi:10.5511/plantbiotechnology.22.81

Rao, B., Raghavendra, M., Nagasampige, H., & Ravikiran, M. (2011). Evaluation of nutraceutical properties of selected small millets. *Journal of Pharmacy & Bioallied Sciences*, *3*(2), 277. doi:10.4103/0975-7406.80775 PMID:21687358

Singh, G. (2023). Food of the future: The business and economy of super grains. *Economic Times*. https://economictimes.indiatimes.com/industry/cons-products/food/food-of-the-future-the-business-and-economy-of-super-grains/articleshow/99336226.cms

KEY TERMS AND DEFINITIONS

Economic Development: Economics development is a process by which policies related to programs and projects which are implemented in order to improve the economic wellbeing and quality of life of the people and their communities. For example, infrastructure development such as road, rial, seaport and airports, which creates jobs and improves transportation of goods and people and are the supply chain lifelines of the nation.

Millets vs. Cereals: Millets are grass crops growing small edible seeds, with high nutritional properties, drought tolerant, grown annually in dry to semi-arid low-rainfall regions of Asia and Africa, and do not require fertilizers, and pesticides. **Millets** are of two types: **major** (Foxtail/Kangni, Pearl/Bajra, Sorghum/Jowar,) and **minor** (Kodo, Proso/Cheena, Finger/Ragi, little/kutki, barnyard/sava). Whereas **Cereals** are large-seeded grains grows in richly nutrient soil, requires plenty of water, fertilizer, and pesticides for example wheat, barley, rice, rye, maize, and corn.

Polycrisis: The World Economic Forum (WEF) uses the term, to explain how, "present and future risks can also interact with each other to form a 'polycrisis. It is defined as a cluster of related global risks with compounding effects, such that the overall impact exceeds the sum of each part" (WEF- Global Risk Report, 2023).

Poverty Alleviation programme: When people do not have sufficient economic and noneconomic resources to meet their basic needs like food, clothing, and shelter, then poverty alleviation programmes have to be planned and implemented by the government and non-government organsiation. The first sustainable development goal of the United Nations is 'No Poverty.' Poverty alleviation programs aim to irradicate poverty and improve the quality of lives of the people living below the poverty line. For example, in India various programs to alleviate poverty are food for work programs, universal free education for the poor children, providing free meals and stationery in school s for poor homeless children.

Sustainable Development: Sustainable development of the economies aims to meet the present needs of the17 human development goals and simultaneously not compromising the needs of the future generations to meet their own basic needs.

Sustainable Societies: Societies are sustainable when they learn to live within the resources available provided by the nature and also maintaining and preserving them for the future generations. That is by not damaging or depleting the environmental resources through overuse (such as logging trees, overfishing, and mining) but enhancing the environment through sustainable practices such as tree planting, recycling, adopting green energy and business production practices.

Women's Empowerment Programme: Women's empowerment programmes organised by the government and non-government organsiation, aims to promote equality, and empower women to participate in all decision-making processes, act, and give them voice, dignity, and respect, by enhancing access to basic health, education, nutritional food, jobs, financial inclusion.

Chapter 13
Millet Cultivation as a Tool for Women's Empowerment:
Lessons From Odisha, India

Dilip Kumar Bagal
https://orcid.org/0000-0002-8105-2616
Government College of Engineering, Kalahandi, India

Lakshmi Prasad Panda
Government College of Engineering, Kalahandi, India

ABSTRACT

The experiences of women who grow millet in Odisha, India, are investigated in this research. The study identifies the constraints and opportunities for women to participate in millet farming, as well as the effects of millet farming on their lives and means of subsistence, using in-depth interviews, focus groups, and surveys. According to the research, although cultural expectations and a lack of resources present substantial obstacles for women, their participation may be facilitated by establishing women's groups and cooperatives, access to training, and support from NGOs and government initiatives. In addition to proposing steps to increase women's access to resources and support gender-sensitive agricultural extension services, the study highlights the potential of millet cultivation for advancing sustainable agriculture and gender equality.

1. INTRODUCTION

In Odisha, India, millet agriculture has a long history, but since high-yielding rice and wheat varieties have been promoted recently, it has diminished. Millets, on the other hand, are resource-efficient, highly nutritious, and climate-resilient crops that are appropriate for smallholder farmers confronting climatic concerns. Additionally, millet planting might increase the influence of women in Odisha, who historically play a significant role in agriculture but encounter obstacles when trying to access resources and authority Figure 1 displays the crop area of "Odisha Millets Mission." This research investigates the

DOI: 10.4018/978-1-6684-9819-4.ch013

Copyright © 2023, IGI Global. Copying or distributing in print or electronic forms without written permission of IGI Global is prohibited.

possibilities for millet farming to empower women in Odisha, India. It specifically looks at how millet farming empowers women in Odisha. Why are women-led millet projects successful? What obstacles and chances do women in millet farming face?

Figure 1. Crop area of "Odisha Millets Mission"
(Eurasiareview.com, 2021)

This research is essential because it helps develop policies and initiatives that promote resilient food systems in Odisha and surrounding areas while empowering women via sustainable agriculture. Designing interventions that support gender parity, women's access to resources, and their authority to make decisions may be more accessible by understanding the variables supporting women-led millet efforts. Resilience is increased by promoting nutrient-rich, climate-resilient crops like millet. According to studies, millet planting gives women in South Asia and India greater power via more money, better nutrition, and increased decision-making. Access to land, financing, markets, high-quality inputs, knowledge/training, and supporting policies/institutions are necessary for success. More resources, market connections, infrastructure, and decision-making authority are needed. Women in Odisha have obstacles to participating in agriculture due to societal conventions, a lack of credit and education, and their limited decision-making ability—women who overcame barriers earned more money and significantly influenced home decisions.

More study is required on the elements that enable success and problems, including women's experiences, resources, and decision-making authority, as well as the effects of millet farming on food security and nutrition.

2. LITERATURE REVIEW

The world's biggest producer of millet in India (Agricultural and Processed Food Products Export Development Authority,). A project in Odisha dubbed "Odisha Millet Mission" attempts to resurrect millets' production and consumption (Odisha Millet Mission,). The many millets that are grown include, among others, ragi, foxtail, pearl, sorghum, kodo, and barnyard. In Odisha, ragi account for almost 86 percent of all millet farming (Mongabay-India, 2022) (India Together,). The Government of Odisha's Department of Agriculture and Farmers Empowerment is responsible for the flagship program, the Odisha Millet Mission (Odisha Millet Mission,). The Government of Odisha introduced it to revitalize millets in fields and on plates in 2017 (Odisha Millet Mission,). The program seeks to encourage household consumption, preserve and develop millet landraces, and establish decentralized processing facilities. The mission has earned praise from a variety of sources. To promote millets, pulses, and oilseeds, the Indian government has requested that other states follow the Odisha Millets Mission model (Disha Odisha,). A millet mission modeled after the "Odisha Millets Mission" has been ordered by the Chhattisgarh State Planning Commission (Millet Mission Odisha,). The Odisha Millets Mission has reached numerous milestones. Odisha was the first state to announce a three-year direct incentive to farmers (Odisha Millet Mission,). The pricing benchmarking for small millet and foxtail millet was also completed in Odisha, making it the first state in the nation (Odisha Millet Mission,).

Additionally, the mission has received commendation from several sites. The Indian government has asked other states to imitate the Odisha Millets Mission to promote millets, pulses, and oilseeds (World Food Programme, 2020). The Chhattisgarh State Planning Commission has asked for a millet mission that is patterned after the "Odisha Millets Mission" (Odisha Millet Mission,)[10]. To revive millet in fields and on plates, the government of Odisha formed the Odisha Millet Mission (OMM) in 2017 (United Nations in India,). The mission wants to bring millets back to its fields and food plates by encouraging farmers to produce the crops that have traditionally played a vital part in the diet and agricultural system in tribal communities (Down to Earth, 2021). By supporting climatically adaptable crops like millet, the Odisha Millet Mission (OMM) has aided in women's empowerment in Odisha (United Nations in India,). In 13 districts with a predominance of tribal communities, the mission has helped women self-help groups (WSHGs) establish 76 "millet tiffin centers" in different community sites (Kalinga, 2022) (Mongabay-India, 2022). Through the Millet Shakti Cafe, the mission has also united OMM and Mission Shakti to empower women through SHGs. On the grounds of the Keonjhar Collectorate, the first one began operations in October 2021 (Odisha Millet Mission,).

The Odisha Millet Mission (OMM) encourages millet consumption through several programs. By incorporating millet into the public distribution system (PDS) and other government nutrition programs, the mission seeks to combat malnutrition (Vikalp Sangam, 2021)[16]. The goal is to raise awareness among urban residents about millets' nutritional value and health advantages via urban campaigns and promotional activities (Vikalp Sangam, 2021). Mission Shakti and the Odisha Millets Mission have teamed together to launch the state's first millet café in the Keonjhar district to encourage millet consumption in the form. The women's self-help group federation that Mission Shakti supports will run the Millet Shakti Café, which will serve a variety of culinary items made from millet (Odisha Millet Mission,). According to Vijayalakshmi et al. (2012) (Vijayalakshmi et al., 2012), finger millet's value addition raised the level of life for Indian rural women. According to Padulosi et al. (2015) (Padulosi et al., 2015), growing minor millets may increase earnings, nutritional status, and women's empowerment. According to Jahan et al. (2020) (Jahan et al., 2020), millet-based value-added goods improved farm women's business chances in India, boosting their income and frequency of millet-based product consumption. Although organic farming programs favor women's health and food security, Altenbuchner et al. (2017) (Altenbuchner et al., 2017) reported that women are excluded from commercial choices and agricultural training, reinforcing gender inequality. Diiro et al. (2018) (Diiro et al., 2018) discovered a positive association between maize production and women's empowerment in agriculture in western Kenya. This indicates that including women's empowerment in rural development, interventions might increase agricultural productivity and lower poverty. Women's participation in Tanzanian sunflower commercialization may improve agency and access to resources, according to Mosha et al. (2021) (Mosha et al., 2021), but cultural norms and power imbalances may still constrain the advantages.

According to Mishra and Sam (2016) (Mishra & Sam, 2016), women who owned property in Nepal had more negotiating leverage inside their households and invested more in the community's human capital. Parveen and Leonhauser (2008) (Parveen & Leonhauser, 2008) discovered a variety of traits in rural houses in Bangladesh, including freedom of mobility, media exposure, education, and training. Murugani and Thamaga-Chitja (2019) (Murugani & Thamaga-Chitja, 2019) examined the link between women's empowerment in farming and food availability in South Africa; crop diversification and market-oriented empowerment may boost food and nutrition security. According to a study by Forsythe et al. (2016) (Forsythe et al., 2016) on women's perceptions of cassava commercialization in Nigeria and Malawi, women may participate in new commercial opportunities while maintaining or enhancing food security. However, this is heavily dependent on gender norms and familial bonds. According to Michaelraj and Shanmugam (2013) (Michaelraj & Shanmugam, 2013), millets are resilient plants with a short growing season that thrive in arid climates. While there was an upward trend in output and area in the 1960s and 1970s, Jena and Mishra (2022) (Jena & Mishra, 2022) analyzed the development of millet production, size, and yield in Odisha from 1960 to 2020 and found that there has been a fall since the 1980s, primarily because of a reduction in area. However, Kothari et al. (2005) (Kothari et al., 2005) point out that there has yet to be much study and development in using biotechnology to enhance millet crops. The relevance of millets in tribal parts of Odisha's diets and cropping systems, as well as the need to revive millet production to treat malnutrition and improve climate change resistance, are highlighted by Parida (2020) (Parida, 2020).

Microenterprises had a considerable positive influence on rural women's social and economic circumstances in Odisha, which led to their empowerment, according to Mohanty et al. (2013) (Mohanty et al., 2013). Singh (2020) (Singh, 2020) highlights the initiatives taken by marginal women farmers in Odisha to revive seed saving and natural farming; this has improved food security and seed sovereignty while empowering women farmers. According to Vijayalakshmi et al. (2010) (Vijayalakshmi et al., 2010), empowering farmers in the production of high-quality seeds and supply conservation using a village seed bank approach and utilizing Self Help Groups (SHGs) of farm women for grain value addition and its marketing can increase farmers' incomes who grow finger millet. According to Baruah et al. (2022) (Baruah et al., 2022), collective action among farmers may have various positive effects, such as increased technological and production practice knowledge, a more robust social network and civic participation, and adequate water management that empowers women. However, research on the precise effect of millet cultivation on women's empowerment in Odisha still needs to be filled in since there are gaps in the literature.

3. METHODOLOGY

This research uses a qualitative case study technique to examine the potential of millet farming for women's empowerment in Odisha, India. A case study technique was selected because it enables a thorough analysis of the intricate problems related to women's involvement in millet farming. Case studies may provide deep insights into women farmers' lives, the conditions that support or obstruct their empowerment, and the difficulties they encounter while attempting millet farming. Data from women engaging in millet growing efforts around Odisha will be gathered via semi-structured interviews, focus groups, and field observations. Women's perceptions of how millet farming has affected their access to resources, income, nutrition, decision-making ability, and well-being will be captured via interviews and discussions. Participant observation and a fuller comprehension of women's roles and responsibilities in millet production will be made possible via field observations.

Participants will be chosen using targeted and snowball sampling strategies from various geographical locations, socioeconomic classes, and millet farming expertise levels. Data analysis will uncover important themes, patterns, and insights by classifying, categorizing, and interpreting the results. Triangulation will be made more accessible using many data sources, increasing the findings' reliability and validity. By concentrating exclusively on women's experiences of empowerment via millet farming, the study methodology seeks to overcome several significant shortcomings of prior studies. Greater understandings are possible via a qualitative approach than through strictly quantitative means. High-quality information to support sustainable and empowering practices in agriculture will be ensured through rigorous and open methodologies.

Twenty to thirty women involved in millet growing efforts in various districts of Odisha will be subjected to in-depth interviews. Through in-depth interviews, we want to learn more about how millet farming has affected women's lives, livelihoods, nutrition, income, workload, decision-making ability, and general well-being. To learn more about the potential and difficulties of encouraging women's participation in millet production, key informants, including agricultural officials, NGO workers, and women farmer leaders, will also be interviewed (n=10–15). Focus group talks (n=4-6) will provide chances for experience sharing and acquiring various viewpoints in a group context. In-depth topics will be explored in the FGDs, and group dynamics relating to gender, power, resource, and empowerment in millet farming will be observed. With an emphasis on varying experiences depending on criteria including age, caste, class, and amount of millet farming expertise, discussions will be fostered to express various viewpoints.

Through closed- and open-ended questions, surveys with women millet growers (n=40–60) will gather various experiences. The surveys are designed to reach more women in different places, enhance in-depth qualitative results, and pinpoint the main problems that might serve as the foundation for more study and action. Participants will be chosen from a variety of socioeconomic backgrounds, degrees of millet farming experience (uptake, duration), access to resources (land, credit, markets), and reported effects (income, nutrition, decision-making, and well-being) by purposive and snowball sampling. Data saturation will determine the ultimate sample size and distribution among approaches, ensuring that new themes are well supported before data gathering is finished. Triangulation to confirm and justify conclusions would be made more accessible by collecting data from many sources. To improve the technique and ensure an understanding of the underlying challenges, an interactive and iterative approach will consider new insights as they are discovered throughout the continuous sampling and data gathering. Overall, the richness and variety of the data would increase the validity and applicability of the results in guiding the development of policies, programs, and interventions targeted at empowering women via sustainable agriculture in Odisha and beyond.

The thematic analysis examined qualitative information on women's empowerment via millet growing in Odisha. This required coding interviews, focus groups, field notes, and documentation to uncover significant themes, trends, and insights relating to factors enabling/hindering women's empowerment, effects on their lives and livelihoods, obstacles encountered, and suggestions for supporting their achievement. The six phases of the thematic analysis process—data collection, data familiarisation, code generation, theme identification, grounding, and comparison—are shown in the flowchart (in

Figure 2). The rectangular nodes indicate the many phases in the procedure, and the arrows show how the analytical process flows.

Figure 2. Flowchart of the thematic analysis process for qualitative data on women's experiences with millet cultivation and empowerment in Odisha

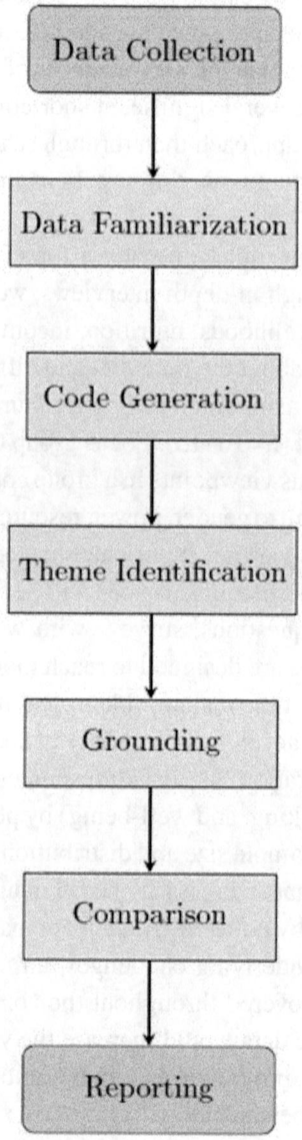

3.1. Data Collection

We interviewed 25 women growing millet in-depth for this research and ten key informant interviews with agricultural officials, NGO employees, and influential women farmers. Additionally, we polled 50 women from various parts of Odisha during four focus group talks with 24 female millet farmers.

3.2. Data Analysis

The information gathered from the interviews, focus groups, and surveys was examined using thematic analysis. We started by becoming acquainted with the knowledge and looking through the transcripts and notes to find the first codes. We created a list of regulations based on the data and categorized them into more important topics.

3.2.1. Theme One: Barriers to Women's Engagement in Millet Cultivation

Women's difficulties growing millet was one of the significant issues from the study. The capacity of many women to engage in agricultural activities has been restricted by societal norms and gender-based discrimination, as well as difficulties with accessing resources like land, loans, and markets. One participant gave the following example: "I have been trying to get a loan to buy seeds and fertilizer for my millet crop, but the bank won't lend to me because I don't have collateral." Another participant said that my spouse and in-laws don't believe it suits me to work in the fields since I'm a woman. They anticipate that I will look after the home and the kids.

3.2.2. Theme Two: Enablers of Women's Engagement in Millet Cultivation

Many women reported encountering facilitators that helped them participate in millet farming despite their obstacles. These include the opportunity to create women's clubs and cooperatives to pool resources and expertise, access to training and assistance from NGOs and government programs, and more. One participant said, "The government has provided us with training on how to grow millet using organic methods, and it has helped me to increase my yields." A different participant said, "I joined a women's cooperative that has given me access to seeds and tools, and we share knowledge and experiences about millet cultivation."

3.2.3. Theme Three: Impacts of Millet Cultivation on Women's Lives and Livelihoods

Many women said that growing millet had improved their life and means of subsistence, including their income, nutrition, and capacity for decision-making. Before I began producing millet, I was having trouble making ends meet, said one participant. Nevertheless, I can sell my harvest and make enough money to sustain my family. Another participant said, "We used to rely on rice and wheat for our meals, but now we have more variety in our diet because we grow and eat millet."

4. DISCUSSION

This research provides crucial new information on the experiences of women working in millet farming in Odisha. The statistics indicate that while there are many obstacles for women to participate in agriculture, there are also enablers that enhance their involvement and empowerment. The advantages of millet farming for women's life and means of subsistence demonstrate the potential of sustainable agriculture to aid in eradicating poverty and promoting gender equality in the area. The report also emphasizes the necessity for initiatives that deal with the social norms and structural constraints that prevent women from participating in agricultural activities.

We advise policymakers and development professionals to prioritize programs that address the underlying factors contributing to gender inequality in agriculture, such as uneven access to resources and discriminatory societal norms. Our is based on the results of our study. This might include programs to improve women's access to loans, land, markets, and campaigns to encourage creating women's organizations, cooperatives, and gender-sensitive agricultural extension services. The summary of the topics and sub-themes from the data on women's experiences with millet cultivation and empowerment in Odisha is shown in Table 1.

Table 1. Summary of themes and sub-themes from thematic analysis of data on women's experiences with millet cultivation and empowerment in Odisha

Theme	Sub-themes
Barriers to Women's Engagement in Millet Cultivation	• Lack of access to resources (e.g., land, credit, markets) • Social norms and discrimination
Enablers of Women's Engagement in Millet Cultivation	• Support from NGOs and government programs • Formation of women's groups and cooperatives
Impacts of Millet Cultivation on Women's Lives and Livelihoods	• Increased income • Improved nutrition • Greater decision-making power

Concepts from the literature on women empowerment, sustainable agriculture, and food safety in India guided the initial codes and frameworks. This enabled the topic to emerge from the participants' views. To verify that emerging themes were based on the data, they were regularly compared to the data while considering linkages. Insights from various areas and socioeconomic classes were contrasted to show the similarities and contrasts in how millet growing has been used as a tool for empowerment.

Techniques for improving rigor, validity, and dependability included:

- Researcher reflexivity: Analysing one's own biases, preconceptions, and positionality critically to provide unbiased, ethical analysis.
- Peer debriefing: Discuss themes and interpretations with peers or mentors to get other viewpoints, explore possible weaknesses, and deepen the analysis.
- Member checking: Discuss results with participants and essential sources to ensure correctness and resonance with intended meanings and clarify misunderstandings.

- Data triangulation: Using various sources to create a thorough, all-encompassing grasp of essential topics. Central themes were supported by consistency among sources.
- Long-term involvement: Spending significant time in the field to understand the contextual factors affecting women's experiences of empowerment via millet farming.
- Thick description: Providing in-depth information on the background, procedures, viewpoints, and connections to enable assessment of applicability and implications in various settings/populations.
- Audit trail: Carefully recording the study processes, procedures, data, analysis, and conclusions. This enables openness and impartial scrutiny and rigor, credibility, and integrity assessment.
- Active consideration of various reasonable explanations for the facts and themes to prevent drawing too many or too incorrect conclusions. The chosen interpretation was the most logical, perceptive, and well-supported.

5. MILLET CULTIVATION AND WOMEN'S EMPOWERMENT

Odisha has a long history of millet agriculture, but as high-yielding rice and wheat varieties have been promoted, millet output has decreased. However, millet farming remains a significant source of income and food security for smallholder communities, particularly for rural women. Women may become more independent, enhance family nutrition and health, acquire skills, and contribute to the community and food system change when provided access to resources, income, and decision-making power through millet production. For instance, with the assistance of a nearby NGO, women farmers in the hamlet of Barapali restored ragi (finger millet) production. They had access to better seeds, training in more effective methods, linkages to government programs for small farmers, and marketplaces where they could sell their goods. Crop yields, revenue, job opportunities, and investments in the education of young people all grew as a result. The ladies said their success as ragi growers and organizers of the community effort gave them more self-assurance, financial clout, and social prestige.

Essential elements that support women's emancipation via millet cultivation include:

- Resources: When women have access to land, seeds, credit, resources, expertise, and training, they may take charge of production and help to increase their living and independence.
- Social support: Support from community members, leaders, and programs/schemes helps people get over challenges, build skills, and open doors to possibilities. Peer support is another feature of group-based projects that improves success and well-being.
- Decision-making participation: Women may direct the rewards of their labor towards objectives and priorities that are most important to them and their communities when they are given a say in how resources and revenue will be utilized. Within the home and outside, this increases authority and confidence.

These elements provide women more financial power and control over their income, which they may use to raise their quality of life, become financially independent, and support dependents.

- Increasing access to wholesome food by increasing production and buying capacity. This improves development potential, productivity, and health, particularly for young children.

- Acquiring expertise and knowledge via training, experience, and hands-on resource access. This may increase possibilities and boost resistance to setbacks.
- Increasing one's sense of self-worth, confidence, and social standing. Women can better fight for their rights and influence laws and programs that help other women farmers when acknowledged for their accomplishments and achievements.
- Having more autonomy and control over one's life and means of subsistence. Women who are assisted in leadership positions develop their independence and the capacity to actively shape a better, more empowered future for their families, communities, and themselves.

The following are the main effects and results of women's empowerment via millet farming:

- Well-being: Women report much greater well-being when they have more influence over the assets, income, and choices that affect their lives. Better health, increased productivity, and interpersonal harmony result from less stress, anxiety, and sensitivity to adversity. Women feel more safe and respected when they can sustain themselves financially and, if necessary, their dependents.
- Food security: Women may guarantee their families' year-round access to inexpensive, high-quality food by growing their healthy food. Additional money from surplus output may be used to pay for other requirements. Higher yield/quality crops result in fewer cases of malnutrition and associated health problems, with positive communal implications.
- Income generation: Women may contribute financially to homes, invest in the welfare of their families and the possibilities for their children, and accumulate assets and savings by selling millet crop produce and value-added goods. This revenue often stays in the hands of women, enhancing their economic independence and empowerment.
- position and decision-making ability: Women's role and influence in family and community issues rise due to millet farming success as they acquire knowledge, skills, money, and confidence. They can disrupt conventional gender stereotypes by having a voice in crucial financial and resource allocation decisions. The high status also gives women more power to advocate for policies that would help all women farmers.

Benefits transcend the individual level to affect communities and systems:

- Households experience improved living conditions, intergenerational development, and health. Children and families prosper when women and carers prosper.
- Communities benefit from a multiplier effect from more productive, well-fed people. Food systems become more sustainable, just, and powerful due to chances for collaboration, skill-sharing, market connections, and fair distribution of activities, resources, and outcomes. Millet production improves food security by supplying a sustainable smallholder farmer livelihood, particularly for women, and climate-resilient crops. It improves poor or marginalized communities regarding accessibility, affordability, and agency over food choices.
- Policies and programs may be improved to more effectively address the diverse needs of women farmers. Sustainable, practical solutions that advance development, rights, justice, and equality arise when women actively engage in the systems that shape their lives.

6. CHALLENGES AND OPPORTUNITIES

6.1. Challenges

- Limited access to resources: Women's capacity to farm millets commercially is constrained by a lack of land, credit, seeds, and skills/training. They rely on men's and families' consent and often battle prejudice. Equal access to resources is crucial for empowering women and their success in millet growing.
- Social and cultural norms: Traditional gender roles strongly emphasize household duties for women, which limits their time for independent farming and their ability to travel or contact others outside the home. To defeat prejudice and establish justice/equality, it is necessary to challenge detrimental norms.
- Despite the benefits of millet farming, women are disproportionately responsible for unpaid care and household tasks, which leaves them with little free time or energy. Promoting an equal distribution of home tasks is essential to the prosperity and well-being of women.
- The effects of climate change: As climatic variability rises, recurrent crop losses put advances in development and food security at risk. Women risk increased marginalization since they have fewer resources to adjust to and recover from disasters. Measures to increase resilience must consider socioeconomic and gender differences in vulnerability and adaptive capability.

6.2. Opportunities and Solutions

- Encouraging access: Giving women access to land, financing, resources, training, and markets so they may build sustainable millet-based livelihoods. Quotas, reserves, legal rights, and lobbying may assist in amplifying the voice of women in agriculture.
- Challenging harmful norms: increasing awareness of injustice and discrimination; supporting movements reimagining gender roles and identities; elevating women's voices in society and decision-making; and funding girls' education and exposure outside the home.
- Equitable duty sharing: Supporting joint/shared responsibility between men and women for caregiving, millet farming, and other tasks. This might promote equality and justice while significantly reducing time poverty and enhancing well-being.
- Increasing women's involvement in resilience planning and decision-making; improving food preservation/storage; diversifying livelihood options; constructing community support systems. • Building context-specific adaptation strategies/techniques based on women's perspectives and vulnerabilities.
- Facilitating connections between women's organizations and aiding cooperative businesses focused on millet value addition, marketing, and distribution. This increases impact, gives women farmers access to more chances and resources, and strengthens their voice and influence in the broader food system and society.

In finality, millet farming offers a compelling chance to empower rural women, but significant obstacles remain, such as prejudice and uneven power structures. We can promote women farmers' success, rights, and flourishing on their terms by encouraging access, addressing harmful practices, dividing duties fairly, and developing resilience with a gender perspective. In the fight for justice, sustainability, and development, scaling up efforts via cooperation and advocacy enables transformative change that benefits people, communities, and systems.

7. CONCLUSION

The chance to empower rural women via millet farming is attractive, but many obstacles remain, including prejudice and unbalanced power structures. We can promote women farmers' success, rights, and flourishing on their terms by encouraging access, combating oppressive practices, allocating duties fairly, and developing resilience with a gender perspective. In the fight for justice, sustainability, and development, scaling up efforts via cooperation and advocacy enables transformative change that benefits people, communities, and systems. The potential for advancement is enormous when indigenous crops, historical knowledge, and women's autonomy are acknowledged and supported. We investigated the dynamics of women's empowerment in millet farming communities in Odisha via this study, uncovered barriers to and drivers of their achievement, and created suggestions for expanding effect. The lessons emphasize how encouraging women to become farmers may enhance well-being, increase food security, create wealthy communities, and influence the development of sustainable food systems and society.

To realize this vision, however, it is necessary to aggressively fight the power relations that support injustice. We can strive towards the egalitarian and empowered future women farmers deserve by amplifying their voices, encouraging their involvement in important choices, and fostering cooperation across differences. Even though there is still a long way to go, the tales of struggle and triumph shared here to give people hope that a different world is possible—one where women reap what they sow, communities prosper, ecosystems are restored, and the most vulnerable people influence our collective destiny. The fairness we establish now will determine how our food systems and planet fare. We can sow the seeds for a fair world and an empowered end by studying and supporting women farmers. Their experiences demonstrate the need to give control of agricultural biodiversity, food sovereignty, and sustainability to people who are most intimately involved with and affected by these issues. Without the women who cultivate the land, the harvest would be dust. Giving them more authority increases our ability as a group to change people's lives, communities, and institutions for freedom, equality, stability, and peace.

In this vein, may the work done here increase opportunities, spur change, and honor current and past farmers for the job that keeps us all alive. Their achievement is our fate; their fight is our joint obligation, and their freedom is our shared freedom. It is time for us to demand their empowerment; for the future, they are farming to become our shared legacy.

REFERENCES

Agricultural and Processed Food Products Export Development Authority. (n.d.). *India Millets Production*. APEDA. https://apeda.gov.in/milletportal/Production.html

Altenbuchner, C., Vogel, S., & Larcher, M. (2017, September). Effects of organic farming on the empowerment of women: A case study on the perception of female farmers in Odisha, India. *Women's Studies International Forum*, *64*, 28–33. doi:10.1016/j.wsif.2017.09.001

Baruah, S., Mohanty, S., & Rola, A. C. (2022). Empowering women farmers through collective action: A case study of Khanizpur Hamlet, Odisha. *Gender, Technology and Development*, *26*(1), 58–73. doi: 10.1080/09718524.2022.2040218

Diiro, G. M., Seymour, G., Kassie, M., Muricho, G., & Muriithi, B. W. (2018). Women's empowerment in agriculture and agricultural productivity: Evidence from rural maize farmer households in western Kenya. *PLoS One*, *13*(5), e0197995. doi:10.1371/journal.pone.0197995 PMID:29852008

Disha Odisha. (n.d.). *Odisha Millets Mission*. Disha Odisha. http://www.dishaodisha.org.in/projects/odisha-millets-mission/

Down to Earth. (2021, August 23). *Millet might: Here is how Odisha can succeed in reviving these cereals lost to the Green Revolution*. Down to Earth. https://www.downtoearth.org.in/news/agriculture/millet-might-here-is-how-odisha-can-succeed-in-reviving-these-cereals-lost-to-the-green-revolution-85263

Eurasiareview.com. (2021, December 30). Odisha Millets Mission: A Step Towards Combating Climate Vulnerability And Promoting Sustainable Food Systems – OpEd. *Eurasia Review*. https://www.eurasiareview.com/30122021-odisha-millets-mission-a-step-towards-combating-climate-vulnerability-and-promoting-sustainable-food-systems-oped/

Forsythe, L., Posthumus, H., & Martin, A. (2016). A crop of one's own? Women's experiences of cassava commercialization in Nigeria and Malawi. Journal of Gender. *Agriculture & Food Security*, *1*(2), 110–128.

India Together. (n.d.). Reviving cultivation and consumption of millets. *India Together*. https://www.indiatogether.org/odisha-millets-agriculture

Jahan, A., Arunjyothi, R., Shankar, A., Reddy, M., Reddy, T. P., Rajasheker, M., Krishna, K. R., & Shravika, L. (2020). Enhancing Entrepreneurship Opportunities through Millet-Based Value Added Products to Farm Women of Nagarkurnool District. *Current Journal of Applied Science and Technology*, *39*(48), 293–299. doi:10.9734/cjast/2020/v39i4831232

Jena, D., & Mishra, S. (2022). Growth and decomposition of millets in Odisha: 1960–61 to 2019–20. *Agricultural Science Digest-A Research Journal*, *42*(3), 272–277. doi:10.18805/ag.D-5447

Kalinga TV. (2022, January 20). *Odisha tribal women spearhead millets mission*. Kalinga TV. https://kalingatv.com/state/odishas-tribal-women-spearhead-mission-to-promote-millets-from-farm-to-plate/

Kothari, S. L., Kumar, S., Kothari, A., & Watanabe, K. N. (2005). Applications of biotechnology for improvement of millet crops: A review of progress and prospects. *Plant Biotechnology (Tsukuba)*, *22*(2), 81–88. doi:10.5511/plantbiotechnology.22.81

Michaelraj, P. S. J., & Shanmugam, A. (2013). A study on millets-based cultivation and consumption in India. International Journal of Marketing. *Financial Services & Management Research*, *2*(4), 49–58.

Millet Mission Odisha. (n.d.). *Home*. Millet Mission. https://milletmission.wordpress.com/

Mishra, K., & Sam, A. G. (2016). Does women's land ownership promote their empowerment? Empirical evidence from Nepal. *World Development*, *78*, 360–371. doi:10.1016/j.worlddev.2015.10.003

Mohanty, S. R., Das, B., & Mohanty, T. (2013). Empowerment of women in rural Odisha through micro-enterprises. *IOSR Journal of Humanities and Social Sciences*, *12*(1), 1–8. doi:10.9790/0837-1210108

Mongabay-India. (2022, August 10). *Climate resilient millets boost livelihood opportunities for Odisha's tribal women*. Mongabay-India.https://india.mongabay.com/2022/08/climate-resilient-millets-boost-livelihood-opportunities-for-odishas-tribal-women/

Mosha, D. B., Jeckoniah, J., Isinika, A., & Boniface, G. (2021). *The Influence of Sunflower Commercialisation and Diversity on Women's Empowerment: The Case of Iramba and Mkalama Districts*. Singida Region.

Murugani, V. G., & Thamaga-Chitja, J. M. (2019). How does women's empowerment in agriculture affect household food security and dietary diversity? The case of rural irrigation schemes in Limpopo Province, South Africa. *Agrekon*, *58*(3), 308–323. doi:10.1080/03031853.2019.1610976

Odisha Millet Mission. (n.d.). *Home*. Odisha Millet Mission. https://milletsodisha.com/

Odisha Millet Mission. (n.d.). *About Program*. Odisha Millet Mission. https://milletsodisha.com/about-program

Padulosi, S., Mal, B., King, O. I., & Gotor, E. (2015). Minor millets as a central element for sustainably enhanced incomes, empowerment, and nutrition in rural India. *Sustainability (Basel)*, *7*(7), 8904–8933. doi:10.3390u7078904

Parida, P. K. (2020). Magical Hands of a Tribal Farmer Set a Milestone in Finger Millet Cultivation-A Case from Koraput, Odisha, India. *Scholarly Journal of Food and Nutrition*, *2*(4). Advance online publication. doi:10.32474/SJFN.2020.02.000144

Parveen, S., & Leonhauser, I. U. (2008). Factors affecting the extent of economic empowerment of women in farm households: Experiences from rural Bangladesh. *International Journal of Human Ecology*, *9*(2), 117–126.

Singh, P. (2020). Capturing the Narratives of Sustainable Farming: Study of Marginal Women Farmers in Five Districts of Odisha. *The Indian Journal of Public Administration*, *66*(4), 455–465. doi:10.1177/0019556120982199

The Indian Express. (2021, November 16). Millet's cafe was inaugurated in Odisha's Keonjhar. *The India Express*. https://indianexpress.com/article/cities/bhubaneswar/millets-cafe-inaugurated-in-odishas-keonjhar-7601228/

The Locavore. (2023, March 22). *Odisha Millets Mission is a People's Programme*. The Locavore. https://thelocavore.in/2023/03/22/odisha-millets-mission-is-a-peoples-programme/

United Nations in India. (n.d.). *Odisha Millets Mission – Sowing Seeds of Change*. UN. https://india.un.org/en/192879-odisha-millets-mission-sowing-seeds-change

United Nations in India. (n.d.). *Odisha Millets Mission – Sowing Seedsof Change*. UN. https://india.un.org/en/192879-odisha-millets-mission-sowing-seeds-change

Vijayalakshmi, D., Geetha, K., Gowda, J., Ravi, S. B., Padulosi, S., & Mal, B. (2010). Empowerment of women farmersthrough value addition on minor millet genetic resources: A case study in Karnataka. *Indian Journal of Plant Genetic Resources*, *23*(1), 132–135.

Vijayalakshmi, D., Gowda, K. N., Jamuna, K. V., Ray, B. R., & Sajjan, J. T. (2012). Empowerment of self-help group women through value addition of finger millet. *Journal of Dairying, Foods & Home Sciences*, *31*(3).

Vikalp Sangam. (2021, December 26). *Odisha Millet Mission: The Successes and the Challenges*. Vikalp Sangam. https://vikalpsangam.org/article/odisha-millet-mission-the-successes-and-the-challenges/

World Food Programme. (2020). *Lessons from Odisha Millets Mission: A Case for Mainstreaming Millets in India's Food Programs*. WFP. https://www.wfp.org/publications/lessons-odisha-millets-mission-case-mainstreaming-millets

Compilation of References

A brief history of millets. (2022). The Millet Foundation. https://themillet.org/a-brief-history-of-millets/

Abdoulaye, T., & Sanders, J. H. (2006). New technologies, marketing strategies and public policy for traditional food crops: Millet in Niger. *Agricultural Systems*, *90*(1-3), 272–292. doi:10.1016/j.agsy.2005.12.008

Abdullah, K., Abukari, A. B. T., & Abdulai, A. M. (2022). Testing the climate resilience of sorghum and millet with time series data. *Cogent Food & Agriculture*, *8*(1), 2088459. doi:10.1080/23311932.2022.2088459

About Millets. (2022). SV Millets. https://svmillets.com/millets/#:~:text=Millets%20are%20one%20of%20the,became%20a%20staple%20food%20worldwide

Accenture. (2020). How COVID-19 will permanently change consumer behavior. *Accenture*. https://www.accenture.com/us-en/insights/consumer-goods-services/coronavirus-consumer-behavior-research

Adedara, O. A., & Taylor, J. R. N. (2021). Roles of protein, starch and sugar in the texture of sorghum biscuits. *Lebensmittel-Wissenschaft + Technologie*, *136*, 110323. doi:10.1016/j.lwt.2020.110323

Adimassu, Z., Mekonnen, K., Yirga, C., & Kessler, A. (2014). Effect of soil bunds on runoff, soil and nutrient losses, and crop yield in the central highlands of Ethiopia. *Land Degradation & Development*, *25*(6), 554–564. doi:10.1002/ldr.2182

Adweb, K. M. A., Cavus, N., & Sekeroglu, B. (2021). Cervical cancer diagnosis using very deep networks over different activation functions. *IEEE Access : Practical Innovations, Open Solutions*, *9*, 46612–46625. doi:10.1109/ACCESS.2021.3067195

Agarwal, B. (2000). Conceptualising environmental collective action: Why gender matters. *Cambridge Journal of Economics*, *24*(3), 283–310. doi:10.1093/cje/24.3.283

Agricultural and Processed Food Products Export Development Authority (APEDA). (2023). *Indian Millets Production*. APEDA. https://apeda.gov.in/milletportal/Production.html

Agricultural and Processed Food Products Export Development Authority. (n.d.). *India Millets Production*. APEDA. https://apeda.gov.in/milletportal/Production.html

Aidoo, M. K., Bdolach, E., Fait, A., Lazarovitch, N., & Rachmilevitch, S. (2016). Tolerance to high soil temperature in foxtail millet (Setaria italica L.) is related to shoot and root growth and metabolism. *Plant Physiology and Biochemistry*, *106*, 73–81. doi:10.1016/j.plaphy.2016.04.038 PMID:27149034

Akesowan, A. (2016). Influence of konjac flour on foaming properties of milk protein concentrate and quality characteristics of gluten-free cookie. *International Journal of Food Science & Technology*, *51*(7), 1560–1569. doi:10.1111/ijfs.13125

Al Saidi, M. (2023). Caught off guard and beaten: Ukraine war and food security in the Middle East. *Frontiers in Nutrition*, *10*, 983346. doi:10.3389/fnut.2023.983346 PMID:36895270

Alavi, S., Mazumdar, S. D., & Taylor, J. R. (2019). Modern convenient sorghum and millet food, beverage and animal feed products, and their technologies. In *Sorghum and millets* (pp. 293–329). AACC International Press. doi:10.1016/B978-0-12-811527-5.00010-1

Altenbuchner, C., Vogel, S., & Larcher, M. (2017, September). Effects of organic farming on the empowerment of women: A case study on the perception of female farmers in Odisha, India. *Women's Studies International Forum, 64*, 28–33. doi:10.1016/j.wsif.2017.09.001

Altindag, G., Certel, M., Erem, F., & Ilknur Konak, Ü. (2015). Quality characteristics of gluten-free cookies made of buckwheat, corn, and rice flour with/without transglutaminase. *Food Science & Technology International, 21*(3), 213–220. doi:10.1177/1082013214525428 PMID:24613830

Amadou, I. (2022). Millet Based Functional Foods: Bio-Chemical and Bio-Functional Properties. *Functional Foods*, 303-329. doi:10.1002/9781119776345.ch9

Amadou, I., Gounga, M. E., & Le, G.-W. (2013). Millets: Nutritional composition, some health benefits and processing-A review. *Emirates Journal of Food and Agriculture, 25*(7), 501–508. doi:10.9755/ejfa.v25i7.12045

Amritha, G. K., Dharmaraj, U., Halami, P. M., & Venkateswaran, G. (2018). Dephytinization of seed coat matter of finger millet (Eleusine coracana) by Lactobacillus pentosus CFR3 to improve zinc bioavailability. *Lebensmittel-Wissenschaft + Technologie, 87*, 562–566. doi:10.1016/j.lwt.2017.09.024

Anderson, J. W., & Hanna, T. J. (1999). Whole grains and protection against coronary heart disease: What are the active components and mechanisms? *The American Journal of Clinical Nutrition, 70*(3), 307–308. doi:10.1093/ajcn/70.3.307 PMID:10479191

Annor, G. A., Tyl, C., Marcone, M., Ragaee, S., & Marti, A. (2017). Why do millets have slower starch and protein digestibility than other cereals? *Trends in Food Science & Technology, 66*, 73–83. doi:10.1016/j.tifs.2017.05.012

Ansari, A., Pratiksha, B., Tejal, M., & Yelmame, M. (2017). Women Security System using GSM & GPS. *IJIRSET, 6*(3).

Aune, J. B., Doumbia, M., & Berthe, A. (2007). Microfertilizing sorghum and pearl millet in Mali: Agronomic, economic and social feasibility. *Outlook on Agriculture, 36*(3), 199–203. doi:10.5367/000000007781891504

Ávila, B. P., Braganca, G. C. M., Rockenbach, R., Alves, G. D., Monks, J., Gularte, M. A., & Elias, M. C. (2017). Physical and sensory characteristics of cake prepared with six whole-grain flours. *Journal of Food Measurement and Characterization, 11*(3), 1486–1492. doi:10.100711694-017-9527-0

Azare, I.M., Dantata, I.J., Abdullahi, M.S., Adebayo, A.A. & Aliyu, M. (2020). Effects of Climate Change on Pearl Millet (Pennisetum glaucum [L. R. Br.]) Production in Nigeria. *Journal of Applied Sciences and Environmental Management*, 157-162. doi:10.4314/jasem.v24i1.23

Babele, P. K., Kudapa, H., Singh, Y., Varshney, R. K., & Kumar, A. (2022). Mainstreaming orphan millets for advancing climate smart agriculture to secure nutrition and health: Systematic Review. *Frontiers in Plant Science, 13*, 902536. doi:10.3389/fpls.2022.902536 PMID:36035707

Bandyopadhyay, R., Atehnkeng, J., Ortega-Beltran, A., Akande, A., Falade, T. D., & Cotty, P. J. (2019). "Ground-truthing" efficacy of biological control for aflatoxin mitigation in farmers' fields in Nigeria: From field trials to commercial usage, a 10-year study. *Frontiers in Microbiology, 10*, 2528. doi:10.3389/fmicb.2019.02528 PMID:31824438

Bandyopadhyay, T., Muthamilarasan, M., & Prasad, M. (2017). Millets for next generation climate-smart agriculture. *Frontiers in Plant Science, 8*, 1266. doi:10.3389/fpls.2017.01266 PMID:28769966

Banerjee, S. (2022). *International Year of Millets: Unleashing the potential of millets for the well-being of people and the environment*. Food and Agriculture Organization of the United Nations (FAO). https://www.fao.org/newsroom/detail/international-year-of-millets-unleashing-the-potential-of-millets-for-the-well-being-of-people-and-the-environment/en

Bangar, S. P., Ashogbon, A. O., Dhull, S. B., Thirumdas, R., Kumar, M., Hasan, M., Chaudhary, V., & Pathem, S. (2021). Proso-millet starch: Properties, functionality, and applications. *International Journal of Biological Macromolecules*, *190*, 960–968. doi:10.1016/j.ijbiomac.2021.09.064 PMID:34536472

Baruah, S., Mohanty, S., & Rola, A. C. (2022). Empowering women farmers through collective action: A case study of Khanizpur Hamlet, Odisha. *Gender, Technology and Development*, *26*(1), 58–73. doi:10.1080/09718524.2022.2040218

Basavaraj, G., Rao, P., Bhagavatula, S., & Ahmed, W. (2010). Availability and utilization of pearl millet in India. *Journal of SAT Agricultural Research*, 8.

Behera, M. K. (2017). Assessment of the State of Millets Farming in India. *MOJ Ecology and Environmental Sciences*, *2*(1), 16–20. doi:10.15406/mojes.2017.02.00013

Bhardwaj, N., & Aggarwal, N. (2014). Design development of "Suraksha". A women safety device. *Int J Inf Computer Technology 4*(8), 787–792.

Bhat, B. V., Rao, B. D., & Tonapi, V. A. (2018). *The Story of Millets*. ICAR-Indian Institute of Millets Research, Hyderabad. https://www.millets.res.in/pub/2018/The_Story_of_Millets.pdf

Bhattacharjee, R., Khairwal, I. S., Bramel, P. J., & Reddy, K. N. (2007). Establishment of a pearl millet [Pennisetum glaucum (L.) R. Br.] core collection based on geographical distribution and quantitative traits. *Euphytica*, *155*(1-2), 35–45. doi:10.100710681-006-9298-x

Bhatt, D., Fairos, M., & Mazumdar, A. (2022). Millets: Nutritional composition, production and significance: A review. ~ 1577 ~ Pharma. *The Innovation Journal*, 1577–1582.

Bilgiçli, N., İbanogˇlu, Ş., & Herken, E. N. (2007). Effect of dietary fiber addition on the selected nutritional properties of cookies. *Journal of Food Engineering*, *78*(1), 86–89. doi:10.1016/j.jfoodeng.2005.09.009

Bopp, J. (2020). Local notions of alternative practices: Organic food movements in Bangkok, Thailand and Chennai, India. *Sustainability (Basel)*, *12*(5), 1952. doi:10.3390u12051952

Bunkar, D. S., Goyal, S. K., Meena, K. K., & Kamalvanshi, V. (2021). Nutritional, Functional Role of Kodo Millet and its Processing: A Review. *International Journal of Current Microbiology and Applied Sciences*, *10*(1), 1972–1985. doi:10.20546/ijcmas.2021.1001.229

Butt, M. S., Arshad, M. U., Alam, M. S., & Nadeem, M. T.BUTT. (2007). Bioavailability and storage stability of vitamin A fortificant (retinyl acetate) in fortified cookies. *Food Research International*, *40*(10), 1212–1219. doi:10.1016/j.foodres.2007.07.002

Campbell, R., Ahrens, C. E., Sefi, T., Wasco, S. M., & Barner, H. E. (2001). Social reactions to rape victims: Healing and hurtfull effects on psychological and physical health outcomes. *Violence and Victims*, *16*(3), 310–367. doi:10.1891/0886-6708.16.3.287 PMID:11437118

Cauvain, S. P. (2013). Evolution in the manufacture of baked products. *Cereal Foods World*, *58*(5), 277–278. doi:10.1094/CFW-58-5-277

Cauvain, S. P. (2016). *Bread and other bakery products, The Stability and Shelf Life of Food*. Elsevier Ltd., doi:10.1016/B978-0-08-100435-7.00015-0

Ceasar, S.A., Maharajan, T., Krishna, T.P.A., & Ignacimuthu, S. (2023). Finger millet (Eleusine coracana (L.) Gaertn). *Neglected Underutilized Crop,* 137–149. doi:10.1016/B978-0-323-90537-4.00031-4

Ceasar, S. A., & Maharajan, T. (2022). The role of millets in attaining United Nation's sustainable developmental goals. *Plants, People, Planet, 345-349*(4), 345–349. Advance online publication. wileyonlinelibrary.com/journal/ppp3345. doi:10.1002/ppp3.10254

Chand, D., Nayak, S., Bhat, K. S., & Parikh, S. (2015). A mobile application for Women's Safety: WoS App. *IEEE Region Conference TENCON*. IEEE.

Chang, J. H. (2017). *Climate and agriculture: An ecological survey*. Routledge. doi:10.4324/9781315081069

Chaudhury, R. (2010). Determinants of Consumer Behavior in Buying RTE Foods. *The Journal of Business and Retail Management Research, 5*(1), 76–86.

Choudhary, S., Guha, A., Kholova, J., Pandravada, A., Messina, C. D., Cooper, M., & Vadez, V. (2020). Maize, sorghum, and pearl millet have highly contrasting species strategies to adapt to water stress and climate change-like conditions. *Plant Science, 295*, 110297. doi:10.1016/j.plantsci.2019.110297 PMID:32534623

Coulibaly, A., Kouakou, B., & Chen, J. (2011). Phytic acid in cereal grains: structure, healthy or harmful ways to reduce phytic acid in cereal grains and their effects on nutritional quality. *American Journal of Plant Nutrition and fertilization technology, 1*(1), 1-22.

Credidio, E. (2005). Alimentos Fortificados. *Boletim Brasileiro de Nutrilogia, 3*(1), 1–3.

Dal Corso, M., Pashkevych, G., Filipovic, D., Liu, X., & Matuzeviciute, G. M. (2022). Between Cereal Agriculture and Animal Husbandry: Millet in the Early Economy of the North Pontic Region. *Journal of World Prehistory, 35*(3-4), 321–374. doi:10.100710963-022-09171-1

Dangi, N., Yadav, B. S., & Yadav, R. B. (2019). Pasting, rheological, thermal and gel textural properties of pearl millet starch as modified by guar gum and its acid hydrolysate. *International Journal of Biological Macromolecules, 139*, 387–396. doi:10.1016/j.ijbiomac.2019.08.012 PMID:31381921

Dapčević Hadnadev, T. R., Torbica, A. M., & Hadnadev, M. S. (2013). Influence of Buckwheat Flour and Carboxymethyl Cellulose on Rheological Behaviour and Baking Performance of Gluten-Free Cookie Dough. *Food and Bioprocess Technology, 6*(7), 1770–1781. doi:10.100711947-012-0841-6

Das, K., & Ismail, S. (2023). Millet Mania: The Resurgence of Ancient Grains in Indian Agriculture. *Just Agriculture, Multidisciplinary e- Newsletter, 3*(6).

Dayakar Rao, B., Bhat, V., Niranjan, T., Sujatha, M., & Tonapi, V. A. (2021). Demand creation measures and value chain model on millets in India. *Millets and millet technology*, 381-411.

Dayakar Rao, B., Kandlakunta, B., Christina, A., Golla, S. D., & Tonapi, V. (2017). *Nutritional and Health Benefits of Millets [WWW Document]*. ICAR_Indian Inst. Millets Res. Rajendranagar. https://www.researchgate.net/publication/318921493_Nutritional_and_Health_Benefits_of_Millets (accessed 3.29.23).

Dayakar Rao, B., Patil, J. V., & Hemavati, T. V. (2014). *Terminal report of NAIP sub-project on Creation of Demand for Millet Foods through PCS Value Chain*. Directorate of Sorghum Research, Rajendra nagar, Hyderabad. https://naip.icar.gov.in/download/79347/207601.pdf/207601.pdf

De Groote, H., Kariuki, S. W., Traore, D., Taylor, J. R., Ferruzzi, M. G., & Hamaker, B. R. (2018). Measuring consumers' interest in instant fortified pearl millet products: A field experiment in Touba, Senegal. *Journal of the Science of Food and Agriculture, 98*(6), 2320–2331. doi:10.1002/jsfa.8722 PMID:28990670

De la Hera, E., Martinez, M., Oliete, B., & Gómez, M. (2013). Influence of Flour Particle Size on Quality of Gluten-Free Rice Cakes. *Food and Bioprocess Technology*, *6*(9), 2280–2288. doi:10.100711947-012-0922-6

De la Hera, E., Ruiz-París, E., Oliete, B., & Gómez, M. (2012). Studies of the quality of cakes made with wheat-lentil composite flours. *Lebensmittel-Wissenschaft + Technologie*, *49*(1), 48–54. doi:10.1016/j.lwt.2012.05.009

Delind, L. B. (2006). Of bodies, place, and culture: Re-situating local food. *Journal of Agricultural & Environmental Ethics*, *19*(2), 121–146. doi:10.100710806-005-1803-z

Descamps, M. J., Rothblum, E., Bradford, J., & Ryan, C. (2000). Mental health impact of child sexual abuse, rape, intimate partner violence, and hate crimes. National lesbian health care survey, 9(5).

Devi, K., Kumar, V., Kumar, V., Mahajan, N., Kaur, J., Sharma, S., Kumar, A., Khan, R., Bishnoi, M., & Kondepudi, K. K. (2023). Modified cereal bran (MCB) from finger millet, kodo millet, and rice bran prevents high-fat diet-induced metabolic derangements. *Food & Function*, *14*(3), 1459–1475. doi:10.1039/D2FO02095E PMID:36648164

Devi, P. B., Vijayabharathi, R., Sathyabama, S., Malleshi, N. G., & Priyadarisini, V. B. (2014). Health benefits of finger millet (Eleusine coracana L.) polyphenols and dietary fiber: A review. *Journal of Food Science and Technology*, *51*(6), 1021–1040. doi:10.100713197-011-0584-9 PMID:24876635

Di Cairano, M., Galgano, F., Tolve, R., Caruso, M. C., & Condelli, N. (2018). Focus on gluten free biscuits: Ingredients and issues. *Trends in Food Science & Technology*, *81*, 203–212. doi:10.1016/j.tifs.2018.09.006

Di Stefano, E., White, J., Seney, S., Heckmat, S., Mc Dowell, T., Sumarah, M., & Reid, G. (2017). A Novel Millet-Based Probiotic Fermented Food for the Developing World. *Nutrients*, *9*(5), 529. doi:10.3390/nu9050529 PMID:28531168

Diiro, G. M., Seymour, G., Kassie, M., Muricho, G., & Muriithi, B. W. (2018). Women's empowerment in agriculture and agricultural productivity: Evidence from rural maize farmer households in western Kenya. *PLoS One*, *13*(5), e0197995. doi:10.1371/journal.pone.0197995 PMID:29852008

Disha Odisha. (n.d.). *Odisha Millets Mission*. Disha Odisha. http://www.dishaodisha.org.in/projects/odisha-millets-mission/

Down To Earth. (2020). *Daughters of the soil: Meet the fiery Dalit women working to preserve millet biodiversity*. Video. [Youtube]. https://www.youtube.com/watch?v=E_HHEF1guH8

Down to Earth. (2021, August 23). *Millet might: Here is how Odisha can succeed in reviving these cereals lost to the Green Revolution*. Down to Earth. https://www.downtoearth.org.in/news/agriculture/millet-might-here-is-how-odisha-can-succeed-in-reviving-these-cereals-lost-to-the-green-revolution-85263

Durairaj, M., Gurumurthy, G., Nachimuthu, V., Muniappan, K., & Balasubramanian, S. (2019). Dehulled small millets: The promising nutricereals for improving the nutrition of children. *Maternal and Child Nutrition*, *15*(S3), e12791. doi:10.1111/mcn.12791 PMID:31148399

Duta, D. E., & Culetu, A. (2015). Evaluation of rheological, physicochemical, thermal, mechanical and sensory properties of oat-based gluten free cookies. *Journal of Food Engineering*, *162*, 1–8. doi:10.1016/j.jfoodeng.2015.04.002

Eurasiareview.com. (2021, December 30). Odisha Millets Mission: A Step Towards Combating Climate Vulnerability And Promoting Sustainable Food Systems – OpEd. *Eurasia Review*. https://www.eurasiareview.com/30122021-odisha-millets-mission-a-step-towards-combating-climate-vulnerability-and-promoting-sustainable-food-systems-oped/

Ezeagu, I. E. (2015). Nutritional evaluation of malted millet fortified with defatted sesame seed flour. *African Journal of Food Science*, *9*(2), 70–75.

Falconí, L. G., Pérez, M., & Aguilar, W. G. (2019, June). Transfer learning in breast mammogram abnormalities classification with mobilenet and nasnet. In 2019 international conference on systems, signals and image processing (IWSSIP) (pp. 109-114). IEEE. doi:10.1109/IWSSIP.2019.8787295

Fan, D., Ruiz, L., & Gong, J. (2018). An Energy Harvesting Modeling and Profiling Platform for Sensor Networks. *IEEE Journal of biomedical conference in health informatics, 22*(1).

FAO, UN. (2023). *International Year of Millets 2023*. Food and Agriculture Organization of the United Nations. https://www.fao.org/millets-2023/en

FAO. (2021). *FAO Statistical Yearbook 2021 - World Food and Agriculture*. Food and Agriculture Organization of the United Nations. https://reliefweb.int/report/world/fao-statistical-yearbook-2021-world-food-and-agriculture#:~:text=In%20fact%2C%20the%20global%20value,percent%20of%20the%20global%20workforce

FAO/WHO. (1998). Carbohydrates in human nutrition: Report of joint FAO/WHO expert consultation. *FAO Food and Nutrition, 66*, 1–140. PMID:9743703

Farnworth, C. R. (2003). Tapping indigenous knowledge to save millet and sorghum biodiversity. *LEISA Magazine, 19*(1), 10–11.

Fischer, H. W., Reddy, N. L., & Rao, M. L. S. (2016). Can more drought resistant crops promote more climate secure agriculture? Prospects and challenges of millet cultivation in Ananthapur, Andhra Pradesh. *World Development Perspectives, 2*, 5–10. doi:10.1016/j.wdp.2016.06.005

Foley, J. K., Michaux, K. D., Mudyahoto, B., Kyazike, L., Cherian, B., Kalejaiye, O., Ifeoma, O., Ilona, P., Reinberg, C., Mavindidze, D., & Boy, E. (2021). Scaling up delivery of biofortified staple food crops globally: Paths to nourishing millions. *Food and Nutrition Bulletin, 42*(1), 116–132. doi:10.1177/0379572120982501 PMID:33593095

Food and Agriculture Organisation. (2023). *International Year of Millets*. FAO. https://www.fao.org/millets-2023/en

Food and Drug Administration- FDA. (2013). *'Gluten-Free' Means What It Says*. FDA.

Forsythe, L., Posthumus, H., & Martin, A. (2016). A crop of one's own? Women's experiences of cassava commercialization in Nigeria and Malawi. Journal of Gender. *Agriculture & Food Security, 1*(2), 110–128.

Gahukar, R. T. (1988). *Problems and perspectives of pest management in the Sahel: A case study of pearl millet*. Taylor & Francis. doi:10.1080/09670878809371202

Gaikwad, P. S., Reddy, N. B. P., Sandhya, K., & Yadav, B. K. (2023). Traditional vs. Modern Usage of Nutri-Cereals. In *Nutri-Cereals* (pp. 27–56). CRC Press. doi:10.1201/9781003251279-2

Gassner, A., Harris, D., Mausch, K., Terheggen, A., Lopes, C., Finlayson, R. F., & Dobie, P. (2019). Poverty eradication and food security through agriculture in Africa: Rethinking objectives and entry points. *Outlook on Agriculture, 48*(4), 309–315. doi:10.1177/0030727019888513 PMID:33281227

Gatti, M., Garavani, A., Vercesi, A., & Poni, S. (2017). Ground-truthing of remotely sensed within-field variability in a cv. Barbera plot for improving vineyard management. *Australian Journal of Grape and Wine Research, 23*(3), 399–408. doi:10.1111/ajgw.12286

Geetha, K., Yankanchi, G. M., Hulamani, S., & Hiremath, N. (2020). Glycemic index of millet based food mix and its effect on pre diabetic subjects. *Journal of Food Science and Technology, 57*(7), 2732–2738. doi:10.100713197-020-04309-5 PMID:32549623

George, R., Anjaly Cherian, V., & Antony, A. (2014, April). An intelligent security system for violence against women in public places. *IJEAT*, *3*(4), 64–68.

Gerzhova, A., Mondor, M., Benali, M., & Aider, M. (2016). Incorporation of canola proteins extracted by electroactivated solutions in gluten-free biscuit formulation of rice–buckwheat flour blend: Assessment of quality characteristics and textural properties of the product. *International Journal of Food Science & Technology*, *51*(3), 814–827. doi:10.1111/ijfs.13034

Glew, R. H., Ayaz, F. A., Sanz, C., Huang, H. S., Chuang, L. T., & VanderJagt, D. J. (2006). Changes in the amino acid and mineral content of cereals, legumes, and seeds during sous vide processing. *Journal of Agricultural and Food Chemistry*, *54*(22), 8397–8402. PMID:17061812

Godefroidt, T., Ooms, N., Pareyt, B., Brijs, K., & Delcour, J. A. (2019). Ingredient Functionality During Foam-Type Cake Making: A Review. *Comprehensive Reviews in Food Science and Food Safety*, *18*(5), 1550–1562. doi:10.1111/1541-4337.12488 PMID:33336911

Government of India Budget. (2023-24). *Government of India Budget Speech 2023-24*. Ministry of Finance. https://www.indiabudget.gov.in/doc/budget_speech.pdf

Government of India. (2023). *Millet Portal*. Indian Millets Production, APEDA. https://apeda.gov.in/milletportal/Production.html

Govindaraj, M., Rai, K. N., Cherian, B., Pfeiffer, W. H., Kanatti, A., & Shivade, H. (2019). Breeding biofortified pearl millet varieties and hybrids to enhance millet markets for human nutrition. *Agriculture*, *9*(5), 106. doi:10.3390/agriculture9050106

Gowda, N. N., Siliveru, K., Prasad, P. V., Bhatt, Y., Netravati, B. P., & Gurikar, C. (2022). Modern processing of Indian millets: A perspective on changes in nutritional properties. *Foods*, *11*(4), 499. doi:10.3390/foods11040499 PMID:35205975

Gowri, S., & Anandha Mala, G. S. (2015, June). IR system for investigation in textual data. *Indian Journal of Science and Technology*, *8*(12), 1–7.

Gujral, S. K. (2017). Millets-the Climate Resilient Crop. *Journal of Emerging Technologies and Innovative Research*, *4*(6), 663–667.

Gupta, N., Srivastava, A., & Pandey, V. (2012). Biodiversity and nutraceutical quality of some Indian millets. *Proceedings of the National Academy of Sciences. India. Section B, Biological Sciences*, *82*(2), 265–273. doi:10.100740011-012-0035-z

Gupta, R. K., Gangoliya, S. S., & Singh, N. K. (2015). Reduction of phytic acid and enhancement of bioavailable micronutrients in food grains. *Journal of Food Science and Technology*, *52*(2), 676–684. doi:10.100713197-013-0978-y PMID:25694676

Gupta, R., Singh, H., Kumar, R. P., & Meghwal, M. (2023). Economic, Social, and Market Feasibility of Nutri-Cereals. In *Nutri-Cereals* (pp. 345–360). CRC Press. doi:10.1201/9781003251279-16

Handoo, S. (2023). *What Is Glycemic Index & List Of Foods With Their GI*.

Hassen, T. B., & ElBilali, H. (2022). Impacts of the Russia-Ukraine War on Global Food Security: Towards More Sustainable and Resilient Food Systems? *Foods*, *11*(15), 2301. doi:10.3390/foods11152301 PMID:35954068

Hatfield, J. L., Boote, K. J., Kimball, B. A., Ziska, L. H., Izaurralde, R. C., Ort, D., Thomson, A. M., & Wolfe, D. (2011). Climate impacts on agriculture: Implications for crop production. *Agronomy Journal*, *103*(2), 351–370. doi:10.2134/agronj2010.0303

Haussmann, B. I., Fred Rattunde, H., Weltzien-Rattunde, E., Traoré, P. S., Vom Brocke, K., & Parzies, H. K. (2012). Breeding strategies for adaptation of pearl millet and sorghum to climate variability and change in West Africa. *Journal Agronomy & Crop Science, 198*(5), 327–339. doi:10.1111/j.1439-037X.2012.00526.x

Hazaresingh, H. (2020). 'Our Grandmother Used to Sing Whilst Weeding': Oral histories, millet food culture, and farming rituals among women smallholders in Ramanagaram district, Karnataka. *Modern Asian Studies*, 1–35.

Heenan, S. P., Hamid, N., Dufour, J. P., Harvey, W., & Delahunty, C. M. (2009). Consumer freshness perceptions of breads, biscuits and cakes. *Food Quality and Preference, 20*(5), 380–390. doi:10.1016/j.foodqual.2009.02.008

Hegarty-Craver, M., Polly, J., O'Neil, M., Ujeneza, N., Rineer, J., Beach, R. H., Lapidus, D., & Temple, D. S. (2020). Remote crop mapping at scale: Using satellite imagery and UAV-acquired data as ground truth. *Remote Sensing (Basel), 12*(12), 1984. doi:10.3390/rs12121984

He, L., Zhang, B., Wang, X., Li, H., & Han, Y. (2015). Foxtail millet: Nutritional and eating quality, and prospects for genetic improvement. *Frontiers of Agricultural Science and Engineering, 2*(2), 124–133. doi:10.15302/J-FASE-2015054

Hoffpauer. (1994). Enrichment of rice. In W. E. Marshall & A. Wadsworth (Eds.), *Rice science and Technology*. Marshall Rekker.

Howard, E., & LeWine, M. D. (2021). *Glycemic index for 60+ foods*. Harvard Medical School.

Huebner, F. R., & Wall, J. S. (1981). Fractionation and quantitative differences of glutenin from wheat varieties varying in baking quality. *Cereal Chemistry, 53*, 258–268.

Hussain, S. M., Nizamuddin, S. A., Asunico, R., Ramaiah, C., & Vikram Singh, A. (2016) Prototype of an intelligent system based on RFID and GPS technologies for women safety. In: *IEEE 5th international conference on reliability, infocom technologies and optimization (ICRITO), 5.* 10.1109/ICRITO.2016.7784986

Hussain, S., Mohamed, A. A., Alamri, M. S., Ibraheem, M. A., Abdo Qasem, A. A., Serag El-Din, M. F., & Almaiman, S. A. M. (2004). Wheat–millet flour cookies: Physical, textural, sensory attributes and antioxidant potential. *Food Science & Technology International, 26*(4), 311–320. doi:10.1177/1082013219894127 PMID:31818128

India Together. (n.d.). Reviving cultivation and consumption of millets. *India Together.* https://www.indiatogether.org/odisha-millets-agriculture

International Nutritional Anemia Consultative Group - INACG. (1977). *Guidelines for the eradication of iron deficiency anemia Washington: International Nutritional Anemia Consultative Group*. Nutrition Foundation.

Jahan, A., Arunjyothi, R., Shankar, A., Reddy, M., Reddy, T. P., Rajasheker, M., Krishna, K. R., & Shravika, L. (2020). Enhancing Entrepreneurship Opportunities through Millet-Based Value Added Products to Farm Women of Nagarkurnool District. *Current Journal of Applied Science and Technology, 39*(48), 293–299. doi:10.9734/cjast/2020/v39i4831232

Jeeva, J. C. (2019). Engendering finger millet-based value chains for livelihood and nutritional security of women in agriculture. *Current Science, 116*(11), 1893–1896. https://www.currentscience.ac.in/Volumes/116/11/1893.pdf. doi:10.18520/cs/v116/i11/1893-1896

Jena, D., & Mishra, S. (2022). Growth and decomposition of millets in Odisha: 1960–61 to 2019–20. *Agricultural Science Digest-A Research Journal, 42*(3), 272–277. doi:10.18805/ag.D-5447

Jia, X., Mai, X., Cui, Y., Yuan, Y., Xing, X., Seo, H., Xing, L., & Meng, M. Q.-H. (2020). Automatic polyp recognition in colonoscopy images using deep learning and two-stage pyramidal feature prediction. *IEEE Transactions on Automation Science and Engineering, 17*(3), 1570–1584. doi:10.1109/TASE.2020.2964827

Jindal, P., & Nikhanj, P. (2023). A review on processing technologies for value added millet products. *Journal of Food Process Engineering, 14419*. doi:10.1111/jfpe.14419

John, M. (2020). The coronavirus pandemic and US consumption. *Vox EU& CEPR Coverage of the Covid-19 Global Pandemic*. https://voxeu.org/article/coronavirus-pandemic-and-us- consumption

Kalinga TV. (2022, January 20). *Odisha tribal women spearhead millets mission*. Kalinga TV. https://kalingatv.com/state/odishas-tribal-women-spearhead-mission-to-promote-millets-from-farm-to-plate/

Kamali Rousta, L., Bodbodak, S., Nejatian, M., Ghandehari Yazdi, A. P., Rafiee, Z., Xiao, J., & Jafari, S. M. (2021). Use of encapsulation technology to enrich and fortify bakery, pasta, and cereal-based products. *Trends in Food Science & Technology, 118*, 688–710. doi:10.1016/j.tifs.2021.10.029

Kane-Potaka, J., Anitha, S., Tsusaka, T. W., Botha, R., Budumuru, M., Upadhyay, S., Kumar, P., Mallesh, K., Hunasgi, R., Jalagam, A. K., & Nedumaran, S. (2021). Assessing millets and sorghum consumption behavior in urban India: A large-scale survey. *Frontiers in Sustainable Food Systems, 5*, 680777. doi:10.3389/fsufs.2021.680777

Kanwal, S., Joseph, T. P., Aliya, S., Song, S., Saleem, M. Z., Nisar, M. A., Wang, Y., Meyiah, A., Ma, Y., & Xin, Y. (2020). Attenuation of DSS induced colitis by Dictyophora indusiata polysaccharide (DIP) via modulation of gutmicrobiota and inflammatory related signaling pathways. *Journal of Functional Foods, 64*, 103641. doi:10.1016/j.jff.2019.103641

Karpagapandi, L., Kalaiselvan, A., & Baskaran, V. (2023). Nutraceutical and Functional Attributes of Foxtail Millet (Setaria italica). In Nutri-Cereals (pp. 141-177). CRC Press.

Kassie, M., Pender, J., Yesuf, M., Kohlin, G., Bluffstone, R., & Mulugeta, E. (2007). *Impact of soil conservation on crop production in the Northern Ethiopian Highlands*. Intl Food Policy Res Inst.

Kaur, N., Agarwal, A., & Sabharwal, M. (2022). Food fortification strategies to deliver nutrients for the management of iron deficiency anaemia. *Current Research in Food Science, 5*, 2094–2107. doi:10.1016/j.crfs.2022.10.020 PMID:36387591

Kerr, R. B. (2014). Lost and found crops: Agrobiodiversity, indigenous knowledge, and a feminist political ecology of sorghum and finger millet in northern Malawi. *Annals of the Association of American Geographers, 104*(3), 577–593. doi:10.1080/00045608.2014.892346

Kiehl, L., Kuntz, S., Höhn, J., Jutzi, T., Krieghoff-Henning, E., Kather, J. N., Holland-Letz, T., Kopp-Schneider, A., Chang-Claude, J., Brobeil, A., von Kalle, C., Fröhling, S., Alwers, E., Brenner, H., Hoffmeister, M., & Brinker, T. J. (2021). Deep learning can predict lymph node status directly from histology in colorectal cancer. *European Journal of Cancer, 157*, 464–473. doi:10.1016/j.ejca.2021.08.039 PMID:34649117

Koss, M.P., Figuerdo, A.J., Bell, I., Tharan, S. (1996). Traumatic memory characteristics: a cross validate mediational model of response to rape among employed women. *Abnormal Psychol, 105*(3), 421–432.

Kothari, S. L., Kumar, S., Kothari, A., & Watanabe, K. N. (2005). Applications of biotechnology for improvement of millet crops: A review of progress and prospects. *Plant Biotechnology (Tsukuba), 22*(2), 81–88. doi:10.5511/plantbiotechnology.22.81

Krishnamoorthy, S., Kunjithapatham, S., & Manickam, L. (2013). Traditional I ndian breakfast (I dli and D osa) with enhanced nutritional content using millets. *Nutrition & Dietetics: the Journal of the Dietitians Association of Australia, 70*(3), 241–246. doi:10.1111/1747-0080.12020

Krishnan, P., & Sharma, M. (2013). Crop monitoring and yield estimation based on MODIS data for millet cultivation areas of Rajasthan, India. *Journal of Agrometeorology, 15*(2), 190–197.

Krüger, C. C. H., Comassetto, M. C. G., Cândido, L. M. B., Baldini, V. L. S., Santtucci, M. C., & Sgarbieri, V. C. (2003). Biscoitos tipo "cookie" e "snack" enriquecidos, respectivamente com caseína obtida por coagulação enzimática e caseinato de sódio. *Food Science and Technology (Campinas)*, *23*(1), 81–86. doi:10.1590/S0101-20612003000100017

Kulkarni, D. B., Sakhale, B. K., & Chavan, R. F. (2021). Studies on development of low gluten cookies from pearl millet and wheat flour. *Food Research*, *5*(4), 114–119. doi:10.26656/fr.2017.5(4).028

Kumar, A., Tomer, V., Kaur, A., Kumar, V., and Gupta, K. (2018). Millets: A solution to agrarian and nutritional challenges. *Agriculture and Food Security*, *7*(1), 1–15. doi:10.1186/s40066-018-0183-3

Kumar, D., & Joshi, N. L. (2006). Indigenous knowledge of millet cultivation in tribal communities of Uttarakhand, Indian Himalaya. *Indian Journal of Traditional Knowledge*, *5*(2), 189–196.

Kumari, P., Pareek, V., Kajla, P., & Khurana, S. (2024). Composition, structure and functionality of starch isolated from Kodo millet. In *Non-Conventional Starch Sources* (pp. 253–278). Academic Press. doi:10.1016/B978-0-443-18981-4.00009-4

Kumar, S., Sridhar, R., Monika, S., Kumar, A., Bhayankar, Raghavan, M., Tiwari, H., Kumar, A., Singh, S., & Yadav, R. (2023). A comprehensive review on Millets: A potential source of energy and nutrients for health. *International. Journal of Environment and Climate Change*, *13*(9), 2531–2538. doi:10.9734/ijecc/2023/v13i92487

Lenne, J. M., Takan, J. P., Mgonja, M. A., Manyasa, E. O., Kaloki, P., Wanyera, N., Okwadi, J., Muthumeenakshi, S., Brown, A. E., Tamale, M., & Sreenivasaprasad, S. (2007). Finger millet blast disease management: A key entry point for fighting malnutrition and poverty in East Africa. *Outlook on Agriculture*, *36*(2), 101–108. doi:10.5367/000000007781159994

Lester, J. N. (2023). Introduction to special issue: Qualitative research methodologies and methods for theory building in human resource development. *Human Resource Development Review*, *22*(1), 7–14. doi:10.1177/15344843221146871

Li, J., Li, W., Hu, J., Li, C., & Cui, X. (2023). Proso Millet Peroxidase-mediated Degradation and Detoxification of Rhodamine B in Water. *Environmental Technology*, 1-33.

Lightfoot, E., Liu, X., & Jones, M. K. (2013). Why move starchy cereals? A review of the isotopic evidence for prehistoric millet consumption across Eurasia. *World Archaeology*, *45*(4), 574–623. doi:10.1080/00438243.2013.852070

Li, S., Xian, F., Guan, X., Huang, K., Yu, W., & Liu, D. (2020). Neural protective effects of millet and millet polyphenols on high-fat diet-induced oxidative stress in the brain. *Plant Foods for Human Nutrition (Dordrecht, Netherlands)*, *75*(2), 208–214. doi:10.100711130-020-00802-6 PMID:32078112

Liu, S., Willett, W., Stampfer, M., Hu, F. B., Franz, M., Sampson, L., Hennekens, C. H., & Manson, J. A. E. (2000). A prospective study of dietary glycemic load, carbohydrate intake, and risk of coronary heart dis- ease in US women. *The American Journal of Clinical Nutrition*, *71*(6), 1455–1461. doi:10.1093/ajcn/71.6.1455 PMID:10837285

Lobell, D. B., & Burke, M. (Eds.). (2009). *Climate change and food security: Adapting agriculture to a warmer world* (Vol. 37). Springer Science & Business Media.

Lopez-Ridaura, S., Barba-Escoto, L., Reyna, C., Hellin, J., Gerard, B., & Van Wijk, M. (2019). Food security and agriculture in the Western Highlands of Guatemala. *Food Security*, *11*(4), 817–833. doi:10.100712571-019-00940-z

Maghaydah, S., Abdul-Hussain, S., Ajo, R., Obeidat, B., & Tawalbeh, Y. (2013). Enhancing the Nutritional Value of Gluten-Free Cookies with Inulin. *Advance Journal of Food Science and Technology : AJFST*, *5*(7), 866–870. doi:10.19026/ajfst.5.3174

Mal, B., Padulosi, S., & Bala Ravi, S. (2010). *Minor millets in South Asia: learnings from IFAD-NUS Project in India and Nepal.*

Maningat, C., Bassi, S., Woo, K., Dohl, C., Gaul, J., & Stempien, G. (2005). Formulation of High-Protein, High-Fiber (Low-Carbohydrate), Reduced Calorie Bread. *Technical Bulletin.*, *27*(4), 1–16.

Manley, D. (2011). Product development in the biscuit industry. Manley's Technol. Biscuits, Crackers Cookies. Woodhead Publishing. doi:10.1533/9780857093646.1.69

Marteau, R., Sultan, B., Moron, V., Alhassane, A., Baron, C., & Traoré, S. B. (2011). The onset of the rainy season and farmers' sowing strategy for pearl millet cultivation in Southwest Niger. *Agricultural and Forest Meteorology*, *151*(10), 1356–1369. doi:10.1016/j.agrformet.2011.05.018

Martin-Diana, A. B., Izquierdo, N., Albertos, I., Sanchez, M. S., Herrero, A., Sanz, M. A., & Rico, D. (2017). Valorization of carob's germ and seed peel as natural antioxidant ingredients in gluten-free crackers: Carob antioxidant by-products in gluten-free snack. *Journal of Food Processing and Preservation*, *41*(2), e12770. doi:10.1111/jfpp.12770

Mason. J. B. (2001). The Micronutrient Report: Current Progress and Trends in the Control of Vitamin A, Iodine, and Iron Deficiencies. Ottawa Micronutrient Initiative

Matias, M. F. O., Oliveira, E. L., Gertrudes, E., & Magalhães, M. M. A. (2005). Use of fibers obtained from the cashew (Anacardium ocidentale) and guava (Psidium guayava) fruits for enrichment of food products. *Brazilian Archives of Biology and Technology*, *48*(spe), 143–150. doi:10.1590/S1516-89132005000400018

Mbinda, W., & Masaki, H. (2021). Breeding strategies and challenges in the improvement of blast disease resistance in finger millet. A current review. *Frontiers in Plant Science*, *11*, 602882. doi:10.3389/fpls.2020.602882 PMID:33488650

Meek, D. (2022). From marginalized to miracle: Critical bioregionalism, jungle farming and the move to millets in Karnataka, India. *Agriculture and Human Values*, *39*(3), 871–883. doi:10.100710460-021-10287-x

Meena, R. P., Joshi, D., Bisht, J. K., & Kant, L. (2021). Global scenario of millets cultivation. *Millets and millet technology*, 33-50.

Meena, R.J., Joshi, D.S., Bisht, J.K., & Kant, L. (2021). Global scenario of Millets Cultivation. *Millets and Millet Technology*, (pp. 35-50). Springer. doi:10.1007/978-981-16-0676-2_2

Meldrum, G., Roy, S., Lauridsen, N., & King, O. (2020). *India impact brief: Promoting kodo and kutki millets for improved incomes, climate resilience and nutrition in Madhya Pradesh.* Bioversity International. cgiar.org/bitstream/handle/10568/109361/India%20Impact%20Brief%20A4.pdf

Michaelraj, P. S. J., & Shanmugam. (2013). A Study on Millets Based Cultivation and Consumption in India. *International Journal of Marketing, Financial Services and Management Research, 2*(4), 49–58. www.semanticscholar.org/paper/A-Study-on-Millets-Based-Cultivation-and-in-India-Michaelraj-Shanmugam/0973456 81c669b828c2fb8d8b-bee94d92414388f

Michaelraj, P. S. J., & Shanmugam, A. (2013). A study on millets-based cultivation and consumption in India. International Journal of Marketing. *Financial Services & Management Research*, *2*(4), 49–58.

Millet Mission Odisha. (n.d.). *Home.* Millet Mission. https://milletmission.wordpress.com/

Mishra, A., Pattnaik, B., Dutta, T., & Baitharu, I. (2022). Nutritional Values and Potential Health Benefits of Millets- A Review. *The Journal of Nutrition*, *8*(1), 9–26. doi:10.18488/87.v8i1.3176

Mishra, K., & Sam, A. G. (2016). Does women's land ownership promote their empowerment? Empirical evidence from Nepal. *World Development*, *78*, 360–371. doi:10.1016/j.worlddev.2015.10.003

Mitelut, A.C., Popa, E.E., Popescu, P.A., & Popa, M.E. (2021). Trends of innovation in bread and bakery production. *Trends Wheat Bread Mak*, 199–226. doi:10.1016/B978-0-12-821048-2.00007-6

Mohanty, S. R., Das, B., & Mohanty, T. (2013). Empowerment of women in rural Odisha through micro-enterprises. *IOSR Journal of Humanities and Social Sciences*, *12*(1), 1–8. doi:10.9790/0837-1210108

Mongabay-India. (2022, August 10). *Climate resilient millets boost livelihood opportunities for Odisha's tribal women.* Mongabay-India.https://india.mongabay.com/2022/08/climate-resilient-millets-boost-livelihood-opportunities-for-odishas-tribal-women/

Morrison, K. (2016). From millets to rice (and back again?): Cuisine, cultivation, and health in southern India. In *A companion to South Asia in the past* (pp. 358–373). Wiley Blackwell. doi:10.1002/9781119055280.ch23

Mosha, D. B., Jeckoniah, J., Isinika, A., & Boniface, G. (2021). *The Influence of Sunflower Commercialisation and Diversity on Women's Empowerment: The Case of Iramba and Mkalama Districts*. Singida Region.

Mulenga, M., Kareem, S. A., Sabri, A. Q. M., & Seera, M. (2021). Stacking and chaining of normalization methods in deep learning-based classification of colorectal cancer using gut microbiome data. *IEEE Access : Practical Innovations, Open Solutions*, *9*, 97296–97319. doi:10.1109/ACCESS.2021.3094529

Mulualem, T., & Melak, A. (2013). A survey on the status and constraints of Finger Millet (Eleusine coracana L.) production in Metekel Zone, North Western Ethiopia. *Direct Research Journal of Agriculture and Food Science, 1* (5), 67-72. https://directresearchpublisher.org/drjafs/abstract/a-survey-on-the-status-and-constraints-of-finger-millet-eleusine-coracana-l-production-in-metekel-zone-north-western-ethiopia/

Murugani, V. G., & Thamaga-Chitja, J. M. (2019). How does women's empowerment in agriculture affect household food security and dietary diversity? The case of rural irrigation schemes in Limpopo Province, South Africa. *Agrekon*, *58*(3), 308–323. doi:10.1080/03031853.2019.1610976

Muthamilarasan, M., Dhaka, A., Yadav, R., & Prasad, M. (2016). Exploration of millet models for developing nutrient rich graminaceous crops. *Plant Science*, *242*, 89–97. doi:10.1016/j.plantsci.2015.08.023 PMID:26566827

Muthamilarasan, M., & Prasad, M. (2021). Small Millets for Enduring Food Security Amidst Pandemics. *Trends in Plant Science*, *26*(1), 33–40. doi:10.1016/j.tplants.2020.08.008 PMID:32900620

Nadeem, M. S., Kumar, V., Al-Abbasi, F. A., Kamal, M. A., & Anwar, F. (2020). Risk of colorectal cancer in inflammatory bowel diseases. *Seminars in Cancer Biology*, *64*, 51–60. doi:10.1016/j.semcancer.2019.05.001 PMID:31112753

Nagarajan, L., Smale, M., & Glewwe, P. (2007). Determinants of millet diversity at the household-farm and village-community levels in the drylands of India: The role of local seed systems. *Agricultural Economics*, *36*(2), 157–167. doi:10.1111/j.1574-0862.2007.00195.x

Nagaraj, N., Basavaraj, G., Rao, P., Bantilan, C., & Haldar, S. (2013). Sorghum and Pearl Millet Economy in India: Future Outlook and Options. *Economic and Political Weekly*, *48*(52), 74–81. https://www.jstor.org/stable/24477899

Nair, S., & Kumar, B. M. (2007). Women's role in biodiversity conservation and sustainable use: A case study of home gardens in Kerala, India. *Biodiversity and Conservation*, *16*(8), 2589–2608.

Naseer, B., Naik, H. R., Hussain, S. Z., Zargar, I., Beenish, Bhat, T. A., & Nazir, N. (2021). Effect of carboxymethyl cellulose and baking conditions on in-vitro starch digestibility and physico-textural characteristics of low glycemic index gluten-free rice cookies. *Lebensmittel-Wissenschaft + Technologie*, *141*, 110885. doi:10.1016/j.lwt.2021.110885

Nath, M., Roy, P., Shukla, A., & Kumar, A. (2013). Spatial distribution and accumulation of calcium in different tissues, developing spikes and seeds of finger millet genotypes. *Journal of Plant Nutrition*, *36*(4), 539–550. doi:10.1080/01904167.2012.748072

NCDS. (2022). *Mid Term Evaluation of Special Programme for Promotion of Millets in Tribal Areas of Odisha*. Phase-I Blocks, Koraput District. https://ncds.nic.in/?q=node/744

Nelsen, J. (2002). Gluten-Sensitive Enteropathy (Celiac Disease). *More Common Than You Think. Am. Fam. Physician*, *66*, 2259–2266. PMID:12507163

Numan, M., Serba, D. D., & Ligaba-Osena, A. (2021). Alternative strategies for multi-stress tolerance and yield improvement in millets. *Genes*, *12*(5), 739. doi:10.3390/genes12050739 PMID:34068886

Obilana, A. (2003). Overview: Importance of millets in Africa. [all cultivated millet species]. *WORLD (Oakland, Calif.)*, *38*(2), 28.

Odisha Millet Mission. (n.d.). *About Program*. Odisha Millet Mission. https://milletsodisha.com/about-program

Odisha Millet Mission. (n.d.). *Home*. Odisha Millet Mission. https://milletsodisha.com/

Okpala, L., Okoli, E., & Udensi, E. (2013). Physico-chemical and sensory properties of cookies made from blends of germinated pigeon pea, fermented sorghum, and cocoyam flours. *Food Science & Nutrition*, *1*(1), 8–14. doi:10.1002/fsn3.2 PMID:24804009

Ongudi, S. O., Ngigi, M. W., & Kimurto, P. K. (2017). Determinants of Consumers' Choice and Willingness to Pay for Biofortified Pearl Millet in Kenya. *East African Agricultural and Forestry Journal*, *82*(2-4), 175–187. doi:10.1080/00128325.2017.1391489

Onyango, C., Mutungi, C., Unbehend, G., & Lindhauer, M. G. (2010). Rheological and baking characteristics of batter and bread prepared from pregelatinised cassava starch and sorghum and modified using microbial transglutaminase. *Journal of Food Engineering*, *97*(4), 465–470. doi:10.1016/j.jfoodeng.2009.11.002

Opole, R. (2019). Opportunities for enhancing production, utilization and marketing of finger millet in Africa. *African Journal of Food, Agriculture, Nutrition and Development*, *19*(1), 13863–13882. doi:10.18697/ajfand.84.BLFB1004

Paciulli, M., Rinaldi, M., Cavazza, A., Ganino, T., Rodolfi, M., Chiancone, B., & Chiavaro, E. (2018). Effect of chestnut flour supplementation on physico-chemical properties and oxidative stability of gluten-free biscuits during storage. *Lebensmittel-Wissenschaft + Technologie*, *98*, 451–457. doi:10.1016/j.lwt.2018.09.002

Padulosi, S., King, I. S., Hunter, D., & Swaminathan, M. S. (Eds.). (2022). *Orphan Crops for Sustainable Food and Nutrition Security: Promoting neglected and Underutilized Species.: Issues in Agricultura Biodiversity*. Routledge.

Padulosi, S., Mal, B., King, O. I., & Gotor, E. (2015). Minor Millets as a Central Element for Sustainably Enhanced Incomes, Empowerment, and Nutrition in Rural India. *Sustainability (Basel)*, *7*(7), 8904–8933. doi:10.3390u7078904

Pandey, N. L., Rath, R., Datta, M., Choudhury, B. R., & Kandpal, R. (2023). *Mainstreaming Millets in G20 Countries Through Integrated Behavioural and Policy Shifts*. ORF.

Pandey, A., & Bolia, N. B. (2023). Millet value chain revolution for sustainability: A proposal for India. *Socio-Economic Planning Sciences*, *101592*, 101592. doi:10.1016/j.seps.2023.101592

Parida, P. K. (2020). Magical Hands of a Tribal Farmer Set a Milestone in Finger Millet Cultivation-A Case from Koraput, Odisha, India. *Scholarly Journal of Food and Nutrition*, *2*(4). Advance online publication. doi:10.32474/SJFN.2020.02.000144

Parveen, S., & Leonhauser, I. U. (2008). Factors affecting the extent of economic empowerment of women in farm households: Experiences from rural Bangladesh. *International Journal of Human Ecology*, *9*(2), 117–126.

Peng, X., & Yao, Y. (2017). Carbohydrates as Fat Replacers. *Annual Review of Food Science and Technology*, *8*(1), 331–351. doi:10.1146/annurev-food-030216-030034 PMID:28245156

Peng, Y., Yan, Y., Wan, P., Chen, D., Ding, Y., Ran, L., Mi, J., Lu, L., Zhang, Z., Li, X., Zeng, X., & Cao, Y. (2019). Gut microbiota modulation and anti-inflammatory properties of anthocyanins from the fruits of Lycium ruthenicum Murrayin dextran sodium sulfate-induced colitis in mice. *Free Radical Biology & Medicine*, *136*, 96–108. doi:10.1016/j.freeradbiomed.2019.04.005 PMID:30959170

Pragya, S., & Rita, S. R. (2012). Finger millet for food and nutritional security. *African Journal of Food Science*, *6*(4), 77–84. doi:10.5897/AJFSX10.010

Priya, V., Verma, R. K., Lakhawat, S., Yadav, V. K., Gacem, A., Abbas, M., Yadav, K. K., Park, H.-K., Jeon, B.-H., & Mishra, S. (2023). Millets: Sustainable treasure house of bioactive components. *International Journal of Food Properties*, *26*(1), 1822–1840. doi:10.1080/10942912.2023.2236317

R M Beattie, J A Walker-Smith, S H Murch (1995). Indications for investigation of chronic gastrointestinal symptoms. *Archives of Disease in Childhood, 73*, 354-355.

Radhika, G., Sathya, R. M., Ganesan, A., Saroja, R., Vijayalakshmi, P., Sudha, V., & Mohan, V. (2011). Dietary profile of urban adult population in South India in the context of chronic disease epidemiology (CURES–68). *Public Health Nutrition*, *14*(4), 591–598. doi:10.1017/S136898001000203X PMID:20701818

Radhika, G., Sumathi, C., Ganesan, A., Sudha, V., Jeya Kumar Henry, C., & Mohan, V. (2010). Glycaemic index of Indian flatbreads (rotis) prepared using whole wheat flour and 'atta mix'-added whole wheat flour. *British Journal of Nutrition*, *103*(11), 1642–1647. doi:10.1017/S0007114509993680 PMID:20100375

Rahman, H., & Gulati, A. (2019). Smart agriculture in India: Current state and future prospects. *Agricultural Economics Research Review*, *32*(2), 155–166.

Rajasekaran, N., Nithya, M., Rose, C., & Chandra, T. (2004). The effect of finger millet feeding on the early responses during the process of wound healing in diabetic rats. *Biochimica et Biophysica Acta (BBA)-. Biochimica et Biophysica Acta. Molecular Basis of Disease*, *1689*(3), 190–201. doi:10.1016/j.bbadis.2004.03.004 PMID:15276645

Raj, S. N., Deepak, S. A., Basavaraju, P., Shetty, H. S., Reddy, M. S., & Kloepper, J. W. (2003). Comparative performance of formulations of plant growth promoting rhizobacteria in growth promotion and suppression of downy mildew in pearl millet. *Crop Protection (Guildford, Surrey)*, *22*(4), 579–588. doi:10.1016/S0261-2194(02)00222-3

Ramashia, S. E., Gwata, E. T., Meddows-Taylor, S., Anyasi, T. A., & Jideani, A. I. O. (2018). Some physical and functional properties of finger millet (Eleusine coracana) obtained in sub-Saharan Africa. *Food Research International*, *104*, 110–118. doi:10.1016/j.foodres.2017.09.065 PMID:29433775

Rana, K., & Sahoo, M. (2023). Millet-based Farming Operations and Women: A Micro-Analysis in Koraput District, Odisha. *Science and Culture*, *89*(9-10), 304–310. doi:10.36094c.v89.2023.Millet_Based_Farming_Operations_and_Women.Rana_Sahoo.304

Ranasalva, N., & Visvanathan, R. (2014). Development of cookies and bread from cooked and fermented pearl millet flour. *African Journal of Food Science*, *8*(6), 330–336. doi:10.5897/AJFS2013.1113

Rao, P. P., & Basavaraj, G. (2015). *Status and prospects of millet utilization in India and global scenario. Millets: Promotion for food, feed, fodder, nutritional and environment security, proceedings of global consultation on millets promotion for health & nutritional security. Society for Millets Research, ICAR*. Indian Institute of Millets Research.

Rao, P. P., Birthal, P. S., & Joshi, P. K. (2006). Changes in consumption of coarse cereals in India during 1983–2000. *Food Policy, 31*(4), 322–332.

Rashmi, S., Dhumketi, K., Chaturvedi, P., Shukla, N., Shukla, R., & Dhumketi, K. (2022). Development and assessment of healthy multigrain biscuits of wheat flour with pearl millet, maize and buckwheat composite flours. *Pharma Innov. J., 11*, 1702–1707.

Reddy, N. R., Pierson, M. D., & Sathe, S. K. (2015). Salient features of millets, with particular reference to their seed proteins. *Food and Nutrition Bulletin, 36*(3), 327–340. PMID:26314732

Rizkalla, S. W., Bellisle, F., & Slama, G. (2002). Health benefits of low glycaemic index foods. Such as pulses, in diabetic petients and health individuals. *British Journal of Nutrition, 88*(S3), 255–262. doi:10.1079/BJN2002715 PMID:12498625

Rizwana, M., Singh, P., Ahalya, N., & Mohanasundaram, T. (2023). Assessing the awareness of nutritional benefits of millets amongst women in Bangalore. *British Food Journal, 125*(6), 2002–2018. doi:10.1108/BFJ-05-2021-0593

Rohrbach, D. D., & Kiriwaggulu, J. A. B. (2001). Commercialization Prospects for Sorghum and Pearl Millet in Tanzania. (Working Paper Series no. 7). *Journal of SAT Agricultural Research, 3*(1), 1–28.

Ronda, F., Oliete, B., Gómez, M., Caballero, P. A., & Pando, V. (2011). Rheological study of layer cake batters made with soybean protein isolate and different starch sources. *Journal of Food Engineering, 102*(3), 272–277. doi:10.1016/j.jfoodeng.2010.09.001

Roudier, P., Alhassane, A., Baron, C., Louvet, S., & Sultan, B. (2016). Assessing the benefits of weather and seasonal forecasts to millet growers in Niger. *Agricultural and Forest Meteorology, 223*, 168–180. doi:10.1016/j.agrformet.2016.04.010

Rowhani, P., Lobell, D. B., Linderman, M., & Ramankutty, N. (2011). Climate variability and crop production in Tanzania. *Agricultural and Forest Meteorology, 151*(4), 449–460. doi:10.1016/j.agrformet.2010.12.002

Roy, T. C. (2023). What are the Opportunities and Challenges in Millet Sector in India? *Millet Advisor.* https://milletadvisor.com/opportunities-and-challenges-in-millet-sector/

Sabuz, A. A., Rana, M. R., Ahmad, T., & Molla, M. M. (2023). Health-Promoting Potential of Millet: A Review. *Separations, 10*(80), 80. doi:10.3390eparations10020080

Sachan, D. S., Kumar, R., Kumar, P., Pal, V., & Yadav, K. K. (2023). Millets Production and Consumption in India. *Just Agriculture, 3*(5), 275–283.

Sahi, S. S., & Alava, J. M. (2003). Functionality of emulsifiers in sponge cake production. *Journal of the Science of Food and Agriculture, 83*(14), 1419–1429. doi:10.1002/jsfa.1557

Sahoo, J. P., & Mahapatra, M. (2023). *Millets towards a sustainable nutritional security national year of millets.* Revitalization.

Saleh, A. S., Zhang, Q., & Chen, J. (2013). Shenquan decortication and grinding: An effective process for milling millets. *International Journal of Food Science & Technology, 48*(5), 1082–1087.

Saleh, A. S., Zhang, Q., Chen, J., & Shen, Q. (2013). Millet grains: Nutritional quality, processing, and potential health benefits. *Comprehensive Reviews in Food Science and Food Safety, 12*(3), 281–295. doi:10.1111/1541-4337.12012

Saloni, S., Sindhu, S., Kumari, S., & Suman, S. (2018). Little millets: Properties, functions and future prospects. *International Journal of Agricultural Engineering*, *11*(Special), 179–181. doi:10.15740/HAS/IJAE/11.Sp.Issue/179-181

Samtiya, M., Aluko, R. E., Dhaka, N., Dhewa, T., & Puniya, A. K. (2023). Nutritional and health-promoting attributes of millet: Current and future perspectives. *Nutrition Reviews*, *81*(6), 684–704. doi:10.1093/nutrit/nuac081 PMID:36219789

Sanodiya, L. K., Sanodiya, P. K., Prajapati, A., Kumawat, P., & Singh, A. (2023). *The sustainability of millets and the advantages they bring for human health*. Research Gate.

Saranraj, P., & Geetha, M. (2012). Microbial Spoilage of Bakery Products and Its Control by Preservatives. *International Journal of Pharmaceutical and Biological Archives*.

Sarita, E. S., & Singh, E. (2016). Potential of millets: Nutrients composition and health benefits. *Journal of Scientific and Innovative Research*, *5*(2), 46–50. doi:10.31254/jsir.2016.5204

Sarkar, P., Kumar, L., Dhumal, C., Panigrahi, S., & Choudhary, R. (2015). Traditional and ayurvedic foods of Indian origin. *Journal of Ethnic Foods*, *2*(3), 97–109. doi:10.1016/j.jef.2015.08.003

Sasmal, P., Paul, A., Bhuyan, M. K., Iwahori, Y., & Kasugai, K. (2021). Extraction of Key-Frames From Endoscopic Videos by Using Depth Information. *IEEE Access : Practical Innovations, Open Solutions*, *9*, 153004–153011. doi:10.1109/ACCESS.2021.3126835

Sato, H., Matsumura, T. and Shibukawa, S. 1995. Apparent heat transfer in a forcedconvection oven and properties of baked food. *J. Food Sci. 52*, 185-188.

Saxena, R. (2019). Feminization of Indian Agriculture. *International Journal of Advanced Scientific Research and Management*, *4*(7), 79–83.

Saxena, R., Vanga, S., Wang, J., Orsat, V., & Raghavan, V. (2018). Millets for Food Security in the Context of Climate Change: A Review. *Sustainability (Basel)*, *10*(7), 2228. doi:10.3390u10072228

Saxena, S., Saini, S., & Sasmal, S. (2022). General Public Awareness Survey Drive on Impact of Indian Culinary Practices on Nutritional Profile of Food: Special Emphasis on Millet Awareness. *Journal of Culinary Science & Technology*, 1–11. doi:10.1080/15428052.2022.2073936

Scabio, A., Granato, D., & Ferrari, R. A. (2005). Aplicação do bagaço de laranja na produção de biscoitos. In: SEMANA DE ENGENHARIA DE ALIMENTOS. Ponta Grossa: Universidade Estadual de Ponta Grossa.

Scherr, S. J., Shames, S., & Friedman, R. (2012). From climate-smart agriculture to climate-smart landscapes. *Agriculture & Food Security*, *1*(1), 1–15. doi:10.1186/2048-7010-1-12

Schlecht, E., & Buerkert, A. (2004). Organic inputs and farmers' management strategies in millet fields of western Niger. *Geoderma*, *121*(3-4), 271–289. doi:10.1016/j.geoderma.2003.11.015

Sebastião, V. G., Batista, D., Rebellato, A. P., Macedo, J. A., & Steel, C. J. (2023). Sustainable Production Of Naturally Colored Extruded Breakfast Cereals From Blends Of Broken Rice And Vegetable Flours. *Food Research International*, *113078*, 113078. doi:10.1016/j.foodres.2023.113078

Selladurai, M., Pulivarthi, M. K., Raj, A. S., Iftikhar, M., Prasad, P. V. V., & Siliveru, K. (2022). *Considerations for gluten free foods - pearl and finger millet processing and market demand*. Grain Oil Sci. Technol. doi:10.1016/j.gaost.2022.11.003

Shahidi, F., & Chandrasekara, A. (2013). Millet grain phenolics and their role in disease risk reduction and health promotion: A review. *Journal of Functional Foods*, *5*(2), 570–581. doi:10.1016/j.jff.2013.02.004

Shah, P., Dhir, A., Joshi, R., & Tripathy, N. (2023). Opportunities and challenges in food entrepreneurship: In-depth qualitative investigation of millet entrepreneurs. *Journal of Business Research*, *155*, 113372. doi:10.1016/j.jbusres.2022.113372

Shan, S., Lu, Y., Zhang, X., Shi, J., Li, H., & Li, Z. (2021). Inhibitory effect of bound polyphenol from foxtail millet bran on miR-149 methylation increases the chemosensitivity of human colorectal cancer HCT-8/Fu cells. *Molecular and Cellular Biochemistry*, *476*(2), 513–523. doi:10.100711010-020-03906-4 PMID:33011952

Sharma, B., & Gujral, H. S. (2020). Modifying the dough mixing behavior, protein & starch digestibility and antinutritional profile of minor millets by sprouting. *International Journal of Biological Macromolecules*, *153*, 962–970. doi:10.1016/j.ijbiomac.2019.10.225 PMID:31759014

Sharma, R., & Sharma, A. (2008). *Assessment of market potential and marketing strategies of Kangra tea. P.G. Dr.* Yashwant Singh Parmar University of Horticulture and Forestry.

Sheahan, C. M. (2014). Plant guide for proso millet (Panicum miliaceum). USDA-Natural Resources Conservation Service, Cape May Plant Materials Center. Cape May, NJ.

Shen, Y., & Ke, J. (2021). Sampling based tumor recognition in whole-slide histology image with deep learning approaches. IEEE/ACM Transactions on Computational Biology and Bioinformatics, 19(4), 2431-2441.

Shobana, S., Krishnaswamy, K., Sudha, V., Malleshi, N. G., Anjana, R. M., Palaniappan, L., & Mohan, V. (2013). Finger millet (Ragi, Eleusine coracana L.): A review of its nutritional properties, processing, and plausible health benefits. *Advances in Food and Nutrition Research*, *69*, 1–39. doi:10.1016/B978-0-12-410540-9.00001-6 PMID:23522794

Shull, J., Chandrashekar, A., Kirleis, A., & Ejeta, G. (1990). Development of sorghum (Sorghum bicolor (L.) Moench) endosperm in varieties of varying hardness. *Food Structure*, *9*(3), 8.

Silva, M. A. (2007). Uso de hidrolisado de pescado como suplemento protéico em biscoito de polvilho. In: Encontro Regional Sul De Ciência E Tecnologia De Alimentos, 9, 475-479.

Silva, M. R. (1997). Caracterização química e nutricional da farinha de jatobá (Hymenaea stigonocarpa Mart.): desenvolvimento e otimização de produtos através de testes sensoriais afetivos. Universidade Estadual de Campinas, São Paulo,.

Singh, N., Kumari, P., & Bassi, J. G. N. (2023). Nutrient rich Kodo millet, importance and value addition: An overview. *The Pharma Innovation Journal.*

Singh, S. (2022). Know about different types of millets. *Krishijagran* https://krishijagran.com/agripedia/know-about-different-types-of-millet/

Singh, S., Sharma, R., Chandra Nayaka, S., Tara Satyavathi, C., & Raj, C. (2021). Understanding pearl millet blast caused by Magnaporthe grisea and strategies for its management. *Blast Disease of Cereal Crops: Evolution and Adaptation in Context of Climate Change*, 151-172.

Singh, K., Mishra, A., & Mishra, H. (2012). Fuzzy analysis of sensory attributes of bread prepared from millet-based composite flours. *Lebensmittel-Wissenschaft + Technologie*, *48*(2), 276–282. doi:10.1016/j.lwt.2012.03.026

Singh, M., & Mohamed, A. (2007). Influence of gluten-soy protein blends on the quality of reduced carbohydrates cookies. *Lebensmittel-Wissenschaft + Technologie*, *40*(2), 353–360. doi:10.1016/j.lwt.2005.09.013

Singh, P. (2020). Capturing the Narratives of Sustainable Farming: Study of Marginal Women Farmers in Five Districts of Odisha. *The Indian Journal of Public Administration*, *66*(4), 455–465. doi:10.1177/0019556120982199

Singh, S., Suri, S., & Singh, R. (2022). Potential and unrealized future possibilities of Browntop millet in the food sector. *Frontiers in Sustainable Food Systems*, *6*, 974126. doi:10.3389/fsufs.2022.974126

Sirisha, S., & Lakshmi, J. (2018*). A study on consumption pattern of millets and millet-based products in Guntur city, Andhra Pradesh PG.* [Thesis, Ranga University, Rajendra nagar, Hyderabad].

SolutionBuggy. (2022). *Millet based business profitable manufacturing msmes solutions.* Linkedin. https://www.linkedin.com/pulse/millet-based-business-profitable-manufacturing-msmes-solutionsbuggy/

Sood, S., Joshi, D. C., Chandra, A. K., & Kumar, A. (2019). Phenomics and genomics of finger millet: Current status and future prospects. *Planta*, *250*(3), 731–751. doi:10.100700425-019-03159-6 PMID:30968267

Srinivasu, P. N., SivaSai, J. G., Ijaz, M. F., Bhoi, A. K., Kim, W., & Kang, J. J. (2021). Classification of skin disease using deep learning neural networks with MobileNet V2 and LSTM. *Sensors (Basel)*, *21*(8), 2852. doi:10.339021082852 PMID:33919583

Stanley, P., Cauvain, & Young, L. (1999). Bakery Food Manufacture and Quality: Water Control and Effects: Second Edition. Wiley.

Stantiall, S.E. & Serventi, L. (2017). *Nutritional and sensory challenges of gluten-free bakery products: a review*. NIH. https://doi.org/ doi:10.1080/09637486.2017.1378626

Stoner, J. L., Felix, R., & Stadler Blank, A. (2023). Best practices for implementing experimental research methods. *International Journal of Consumer Studies*, *47*(4), 1579–1595. doi:10.1111/ijcs.12878

Sun, M., Yan, H., Zhang, A., Jin, Y., Lin, C., Luo, L., Wu, B., Fan, Y., Tian, S., Cao, X., Wang, Z., Luo, J., Yang, Y., Jia, J., Zhou, P., Tang, Q., Jones, C. S., Varshney, R. K., Srivastava, R. K., & Huang, L. (2023). Milletdb: A multi-omics database to accelerate the research of functional genomics and molecular breeding of millets. *Plant Biotechnology Journal*, *21*(11), 1–20. doi:10.1111/pbi.14136 PMID:37530223

Surekha, N., Chimmad, B. V., & Naik, R. S. (2013). Hypoglycaemic and hypolipidemic efficacy of barnyard millet (Echinochloa frumentacea Link) based health food. *Asian Journal of Home Science*, *8*(2), 383–387.

Swamy, B. M., Kumar, T. A., & Rao, K. S. (2012). Buying behavior of consumers towards instant food products. *International Journal of Research and Computational Technology.*, *2*(2).

Taraz, V. (2017). Adaptation to climate change: Historical evidence from the Indian monsoon. *Environment and Development Economics*, *22*(5), 517–545. doi:10.1017/S1355770X17000195

Tavanapong, W., Oh, J., Riegler, M. A., Khaleel, M., Mittal, B., & De Groen, P. C. (2022). Artificial intelligence for colonoscopy: Past, present, and future. *IEEE Journal of Biomedical and Health Informatics*, *26*(8), 3950–3965. doi:10.1109/JBHI.2022.3160098 PMID:35316197

Taylor, J.R.N. & Kruger, J. (2019). *Sorghum and Millets: Food and Beverage Nutritional 36 Attributes.* Sorghum Millets Chem. Technol. Nutr. Attrib. doi:10.1016/B978-0-12-811527-5.00007-1

Taylor, J. R. N., & Duodu, K. G. (2019). *Sorghum and millets: chemistry, technology, and nutritional attributes* (2nd ed.). Woodland Publishing.

Thakur, A., & Vaishnav, K. (2017). Role of millets in human health: A review. *International Journal of Chemical Studies*, *5*(3), 2295–2300.

Thathola, A., & Srivastava, S. (2002). Physicochemical properties and nutritional traits of millet-based weaning food suitable for infants of the Kumaon hills, Northern India. *Asia Pacific Journal of Clinical Nutrition*, *11*(1), 28–32. doi:10.1046/j.1440-6047.2002.00251.x PMID:11890635

The Economic Times. (2023, July 21). India bans export of non-basmati white rice. *The Economic Times.* https://economictimes.indiatimes.com/news/economy/agriculture/govt-bans-export-of-non-basmati-white-rice/articleshow/101989078.cms?from=mdr

The Indian Express. (2021, November 16). Millet's cafe was inaugurated in Odisha's Keonjhar. *The India Express.* https://indianexpress.com/article/cities/bhubaneswar/millets-cafe-inaugurated-in-odishas-keonjhar-7601228/

The Locavore. (2023, March 22). *Odisha Millets Mission is a People's Programme.* The Locavore. https://thelocavore.in/2023/03/22/odisha-millets-mission-is-a-peoples-programme/

The Story of Millets. (2018). Karnataka State Department of Agriculture & ICAR. https://www.millets.res.in/pub/2018/The_Story_of_Millets.pdf

Tiwari, H., Naresh, R. K., Kumar, L., Kataria, S. K., Tewari, S., Saini, A., Yadav, R. K., & Asati, R. (2022). Millets for Food and Nutritional Security for Small and Marginal Farmers of North-West India in the Context of Climate Change: A Review. *International Journal of Plant and Soil Science*, 1694–1705. doi:10.9734/ijpss/2022/v34i232594

Torres, M. A. A., Sato, K., Lobo, N. F., & Queiroz, S. S. (1995). Efeito do uso de leite fortificado com ferro e vitamina C sobre os níveis de hemoglobina e condição nutricional de crianças menores de 2 anos. *Revista de Saude Publica, 29*(4), 301–307. doi:10.1590/S0034-89101995000400008 PMID:8729281

Tsatsaragkou, K., Papantoniou, M., & Mandala, I. (2015). Rheological, Physical, and Sensory Attributes of Gluten-Free Rice Cakes Containing Resistant Starch. *Journal of Food Science, 80*(2), E341–E348. doi:10.1111/1750-3841.12766 PMID:25604540

Ugare, R., Chimmad, B., Naik, R., Bharati, P., & Itagi, S. (2011). Glycemic index and significance of barnyard millet (Echinochloa frumentacae) in type II diabetics. *Journal of Food Science and Technology, 51*(2), pages392–395. doi:10.100713197-011-0516-8 PMID:24493902

Union Budget. (2022-23). *Government of India: Summary of the Union Budget 2022-23.* Government of India. https://static.pib.gov.in/WriteReadData/userfiles/SUMMARY2022.pdf

United Nations in India. (n.d.). *Odisha Millets Mission – Sowing Seeds of Change.* UN. https://india.un.org/en/192879-odisha-millets-mission-sowing-seeds-change

United Nations in India. (n.d.). *Odisha Millets Mission – Sowing Seedsof Change.* UN. https://india.un.org/en/192879-odisha-millets-mission-sowing-seeds-change

United Nations Sustainable Development Goals {UN-SDG}. (2023). *The Sustainable Development Agenda.* UN. https://www.un.org/sustainabledevelopment/development-agenda/

United Nations. (2023). *Millets – good for people, the environment, and farmers.* UN. https://unric.org/en/millets-good-for-people-the-environment-and-farmers/

Upadhyaya, H. D., Ramesh, S., Sharma, S., Singh, S. K., Varshney, S. K., Sarma, N. D. R. K., Ravishankar, C. R., Narasimhudu, Y., Reddy, V. G., Sahrawat, K. L., Dhanalakshmi, T. N., Mgonja, M. A., Parzies, H. K., Gowda, C. L. L., & Singh, S. (2011). Genetic diversity for grain nutrients contents in a core collection of finger millet (Eleusine coracana (L.) Gaertn.) germplasm. *Field Crops Research, 121*(1), 42–52. doi:10.1016/j.fcr.2010.11.017

Uyir-Organic Farmers Market. (2022). *The Several Benefits of Kodo Millet.* Uyir-Organic Farmers Market. https://uyir-organic.farm/benefits-of-the-kodo-millet

Valli, K. (2018). "Siri Dhanya (Millets)"is the Ultimate Solution to Good Health- Dr. Khadar Valli presented at Tech Mahindra campus https://www.youtube.com/watch?v=HtOBAulpR24

Vellakkal, S. (2010). Women's role in food production and agriculture in developing countries. *EAS Journal of Biotechnology and Genetics*, *2*(1), 1–8.

Vijayalakshmi, D., Geetha, K., Gowda, J., Ravi, S. B., Padulosi, S., & Mal, B. (2010). Empowerment of women farmersthrough value addition on minor millet genetic resources: A case study in Karnataka. *Indian Journal of Plant Genetic Resources*, *23*(1), 132–135.

Vijayalakshmi, D., Gowda, K. N., Jamuna, K. V., Ray, B. R., & Sajjan, J. T. (2012). Empowerment of self-help group women through value addition of finger millet. *Journal of Dairying, Foods & Home Sciences*, *31*(3).

Vikalp Sangam. (2021, December 26). *Odisha Millet Mission: The Successes and the Challenges.* Vikalp Sangam. https://vikalpsangam.org/article/odisha-millet-mission-the-successes-and-the-challenges/

Walker, C. E., & Eustace, W. D. (2016). *Milling and Baking, History.* Ref. Modul. Food Sci. doi:10.1016/B978-0-08-100596-5.00158-X

Wickramarachchi, K. S., Sissons, M. J., & Cauvain, S. P. (2015). Puff pastry and trends in fat reduction: An update. *International Journal of Food Science & Technology*, *50*(5), 1065–1075. doi:10.1111/ijfs.12754

Wilderjans, E., Luyts, A., Brijs, K., & Delcour, J. A. (2013). Ingredient functionality in batter type cake making. *Trends in Food Science & Technology*, *30*(1), 6–15. doi:10.1016/j.tifs.2013.01.001

Willett, W., Rockström, J., Loken, B., Springmann, M., Lang, T., Vermeulen, S., Garnett, T., Tilman, D., DeClerck, F., Wood, A., Jonell, M., Clark, M., Gordon, L. J., Fanzo, J., Hawkes, C., Zurayk, R., Rivera, J. A., De Vries, W., Majele Sibanda, L., & Murray, C. J. L. (2019). Food in the Anthropocene: The EAT–Lancet Commission on healthy diets from sustainable food systems. *Lancet*, *393*(10170), 447–492. doi:10.1016/S0140-6736(18)31788-4 PMID:30660336

Wolka, K., Moges, A., & Yimer, F. (2011). Effects of level soil bunds and stone bunds on soil properties and its implications for crop production: The case of Bokole watershed, Dawuro zone, Southern Ethiopia. *Agricultural Sciences*, *2*(03), 357–363. doi:10.4236/as.2011.23047

World Economic Forum. (2023). *We're on a brink of a 'polycrisis'- how worried should we be?* WEF. https://www.weforum.org/agenda/2023/01/polycrisis-global-risks-report-cost-

World Food Programme. (2020). *Lessons from Odisha Millets Mission: A Case for Mainstreaming Millets in India's Food Programs.* WFP. https://www.wfp.org/publications/lessons-odisha-millets-mission-case-mainstreaming-millets

WORLD HEALTH ORGANIZATION – WHO. (2008). Iron deficiency anaemia: assessment, prevention and control. Geneva: World Health Organization.

World Health Organization. (2022). Ageing and health [WWW Document]. World Health Organization. https://www.who.int/news-room/fact-sheets/detail/ageing-and-health

Xu, J., Zhang, Y., Wang, W., & Li, Y. (2020). Advanced properties of gluten-free cookies, cakes, and crackers: A review. *Trends in Food Science & Technology*, *103*, 200–213. doi:10.1016/j.tifs.2020.07.017

Yang, X. (2020). Colon polyp detection and segmentation based on improved MRCNN. *IEEE Transactions on Instrumentation and Measurement*, *70*, 1–10. doi:10.1109/TIM.2020.2986875

Yapi, A. M., Kergna, A. O., Debrah, S. K., Sidibe, A., & Sanogo, O. (2000). *Analysis of the Economic Impact of Sorghum and Millet Research in Mali.* International Crops Research Institute for the Semi-Arid Tropics, Patancheru, Hyderabad, Andhra Pradesh. http://oar.icrisat.org/1084/

Yousaf, L., Hou, D., Liaqat, H., & Shen, Q. (2021). Millet: A review of its nutritional and functional changes during processing. *Food Research International*, *142*, 110197. doi:10.1016/j.foodres.2021.110197 PMID:33773674

Zhang, B., Xu, Y., Liu, S., Lv, H., Hu, Y., Wang, Y., Li, Z., Wang, J., Ji, X., Ma, H., Wang, X., & Wang, S. (2020). Dietary supplementation of foxtail millet ameliorates colitis-associated colorectal cancer in mice via activation of gut receptors and suppression of the STAT3 pathway. *Nutrients*, *12*(8), 2367. doi:10.3390/nu12082367 PMID:32784751

Zhang, C., Di, L., Lin, L., Li, H., Guo, L., Yang, Z., Yu, E. G., Di, Y., & Yang, A. (2022). Towards automation of in-season crop type mapping using spatiotemporal crop information and remote sensing data. *Agricultural Systems*, *201*, 103462. doi:10.1016/j.agsy.2022.103462

About the Contributors

***

Shehbaz Ahmed's research and doctoral work (Ph.D.) are in the areaa of services marketing. He has published several articles in national and International journals. Had been a visiting professor to the leading business schools of the country – IIM(Ahmedabad), ICFAI Business School, Department of Business Management, Osmania University, Symbiosis, Pune.Has made his foray into corporate consultancy and has conducted training programs for multinational companies. His programs on Personality Development and Business Communication always go oversubscribed. He has conducted numerous workshops, seminars, training programs for Managers of govt. owned enterprises BHEL/ DRDL / Sri Ram Group of companies to name a few.

Dilip Bagal is a mechanical engineering academician and researcher Dilip Kumar Bagal is highly accomplished. He graduated from the National Institute of Technology in Rourkela with an M.Tech. (Research) degree. At the Government College of Engineering Kalahandi in Bhawanipatna, Odisha, Dilip is an Assistant Professor in the Department of Mechanical Engineering. He teaches metal machining, unconventional machining, and data analytics classes there. He is also doing cutting-edge research on sophisticated optimization and soft computing in mechanical engineering. Dilip received the Best Research Faculty Award 2021 from the Biju Patnaik University of Technology in Rourkela, Orissa, IN, proving his dedication to excellence in research and education. He also belongs to the Institution of Engineers (India), Kolkata, and the International Association of Engineers, Hong Kong, where he conducts research and engages in peer review to improve the field of mechanical engineering. Dilip is a peer reviewer who regularly contributes to journals such as Cognitive Neurodynamics, Diagnostics, Materials, Petroleum Science and Technology, Proceedings of the Institution of Mechanical Engineers, and Soft Computing. He has examined several research articles for these publications. Several identifiers, including ORCiD, ResearcherID, Scopus Author ID, Loop profile, and SciProfiles, have recognized him as the author of several research publications in prestigious journals. Dilip is a significant addition to any academic institution or research organization due to his knowledge, commitment, and love of mechanical engineering. His work has the potential to spur innovation in mechanical engineering and influence the direction of the sector.

D. Vydeki has received her Doctoral degree from Anna University. She has 16 years of teaching experience. She is currently working as an Associate Professor in the School of Electronics Engineering, VIT Chennai. She has published many papers in international journals and conference proceedings. Her research interests include IOT, wireless networks, network security, pattern recognition, image processing, machine learning and applications.

Saumendra Das is presently working as an Associate Professor at the School of Management Studies, GIET University, Gunupur, Odisha. He has more than 20 years of teaching, research, and industry experience. He has published more than 52 articles in national and international journals, conference proceedings, and book chapters. He also authored one book on advertising effectiveness. Dr Das has participated and presented many papers in seminars, conferences, and workshops in India and abroad. He has organized many FDPs and workshops in his career. He is an academician, author, and editor. He has also published two patents also. He is an active member of various professional bodies such as ICA, ISTE and RFI. In the year 2023, he has been awarded as the best teacher by Research Foundation India.

G. Sethuram Rao is working as an Assistant Professor in Velammal Institute of Technology, Chennai. He has 18 years of teaching experience. He has published many papers in international journals and conference proceedings. His research interest includes image processing, IOT, deep learning, machine learning and embedded systems.

G. Shanmugaraj is working as an Assistant Professor in Velammal Institute of Technology, Chennai. He has 17 years of teaching experience. He has published many papers in international journals and conference proceedings. His research interest includes image processing, IOT, deep learning, machine learning and embedded systems.

Anita Medhekar is a senior lecturer in economics and holds a PhD from Deakin University, Australia. She has completed her double Masters in pure Economics as well as in Education. She has taught at undergraduate and post graduate levels in India, Indonesia, and Australia and has nearly 30 years of teaching experience. Currently, she is teaching at Central Queensland University, Australia. Her research interests are in applied Health Economics, International Trade, Tourism Economics, Development Economics, Asia-Pacific Economies, Public Finance, South-Asia, Australian Economy, Bilateral Trade and Development for Peace, Supply Chains, Public Policy, Gender and Economic Development, Feminisation of Agriculture in Developing Countries, and Public-Private Partnerships. She has presented papers at numerous international conferences and received best research paper wards. She has numerous refereed publications in journals, conference proceedings, edited books, and book chapters to her credit.

Brojo Kishore Mishra awarded Ph.D. in Computer Science from Berhampur University in the year 2012 for his excellent work in the field of Web Mining. He worked in several reputed private Engineering Colleges and state University at a different level for more than last 17 years. Presently, he is a Professor with the Department of Computer Science and Engineering at NIST Institute of Science and Technology (Autonomous), Berhampur, India and also working as Joint Secretary of IEEE Bhubaneswar Subsection. He has published 53 research papers in peer reviewed International Journals, and 44 research papers in proceedings of International conferences, 51 book chapters, 17 edited books and 02 authored books, 05 book series, 02 patents published, 03 copyright and 01 trademark(applied). He has successfully guided 02 Ph.D. research scholar and currently 04 research scholars are continuing. Also he is the Visiting Faculty for the Department of Computer Science at Central University of Odisha. He is also a Life member of ISTE, CSI, and Senior member of IEEE.

Anand Nayyar received a Ph.D (Computer Science) from Desh Bhagat University in 2017 in the area of Wireless Sensor Networks, Swarm Intelligence and Network Simulation. He is currently working in School of Computer Science-Duy Tan University, Da Nang, Vietnam as Professor, Scientist, Vice-Chairman (Research) and Director- IoT and Intelligent Systems Lab. A Certified Professional with 125+ Professional certifications from CISCO, Microsoft, Amazon, EC-Council, Oracle, Google, Beingcert, EXIN, GAQM, Cyberoam and many more. Published more than 175+ Research Papers in various High-Quality ISI-SCI/SCIE/SSCI Impact Factor- Q1, Q2, Q3, Q4 Journals cum Scopus/ESCI indexed Journals, 70+ Papers in International Conferences indexed with Springer, IEEE and ACM Digital Library, 40+ Book Chapters in various SCOPUS/WEB OF SCIENCE Indexed Books with Springer, CRC Press, Wiley, IET, Elsevier with Citations: 10500+, H-Index: 55 and I-Index: 195. Member of more than 60+ Associations as Senior and Life Member like: IEEE (Senior Member) and ACM (Senior Member). He has authored/co-authored cum Edited 50+ Books of Computer Science.

Sasmita Padhy obtained her M. Tech in Computer Science and Engg. from Biju Patnaik University of Technology in 2005 and a Ph.D. degree in Computer Science Engineering from Berhampur University, Odisha, India in 2012. She has served in various colleges as Assistant Professor and Associate Professor in the Department of Computer Science and Engg. At present working as Associate Professor in SCSE, VIT Bhopal University, Bhopal, MP, India. She is having 23 years of teaching experience. Her research interests include Wireless Networks, Information Security, AI and ML, etc. She has a wide range of publications in various national, international journals and conferences. She has 5+ patents. She is a life member of ISTE, and IAENG.

Lakshmi Prasad Panda is a seasoned professor and researcher in business and economics. He earned an MBA in Finance from DDCE at Utkal University in Bhubaneswar and a Ph.D. in Economics from Berhampur University in Orissa. Additionally, he has completed a Faculty Development Programme in Management at the Indian Institute of Management, Ahmadabad, and an NPTEL certification course on Managing Intellectual Property Rights in Universities. Since 2010, Dr. Panda has served as an assistant professor of economics at the Government College of Engineering in Kalahandi. Before this, he had teaching posts at Vignan's Institute of Information Technology in Visakhapatnam, and at Centurion School of Rural Enterprise Management (CSREM) Parlakhemundi. as Dean & Associate Professor of Economics. Dr. Panda's research interests include the effects of financial reforms on the performance of MSMEs, environmental uncertainty, and socioeconomic vulnerability. In addition to publishing research articles in peer-reviewed publications, he has published in Scopus -Indexed Journals. Currently, he is supervising supervised 4nos of Ph.D. scholars. Dr. Panda has actively promoted entrepreneurial and startup initiatives in addition to his academic endeavors. He is the Startup Coordinator and IPR Co-ordinator at GCE Kalahandi and has organized several entrepreneurial awareness courses, seminars, and boot camps. Dr. Panda has held various positions, including dean at CSREM Parlkhemundi, deputy center superintendent of exams at GIET, Gunupur, and hostel superintendent at VHR, GCE Kalahandi. She is an expert user of statistical software, including LIMDEP, Easyreg, SPSS-20, and AMOS. Dr. Panda is a dedicated researcher, a passionate entrepreneur advocate, and an energetic instructor. He is an invaluable addition to any academic institution or research organization due to his experience, knowledge, and commitment.

Narendra Panda pursuing PhD from Berhampur University, Odisha. He joined the School of Management Studies, GIET University in 2010 where he is working as an Assistant Professor. His research interests are focused on operations research, operations management, security analysis and portfolio management, financial decision-making processes and financial services, with a focus on emerging markets. He is the author/co-author of seven articles which are published in international and national journals which were indexed in Scopus, Web of Science and listed in ABDC and UGC-Care.

Karishma Rana is a PhD research scholar of the Department of Economics, Central University of Odisha. She is also working as a Lecture in the Department of Economics in Betnoti College, Mayurbhanj. She is also a motivated, adaptable and responsible fellow with more than five research publications.

Minati Sahoo is working as an Assistant Professor in the Department of Economics since May, 2013. Prior to joining Central University of Odisha, she has worked as a Lecturer in Economics in Ravenshaw University. She did her MA in Economics from Ravenshaw Autonomous College, Cuttack; M Phil in Economics from Utkal University and Ph.D. in Economics from Ravenshaw University. She has also qualified UGC-NET in Economics. She has published various research papers in leading peer-reviewed national and international journals and edited books. She has also presented several research papers in national and international conferences/seminars. Her research pursuits include Economics of Natural Resources, Gender Economics, Tribal Economics and Agricultural Economics. She is currently teaching Macroeconomic Analysis, Agricultural Economics, Environmental Economics, International Trade, Industrial Economics, and Indian Economic Analysis.

Pradeep Kumar Samantray, presently working as Security Officer in Central University of Odisha, has served in Central Reserve Police Force for 11 long Years and National Disaster Response Force (NDRF), MHA for seven years and has done a lot of training on Basic/Advanced/Specialized Disaster Management. He is a Master Trainer of MFR/CSSR (NDRF Academy Nagpur), Master Instructor of Urban Search and Rescue (Swiss Methodology-SDC), School Safety trainer (NIDM New Delhi), Disaster Risk Communication through HAM- trainer (NIDM New Delhi), CBRN Emergency Management trainer (INMAS New Delhi) Chemical and Biological Emergency Management Trainer (DRDE Gwalior), Radiological Safety Officer (BARC Mumbai), Chemical Industrial Emergency Management Trainer (DMI Bhopal). He has published various research papers in leading peer-reviewed national and international journals and edited books. He has also presented several research papers in national and international conferences/seminars. His research pursuits include Environment & Natural Resources, Disaster Management.

Ashita Waghray is a research Scholar in the area of marketing in GIET University, Gunupur.Over twenty years of professional experience in industry handling Human Resource function as well class room teaching / training to MBA students in areas of Human resources and Marketing Relevant work experience in HR function with Big 4 Consulting firm driving the recruitment function and supporting the overall HR Operations Training and resource sharing with various academic bodies like AIMA, indepth knowledge of the concerned subjects and student behavior.Excellent interpersonal skills, multilingual and enjoy working in teams, excellent communication skills.

Index

A

B

C

D

E

F

G

H

I

L

M

N

P

R

S

W

Recommended Reference Books

IGI Global's reference books are available in three unique pricing formats: Print Only, E-Book Only, or Print + E-Book.

Order direct through IGI Global's Online Bookstore at **www.igi-global.com** or through your preferred provider.

ISBN: 9781668445488
EISBN: 9781668445501
© 2022; 334 pp.
List Price: US$ **240**

ISBN: 9781668467503
EISBN: 9781668467527
© 2022; 296 pp.
List Price: US$ **240**

ISBN: 9781799889007
EISBN: 9781799889021
© 2022; 340 pp.
List Price: US$ **215**

ISBN: 9781799898511
EISBN: 9781799898535
© 2022; 276 pp.
List Price: US$ **250**

ISBN: 9781799896913
EISBN: 9781799896937
© 2022; 219 pp.
List Price: US$ **215**

ISBN: 9781668436868
EISBN: 9781668436875
© 2022; 2,064 pp.
List Price: US$ **2,085**

Do you want to stay current on the latest research trends, product announcements, news, and special offers?
Join IGI Global's mailing list to receive customized recommendations, exclusive discounts, and more.
Sign up at: **www.igi-global.com/newsletters.**

Publisher of Timely, Peer-Reviewed Inclusive Research Since 1988

IGI Global
PUBLISHER of TIMELY KNOWLEDGE

www.igi-global.com Sign up at www.igi-global.com/newsletters facebook.com/igiglobal twitter.com/igiglobal linkedin.com/igiglobal

Ensure Quality Research is Introduced to the Academic Community

Become an Evaluator for IGI Global Authored Book Projects

The overall success of an authored book project is dependent on quality and timely manuscript evaluations.

Applications and Inquiries may be sent to:
development@igi-global.com

Applicants must have a doctorate (or equivalent degree) as well as publishing, research, and reviewing experience. Authored Book Evaluators are appointed for one-year terms and are expected to complete at least three evaluations per term. Upon successful completion of this term, evaluators can be considered for an additional term.

If you have a colleague that may be interested in this opportunity, we encourage you to share this information with them.

Easily Identify, Acquire, and Utilize Published Peer-Reviewed Findings in Support of Your Current Research

IGI Global OnDemand

Purchase Individual IGI Global OnDemand Book Chapters and Journal Articles

For More Information:

www.igi-global.com/e-resources/ondemand/

Browse through 150,000+ Articles and Chapters!

Find specific research related to your current studies and projects that have been contributed by international researchers from prestigious institutions, including:

Massachusetts Institute of Technology · HARVARD UNIVERSITY · COLUMBIA UNIVERSITY IN THE CITY OF NEW YORK · Australian National University

- Accurate and Advanced Search
- Affordably Acquire Research
- Instantly Access Your Content
- Benefit from the InfoSci Platform Features

"*It really provides* **an excellent entry into the research literature of the field.** *It presents a manageable number of* **highly relevant sources** *on topics of interest to a wide range of researchers. The sources are* **scholarly, but also accessible** *to 'practitioners'.*"

- Ms. Lisa Stimatz, MLS, University of North Carolina at Chapel Hill, USA

Interested in Additional Savings?

Subscribe to

IGI Global OnDemand *Plus*

Learn More

Acquire content from over 128,000+ research-focused book chapters and 33,000+ scholarly journal articles for as low as US$ 5 per article/chapter (original retail price for an article/chapter: US$ 37.50).

,300+ E-BOOKS. DVANCED RESEARCH. NCLUSIVE & AFFORDABLE.

IGI Global
PUBLISHER of TIMELY KNOWLEDGE

GI Global e-Book Collection

- **Flexible Purchasing Options** (Perpetual, Subscription, EBA, etc.)
- Multi-Year Agreements with **No Price Increases** Guaranteed
- **No Additional Charge** for Multi-User Licensing
- No Maintenance, Hosting, or Archiving Fees
- Continually Enhanced & Innovated **Accessibility Compliance Features** (WCAG)

Handbook of Research n Digital Transformation, ndustry Use Cases, and the Impact of Disruptive Technologies
ISBN: 9781799877127
EISBN: 9781799877141

Handbook of Research on New Investigations in Artificial Life, AI, and Machine Learning
ISBN: 9781799886860
EISBN: 9781799886877

Handbook of Research on Future of Work and Education
ISBN: 9781799882756
EISBN: 9781799882770

Research Anthology on Physical and Intellectual Disabilities in an Inclusive Society (4 Vols.)
ISBN: 9781668435427
EISBN: 9781668435434

Innovative Economic, ocial, and Environmental ractices for Progressing Future Sustainability
ISBN: 9781799895909
EISBN: 9781799895923

Applied Guide for Event Study Research in Supply Chain Management
ISBN: 9781799889694
EISBN: 9781799889717

Mental Health and Wellness in Healthcare Workers
ISBN: 9781799888130
EISBN: 9781799888147

Clean Technologies and Sustainable Development in Civil Engineering
ISBN: 9781799898108
EISBN: 9781799898122

Request More Information, or Recommend the IGI Global e-Book Collection to Your Institution's Librarian

For More Information or to Request a Free Trial, Contact IGI Global's e-Collections Team: eresources@igi-global.com | 1-866-342-6657 ext. 100 | 717-533-8845 ext. 100

www.ingramcontent.com/pod-product-compliance
Lightning Source LLC
LaVergne TN
LVHW080525300926
849162LV00007B/1539
* 9 7 8 1 6 6 8 4 9 8 1 9 4 *